The difficulty that humans have of perceiving and understanding large-scale phenomena is widely acknowledged. The need to study and understand processes associated with global environmental change has created an environment that has promoted unprecedented scientific co-operation between and within disciplines. The science of global change emphasises the need to understand the complex biophysical systems both in terms of components and as a whole. This has involved research at many levels of organisation, or scales, within each system. The need to link and integrate information between scales underlies the scientific approaches described in this book. The theory, practice and challenges of scaling are discussed using examples from current research, combining biology and geography to address issues at a range of scales, from cellular to global levels.

SOCIETY FOR EXPERIMENTAL BIOLOGY
SEMINAR SERIES: 63

SCALING-UP: FROM CELL TO LANDSCAPE

SOCIETY FOR EXPERIMENTAL BIOLOGY SEMINAR SERIES

A series of multi-author volumes developed from seminars held by the Society for Experimental Biology. Each volume serves not only as an introductory review of a specific topic, but also introduces the reader to experimental evidence to support the theories and principles discussed, and points the way to new research.

SCALING-UP
From Cell to Landscape

Edited by

P.R. van Gardingen

Institute of Ecology and Resource Management,
University of Edinburgh

G.M. Foody

University of Salford

and

P.J. Curran

University of Southampton

CAMBRIDGE UNIVERSITY PRESS
Cambridge, New York, Melbourne, Madrid, Cape Town,
Singapore, São Paulo, Delhi, Tokyo, Mexico City

Cambridge University Press
The Edinburgh Building, Cambridge CB2 8RU, UK

Published in the United States of America by Cambridge University Press, New York

www.cambridge.org
Information on this title: www.cambridge.org/9780521187756

First published 1997
First paperback edition 2011

A catalogue record for this publication is available from the British Library

Library of Congress Cataloguing in Publication data
Scaling-up: from cell to landscape / edited by P.R. van Gardingen,
G.M. Foody and P.J. Curran
     p.   cm. – (Society for Experimental Biology seminar series ; 63)
Includes bibliographical references and index.
ISBN 0–521–47109–5 (hb)
1. Biomathematics. 2. Biology – Simulation methods. 3. Ecology –
Simulation methods. 4. Remote sensing – Mathematics. I. Van
Gardingen, P. R. (Paul Richard) II. Foody, Giles M., 1955–   .
III. Curran, Paul J. IV. Series : Seminar series (Society for
Experimental Biology (Great Britain) : 63.
QH323.5.S32   1997
570′.1′5118–dc21   96–51105 CIP

ISBN 978-0-521-47109-1 Hardback
ISBN 978-0-521-18775-6 Paperback

Additional resources for this publication at www.cambridge.org/9780521187756

Contents

Contributors

ATKINSON, D.
Scottish Agricultural College, West Mains Road, Edinburgh EH9 3JG, UK

ATKINSON, P.M.
Department of Geography, University of Southampton, Highfield, Southampton SO17 1BJ, UK

BAND, L.E.
Department of Geography, University of Toronto, Toronto, Ontario M5S 1A1, Canada

BARNSLEY, M.J.
Department of Geography, University of Wales Swansea, Singleton Park, Swansea SA2 8PP, UK

BARR, S.L.
Department of Geography, University of Manchester, Mansfield Cooper Building, Oxford Road, Manchester M13 9PL, UK

BERRY, J.A.
Department of Plant Biology, Carnegie Institute, Washington, Stanford, CA 94305, USA

BLYTH, E.M.
Institute of Hydrology, Wallingford, Oxfordshire, UK

COLELLO, G.D.
Department of Plant Biology, Carnegie Institute, Washington, Stanford, CA 94305, USA

COLLATZ, G.J.
NASA/Goddard Space Flight Center, 923 Greenbelt, MD 20771, USA

COLLINS, J.B.
Boston University Center for Remote Sensing, 725 Commonwealth Avenue, Boston, MA 02215, USA

CRAWFORD, J.W.
Soil–Plant Dynamics Group, Scottish Crop Research Institute, Invergowrie, Dundee DD2 5DA, UK

CURRAN, P.J.
Department of Geography, University of Southampton, Highfield, Southampton SO17 1BJ, UK
DENNING, S.
Department of Atmospheric Science, Colorado State University, Ft Collins, CO 80523, USA
FOGEL, R.
The Herbarium, University of Michigan, Ann Arbor, MI, USA
FOODY, G.M.
Department of Geography, University of Salford, Salford, Manchester M5 4WT, UK
FU, W.
Department of Plant Biology, Carnegie Institute, Washington, Stanford, CA 94305, USA
GIBSON, G.J.
Biomathematics and Statistics Scotland, University of Edinburgh, King's Buildings, Edinburgh EH9 3JU, UK
GRACE, J.
Institute of Ecology and Resource Management, University of Edinburgh, King's Buildings, Edinburgh EH9 3JU, UK
GRIME, J.P.
Unit of Comparative Plant Ecology, University of Sheffield, Department of Animal and Plant Sciences, Sheffield S10 2TN, UK
GRIVET, C.
Department of Plant Biology, Carnegie Institute, Washington, Stanford, CA 94305, USA
GURNEY, R.J.
Environmental Systems Science Centre, University of Reading, Reading RG6 2AB, UK
HARDING, R.J.
Institute of Hydrology, Wallingford, Oxfordshire, UK
HONZÁK, M.
Department of Geography, University of Wales Swansea, Singleton Park, Swansea SA2 8PP, UK
JARVIS, P.G.
Institute of Ecology and Resource Management, University of Edinburgh, King's Buildings, Edinburgh EH9 3JU, UK
JUPP, D.L.B.
Division of Water Resources, CSIRO, Clunies Ross Street, Canberra, 2601 ACT, Australia

KRUIJT, B.
Institute of Ecology and Resource Management, University of
Edinburgh, King's Buildings, Edinburgh EH9 3JU, UK
LAMMERS, R.B.
Department of Geography, University of Toronto, Toronto, Ontario
M5S 1A1, Canada
LAWSON, T.
Department of Biological Sciences, University of Dundee, Dundee
DD1 4HN, UK
LUAN, J.
Institute of Ecology and Resource Management, University of
Edinburgh, King's Buildings, Edinburgh EH9 3JU, UK
LUCAS, R.M.
Department of Geography, University of Southampton, Highfield,
Southampton SO17 1BJ, UK
MACGILLIVRAY, W.C.
Unit of Comparative Plant Ecology, University of Sheffield,
Department of Animal and Plant Sciences, Sheffield S10 2TN, UK
MARSHALL, B.
Soil–Plant Dynamics Group, Scottish Crop Research Institute,
Invergowrie, Dundee DD2 5DA, UK
ONGERI, S.
Institute of Ecology and Resource Management, University of
Edinburgh, King's Buildings, Edinburgh EH9 3JU, UK
PENG, Z.Y.
Department of Biological Sciences, University of Dundee, Dundee
DD1 4HN, UK
PORTER, J.R.
Environmental Sciences, Long Ashton Research Station, Bristol
BS18 9AF, UK
RANDALL, D.A.
Department of Atmospheric Science, Colorado State University, Ft
Collins, CO 80523, USA
RUSSELL, G.
Institute of Ecology and Resource Management, University of
Edinburgh, King's Buildings, Edinburgh EH9 3JU, UK
SELLERS, P.J.
NASA/Goddard Space Flight Center, 923 Greenbelt, MD 20771, USA
SEWELL, I.J.
Environmental Systems Science Centre, University of Reading,
Whiteknights, Reading RG6 6AB, UK

SQUIRE, G.R.
Scottish Crop Research Institute, Invergowrie, Dundee DD2 5DA, UK
TAGUE, C.L.
Department of Geography, University of Toronto, Toronto, Ontario
M5S 1A1, Canada
TAYLOR, C.M.
Institute of Hydrology, Wallingford, Oxfordshire, UK
THOMPSON, K.
Unit of Comparative Plant Ecology, University of Sheffield,
Department of Animal and Plant Sciences, Sheffield S10 2TN, UK
TSANG, T.
Remote Sensing Unit, Department of Geography, University College
London, 26 Bedford Way, London WC1H 0AP, UK
VAN GARDINGEN, P.R.
Institute of Ecology and Resource Management, University of
Edinburgh, King's Buildings, Edinburgh EH9 3JG, UK
WEYERS, J.D.B.
Department of Biological Sciences, University of Dundee, Dundee
DD1 4HN, UK
WOODCOCK, C.E.
Center for Remote Sensing, Boston University, 725 Commonwealth
Avenue, Boston, MA 02215, USA

Preface

In September 1964, the Society for Experimental Biology held its nineteenth Symposium at the University College of Swansea on *The State and Movement of Water in Living Organisms* (Fogg, 1965). This was organised in the belief that discussion of this topic by scientists from different disciplinary backgrounds was long overdue. The section of the symposium focusing on *Water in the Plant* included the seminal paper *Evaporation and Environment* (Monteith, 1965). Since that time, plant scientists have increasingly applied physical principles and techniques to the understanding of water flow in the soil–plant–atmosphere continuum.

Nearly 30 years later in April 1994, the Society for Experimental Biology (SEB) joined with the British Ecological Society (BES) and returned to Swansea to host a similarly interdisciplinary meeting on the topic of *Scaling-up*. The meeting was organised by the Environmental Physiology Groups of the SEB and BES and received very generous financial support from the BES, SEB and the Joint Research Centre of the European Community.

There had been much interest in using science to address environmental issues at scales from a region to the globe. Much of this work resulted from interest in global environmental change. As a result scientists were needing to use information collected at one scale, for example a single plant, to make predictions over a much larger area. This change in scientific culture created an environment that promoted interdisciplinary research, with biologists needing to join with geographers and climate modellers to address these issues. For the biologist there was the challenge of scaling-up to describe processes or systems acting over a large area. For the geographer and climate modeller, there was the difficult task of linking potentially complex biological information into their information systems or models. With this background, it was an appropriate time to bring together the groups to discuss their approaches.

During the meeting on *Scaling-up*, the need for improved communication between the groups or disciplines became apparent. There were

very significant differences in the usage of terminology, in particular
the definition of scale (Curran, Foody & van Gardingen, this volume).
The importance of this issue required an editorial decision to impose a
common definition of scale on all authors within this volume. It is
noteworthy that similar problems were identified in the Preface to the
volume arising from the 1964 Symposium when bringing together biologists and physicists.

We wish to thank everyone who has helped with the running of the
meeting and the production of the resulting volume. Our work has been
a team effort, firstly supported by the authors themselves and then
through the editorial stage with assistance by Mrs Judith van Gardingen,
Dr Colette Robertson, Dr Graham Russell and Mrs Linda Sharp. Dr
Maria Murphy has been the series editor at Cambridge University Press
and has been supportive and patient as the volume progressed. To these
people and the rest of the production team at Cambridge University
Press, we wish to express our appreciation and thanks.

Fogg, G.E. (ed.) (1965). *The State and Movement of Water in Living Organisms*, Society for Experimental Biology Symposium, Vol. 19. Cambridge: Cambridge University Press.
Monteith, J.L. (1965). Evaporation and environment. In: *The State and Movement of Water in Living Organisms*, Fogg, G.E. (ed.) Society for Experimental Biology Symposium, Vol. 19, pp. 205–234. Cambridge: Cambridge University Press.

Paul R. van Gardingen
Giles M. Foody
Paul J. Curran
Editors
February 1997

P.J. CURRAN, G.M. FOODY
and P.R. VAN GARDINGEN

Scaling-up

Places that are near to each other are more alike than those that are further away and the degree of dissimilarity depends on both the environment and the nature of our observations. This view is one that we need to adopt if we wish to move measurements or understanding from the local to the regional or global scale. True, there are some phenomenon that can sometimes be studied in isolation because they show self-similarity with scale (e.g. drainage patterns) or can be considered spatially homogenous (e.g. fresh snow), but in this diverse world of ours these are the exception rather than the rule.

That every place is unique and space is a variable (rather than a parameter) is the essence of geography (Harvey, 1969). This theme echoes backwards to our understanding of measurement uncertainty (Heisenberg, 1932) and fractals (Mandelbrot, 1982) and forwards to landscape ecology (Meentemeyer, 1989) and the needs of Earth System Science (NASA, 1988). However, this understanding of space, time and, thereby, scale sits uneasily with aspatial deterministic models. Assumptions that a relationship at one scale will be the same at another, untested assumptions of spatial homogeneity and limited information on the land cover of our planet limit our ability to address the pressing need to move from a local to a regional or global scale understanding of our environment (IGBP, 1992; Houghton et al., 1996). This confusion over the fundamental importance of scale has been recognised by a number of observers in recent years (Ehleringer & Field 1993; Foody & Curran, 1994a; Stewart et al., 1996). This book intervenes in this debate by exploring first what we know about scale and second by taking that knowledge to scale our local-scale measurements or understanding to larger, particularly regional or global scales, via a mix of models and spatial extrapolation. The dominant voices in this book are those of biologists and geographers but those with relevant viewpoints from the disciplinary vantage points of statistics, hydrology, meteorology and agriculture feature strongly. Fortunately, the only point of contention was semantic and concerned the long running debate over the definition of the word *scale* (Lam & Quattrochi, 1992). In this context scale has two equally valid definitions. The first is *cartographic* and the second is *colloquial* (Foody & Curran, 1994b). Unfortunately what is small scale by one definition is large scale by the other! The cartographic definition of scale relates the distance on a map to the actual distance on the ground via the equation:

$$scale = distance\ on\ map/actual\ distance\ on\ ground$$

Consequently, a *small-scale* map (e.g. 1:10 000 000) covers a large area, like a continent, with little detail and a *large-scale* map (e.g. 1:1000) covers a small area, like a building plot, in great detail. The colloquial definition of scale is that it is a synonym of words such as size or area. Consequently, *small scale* is small size and thereby small area (or period of time) and large scale is large size and thereby large area (or period of time). Scale by the colloquial definition has no commonly accepted bounds and so is relative to the observer. One observer may use small scale to mean a leaf (relative to a field) while another may use small scale to mean a field (relative to a region). The cartographic definition is precise and has a long and unambiguous history (Maling, 1989) whereas the colloquial definition is imprecise and contains in-built redundancy (why use the word scale when no word is needed or when size or area would suffice?). However, the colloquial has the most common contemporary usage and underpins the very notion of 'scaling-up' from small to large. Therefore, in this book we have adopted, where possible, the colloquial definition of scale and authors have adhered to the following five guidelines:

- scale relates to size, area or time period
- the unqualified term *scale* should be avoided
- terms such as *small scale* or *large scale* should be replaced with more specific phrases such as *leaf scale*, *field scale*, *regional scale* and *global scale*
- specific phrases such as *scaling from regional to global scales* should be used to avoid vague phrases such as *scaling-up to larger scales*
- *scaling-up* relates to moving from *small* to *larger* area, but where possible the terms small and large should be defined precisely.

This hopefully will enable the reader to understand each author's use of the term scale. This is vital since the results of an investigation and inferences drawn from it are scale dependent. Thus observations and theories derived at one scale may not apply at another. Furthermore, the differences observed between locations at different scales may be enormous, with, for instance, large changes in both the strength and direction of relationships noted when the scale of the study changed. This type of problem is well known in geographical research, perhaps most notably in the guise of the ecological fallacy (Johnston, 1981; Wrigley, 1995), but has only relatively recently been recognised within the ecological community.

Given the significance of the effects of scale, there is an urgent need for scaling issues, both in space and time, to be recognised as being fundamental to ecological research. The recognition of the existence of domains of scale and the development of scaling theories that generate testable scientific hypotheses are urgent priorities. To achieve these, data will be required at a range of spatial and temporal scales. With much ecological research restricted to relatively small areas, through practical constraints as well as the scientific culture of the ecological research community (Wiens, 1989), much of our knowledge and understanding of the environment relates only to relatively local scales. With the growing awareness and concern over regional- to global-scale issues, such as greenhouse warming and ozone depletion, much emphasis has been placed on scaling-up this detailed knowledge acquired at the local scales. Since the effects of changing scale are complex and non-linear, this is not an easy task, hence the need for multiscale analyses and scaling theories. This will require data at a range of scales. The opportunity to study and exploit scaling effects is provided through remote sensing (Golley, 1989). Presently, remote sensing systems enable the Earth's environment to be studied at spatial scales ranging from the local to the global. Furthermore, the data may be available at a temporal frequency measured over a period of days or weeks, supplemented by an archive that already extends over a number of decades. Remotely sensed data, therefore, provide the basis for observing the environment at a range of scales to facilitate our understanding of the environment. Scale, however, exerts a strong influence on the ability to extract environmental information from remotely sensed data and so the data sets require careful specification and analysis within appropriate scene models (Woodcock & Strahler, 1987; Woodcock & Harward, 1992). The information extracted from the remotely sensed data may then be integrated with other spatial data sets within a geographic information system (GIS). This does not eliminate the need for further accommodation of scale effects. While the data within the GIS may have been acquired at a suitable scale using appropriate techniques, an analyst may wish to change the scale of an investigation and so methods for scaling will be required. Therefore, if scale is to be accommodated appropriately within environmental research, there is an urgent need for theoretical and methodological advancements across a range of fields of study.

It is hoped that the growing awareness of the effects of scale in environmental research and closer linkage between the various research communities involved will aid a better understanding of the environment. We hope that this book will provide a small step in this direction.

References

Ehleringer, J.R. & Field, C.B. (eds.) (1993). *Scaling Physiological Processes: Leaf to Globe*. San Diego: Academic Press.

Foody, G.M. & Curran, P.J. (eds.) (1994a). *Environmental Remote Sensing from Regional to Global Scales*. Chichester: Wiley.

Foody, G.M. & Curran, P.J. (1994b). Scale and environmental remote sensing. In *Environmental Remote Sensing from Regional to Global Scales*, ed. G.M. Foody & P.J. Curran, pp. 223–232. Chichester: Wiley.

Golley, F.B. (1989). A proper scale. *Landscape Ecology*, **2**, 71–72.

Harvey, D. (1969). *Explanation in Geography*. London: Edward Arnold.

Heisenberg, W. (1932). *Nobel Prize in Physics Award Address: W. Heisenberg*. New York: Nobel Foundation/Elsevier.

Houghton. J.T., Meira Filho, L.G., Callander, B.A., Harris, N., Kattenberg, A. & Maskell, K. (eds.) (1996). *Climate Change 1995: The Science of Climate Change*. Cambridge: Cambridge University Press.

IGBP (1992). *Global Change: Reducing the Uncertainties*. International Geosphere Biosphere Programme. Stockholm: Royal Swedish Academy of Sciences.

Johnston, R.J. (1981). Ecological fallacy. In *The Dictionary of Human Geography*, ed. R.J. Johnston, D. Gregory, P. Haggett, D. Smith & D.R. Stoddart, p. 89. Oxford: Blackwell.

Lam, N. S.-N. & Quattrochi, D.A. (1992). On the issues of scale, resolution and fractal analysis in the mapping sciences. *Professional Geographer*, **44**, 88–98.

Maling, D.H. (1989). *Measurements from Maps*. Oxford: Pergamon.

Mandelbrot, B.B. (1982). *The Fractal Geometry of Nature*. San Francisco: Freeman.

Meentemeyer, V. (1989). Geographical perspectives of space, time and scale. *Landscape Ecology*, **3**, 163–173.

NASA (1988). *Earth System Science: A Closer View*. Report of the Earth System Sciences Committee, NASA Advisory Council, Washington DC: NASA.

Stewart, J.B., Engman, E.T., Feddes, R.A. & Kerr, Y. (eds.) (1996). *Scaling-up in Hydrology using Remote Sensing*. Chichester: Wiley.

Wiens, J.A. (1989). Spatial scaling in ecology. *Functional Ecology*, **3**, 385–397.

Woodcock, C.E. & Strahler, A.H. (1987). The factor of scale in remote sensing. *Remote Sensing of Environment*, **21**, 311–332.

Woodcock, C.E. & Harward, V.J. (1992). Nested-hierarchical scene models and image segmentation. *International Journal of Remote Sensing*, **16**, 3167–3187.

Wrigley, N. (1995). Revisiting the modifiable areal unit problem and the ecological fallacy. In *Diffusing Geography*, ed. A. Cliff, P.R. Gould, A.G. Hoare & N.J. Thrift, pp. 49–71. Oxford: Blackwell.

J. GRACE, P.R. VAN GARDINGEN and J. LUAN

Tackling large-scale problems by scaling-up

Introduction

Much of the activity of scientists is described as *analysis*, by which we mean inspection of the whole by examination of the constituent parts and processes. This is evident in the history of science, and the tools of science. Leeuwenhoek (1632–1723) developed the microscope to see small things clearly, but until the development of remote sensing there was no macroscope or equivalent tool to see very large things in their entirety. Yet the difficulty that humans have of perceiving and understanding larger-scale phenomena is widely acknowledged. In the English language we often accuse people of 'being unable to see the wood for the trees' and 'shortsightedness'. It seems that an act of imagination is required to step beyond the scales of time and space that are imposed by human senses, and that the capacity to do this is much cherished (we speak of 'a man of vision' to describe remarkable people who have this capacity). Yet scientists addressing the larger-scale problems have never found it easy to have their ideas accepted, either by fellow scientists or society as a whole. Examples of such ideas include continental drift, evolution by natural selection, population homeostasis, biospheric homeostasis and even climate change. This is because the scientific method works best when the object of study is well defined, can be isolated from extraneous influences and small changes in the conditions can be made at will. Then the classical scientific method of hypothesis testing by experiment is relatively easy. Unfortunately, many of the urgent problems that we face in the world are large-scale problems to do with land and resource utilisation, which, as Waddington (1977) and many others since have pointed out, are not especially amenable to this approach. The recent upsurge of interest in scales of organisation, and scaling-up, is partly a response to the challenge of using science to solve large problems, and partly the need to reconcile the differing world views that we find between scientific disciplines, for example, between population biology and ecosystematics (Levin, 1992; Rowe, 1992; Ehleringer & Field, 1993).

The propositions

The problem of scale is best discussed as a series of propositions, all of which are widely espoused. The first of these represents the reductionist view.

1. The small things are the ones that determine the characteristics of the living world

Examples of small things in this context are genes, ion pumps, stomata and leaves. At first, this proposition seems self-evident and is implicit in many branches of science. However, on reflection, it is clear that matters are less straightforward. For example, taking the case of genes: although each gene codes for a protein, whether or not any specific gene spreads or is extinguished in the species as a whole depends on the process of natural selection. This operates at the population and individual level, and a complete genetic map of, say, *Homo sapiens* would not in itself explain how the species evolved and why differences between individuals occur. Thus, over long (evolutionary) periods the genome is controlled by events operating at a higher level of organisation, and which are subject to their own laws. There are many such examples. A mechanistic understanding of stomatal movement is an important scientific goal but might not help us to model regional scale hydrology, because properties of the leaves, the land cover and the atmosphere have a more dominant role to play than the movement of potassium ions across the plasmalemma of the guard cells. This type of division may be a quite general problem, acute in some cases but less important in others.

2. The small things are the ones most amenable to study by the methods of science

'Small things' fit into test tubes or observation chambers, and they may be subject to experimental manipulations. Not only are they small, but they perform their function over fairly short periods of time and, therefore, may be the subject of experiments several times in a day. The small things are simple, basic, elemental, so we should be able to find out a lot about them in a human lifetime.

For this reason, leaves, stomata and chloroplasts have been studied more than canopies of leaves, which in turn have been studied more than ecosystems. However, even these simple things, in the hands of an experimental scientist with an inquiring spirit, are revealed to be far from simple. Worlds are found to exist inside worlds, and research often generates more questions than answers. Nevertheless, it has been possible to describe their responses to their immediate environment in considerable detail.

3. The large things are the ones that have the most profound effect on humans

Examples of 'large things' are drought, famine and climate change. It is impossible, or at least difficult, to carry out meaningful experiments on them, because the experiments would take too long, or be too expensive or not be ethically acceptable. The best we can do is to gather statistics about them and to construct mathematical models that represent our best guesses about how they work. Small-scale experiments in 'model systems' or 'microcosms' are often attempted but may not be realistic because larger systems have properties of their own, which small-scale experimental systems cannot capture.

4. There is a feeling that we should be able to use our knowledge of small things to predict and manage these large-scale phenomena

The optimist believes that we should be able to use knowledge derived from small things and this is the motivation for scaling-up. Certainly, we need to be able to predict outside our experience, not just by extrapolation but also from a knowledge of processes and mechanisms embodied in physical, chemical and biological disciplines. We may not agree on how far this prediction is possible, but most people would concede that prediction is perfectly possible in principle, although in practice we are limited by lack of knowledge and a shortage of computing power. Here it is useful to distinguish two sorts of scaling-up. The first, we may call it *simple scaling*, is where we want to multiply up a phenomenon observed in a small plot or sample to make a statement about the landscape, the region or the world. An extreme example is the work of Zimmerman *et al.* (1982) on the evolution of methane, a greenhouse gas, by termites. From a study of two termite species *in vitro*, the authors calculated that termites of the world produce 1.5×10^{14} g methane by consuming 33.0×10^{15} g phytomass. However, there are many life forms and species of termite, each with a characteristic metabolism and methane-producing gut flora. They feed at quite different rates. Moreover, much of the methane produced in soil is now known to be oxidised to CO_2 on passage through the soil profile. It is not surprising that the estimate was controversial and has been adjusted downwards (e.g. Collins & Wood (1984) thought the consumption was a mere 3.36×10^{15} g carbon). Analogous scaling-up procedures are inevitably widespread in work on global geochemistry and climate change, though caveats are usually supplied (Grace *et al.*, 1995). The second type of scaling-up is where it becomes necessary or desirable to move between levels of organisation, for example from a knowledge of how stomata behave in relation to light, humidity and CO_2 to a prediction of the hydrology of a region and how this may be affected by climate change (Norman, 1993). We may call this *hierarchical scaling*.

5. In hierarchical systems, information is passed upscale and downscale

Passage of information in a hierarchy is best explained by a diagram (Fig. 1). Hierarchical structure has a horizontal separation that isolates each level from levels above and below, and a vertical separation that segregates the components of any level into groups. A single process operating at any level is the outcome of several lower-level processes that operate more frequently. It should be clear from this that the processes and mechanisms at any one scale, which determine the characteristics of the next largest scale, are subject to feedbacks (Allen & Star,

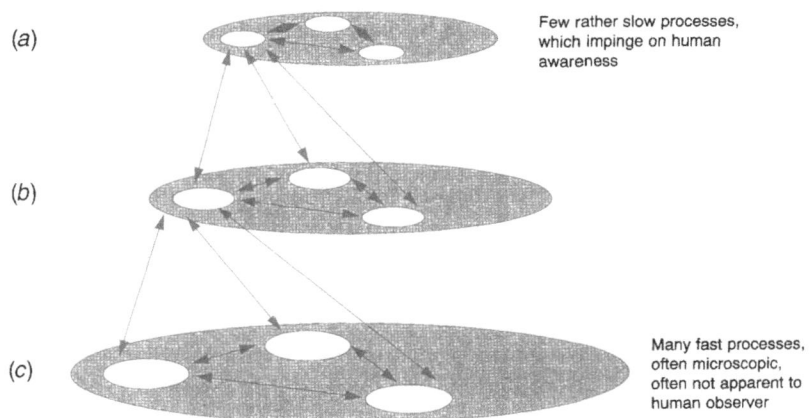

Fig. 1. Scales of organisation, showing feedbacks and feedforwards from the topmost scale (the domain most accessible and familiar to human experience) and the lowermost scale (where structures are microscopic and numerous and processes are fast).

1982). This is the general case of proposition 1. There are several ways this feedback may operate. It may be that the higher level sets the limits for the operation of the processes by a negative feedback. For example, when scaling-up from stomata to a forest region, it is important to realise that the water lost through billions of stomatal pores accumulates in the regional boundary layer, thus decreasing the driving gradient for evaporation (McNaughton & Jarvis, 1991). Therefore, simple scaling of knowledge of evaporation from a population of stomata held at controlled humidity would give the wrong answer, and hierarchical scaling is required. This may be difficult to achieve in practice, as scientific disciplines have evolved to study discrete scales, and scientists may be unaware, or reluctant to assimilate, the wisdom of other disciplines.

A related proposition is that *behaviour at one level cannot always be explained in terms of lower level behaviour simply by adding up the lower level parts*. This is because the parts interact. For example the rate of nutrient supply to a plant cannot be readily predicted from data of nutrient concentrations in the soil; a knowledge of physical chemistry is required to understand how nutrients and toxic ions are made available as the pH and redox potential of the soil changes. The rate will also depend on the distribution of roots and mycorrhizas, as well as on the flux of water through the system, itself a complex function of the vegetational structure and the environment. An example is provided by a study of the impact of elevated CO_2 upon vegetation. Luan (1995) and Luan, Grace & Muetzeldeldt (1996) used a hierarchical model to scale-up the response of leaves to CO_2 in order to predict the response of forest ecosystems to CO_2. The model incorporated many (but by no means all) of the known relationships that act between hierarchical levels. For example, when plants are stimulated to grow fast by elevated supplies of CO_2, then they may become limited by the availability of soil water and nutrients. The study revealed a tendency of the responsiveness to CO_2 to diminish on progressing from leaf scale, through whole plant scale to forest scale. It is too early to say whether this is a general homeostatic principle, essentially the statement of Le Châtelier, which applies to all ecosystems and all perturbations. It seems safer to assume that it is not.

6. Biological systems have evolved bottom-up and tend toward structural stability

In evolution, the lower levels of organisation came first and have been selected for the wider range of environments that they have experienced. This is likely to apply to species and communities. To endure for so long they must be smart, and it may be hard to improve on them. It is not a coincidence that major increases in agricultural production have come not from improving the photosystems or the carboxylating enzymes but from management practices at the farm level, the introduction of fertilisers and pesticides, the use of machinery and improvement of the harvest index.

7. Although the small things are easier to study and understand, they are more numerous

This assertion is obvious from Fig. 1 and represents a stumbling block. If there are so many processes at the lower level, how can we decide which ones are the most important to study; are they all important or can some be neglected altogether? If one is trying to use an ecophysiological approach to predict future distribution of a species, it would not be fruitful to examine all the known physiological processes as a function of environmental variables, as there would be simply too much work. It is normal to seek refuge in the notion of rate-limiting factors, whereby one factor is rate limiting at any time. It may be possible to pin-point this rate-limiting factor by a series of preliminary experiments or observations conducted at the higher level.

8. The large scale is likely to have at least some characteristics that we cannot predict at all from a knowledge of the small scale

This may seem like an admission of defeat, but it is important to realise that prediction is a greater challenge than explanation. The latter, equivalent to 'being wise after the event' is what most of science is about. We often *believe* we understand something, but prediction is the 'acid test' of whether our understanding is complete. One task for the ecologist is to identify processes that feed forward, where a small initial displacement of the system is not suppressed but stimulated. When ecosystems are warmed, the rate of microbial respiration is stimulated more than photosynthesis, and this results in loss of carbon from the soil to the atmosphere. Currently, there is compelling evidence that such a process is occurring in the arctic tundra, which is becoming a substantial source of CO_2 to the atmosphere and may thus be increasing the absorption of outgoing longwave radiation and hence accelerating global warming (Oechel *et al.*, 1993).

9. Scaling-up is not part of our scientific tradition

Although scaling-up is currently fashionable, we find little evidence of it having contributed to scientific progress in the past. There are some examples from physics. The gas laws provide a description of the relationship between temperature, pressure and volume of a gas, and they remain the operational basis for perfectly adequate calculations in many branches of science. The *explanation* of the gas laws, at a lower level of organisation, comes from the kinetic theory of gases but *this came later*. There are many examples like this. The theory of evolution was not derived from Mendel's laws but from a consideration of the patterns of variation at a large scale. It is possible to think of a few examples of scaling-up from other disciplines. Civil engineers make physical models of bridges and buildings and test them in wind tunnels: to scale-up to full-size it is necessary to use dimensionless numbers and sometimes 'rules of thumb', and finally to build in a generous safety margin. This is not hierarchical scaling, as the model bridges are simply small bridges. Civil engineers have been known to copy the hierarchical principles that organisms reveal: large materials are assembled from small material units, which themselves are made of even smaller units (Lakes, 1993). But it is only quite recently that they have tried to design complex structures by scaling-up from a knowledge of the component parts; they prefer a more empirical approach based on accumulated experience because it is regarded as the safer option. The improvement of crops by plant breeding has been an analogous situation (Passioura, 1981).

Conclusions

The propositions above reflect a diversity in the style of scientists, and perhaps in their intellectual processes. Long ago, the psychologist Carl Gustav Jung recognised extreme intellectual types, the *convergent* and the *divergent*. The former works best when focused on a limited field through analysis; the latter prefers the broad brush and the large canvas and can take a broader view. *Both* are required, because analysis and synthesis must proceed hand-in-hand if we are to apply the results of science to the problems of the world.

We are beginning to see in the organisation of scientific endeavour recognition of natural hierarchy, with appropriate structuring of research to reflect the need for both analysis and synthesis. An example of this is given in Fig. 2. The problem here is the impact of forest on the regional and global balance of heat, water and CO_2. The major forested areas of the world are large enough to have an impact on the global climate system. But the system is large and complex, and sensitivity to change exists at several levels of complexity, corresponding to different scientific disciplines (e.g. satellite remote sensing, meteorology, micrometeorology, physiology and biochemistry). Moreover, these

Scales of observation   **Strategy**

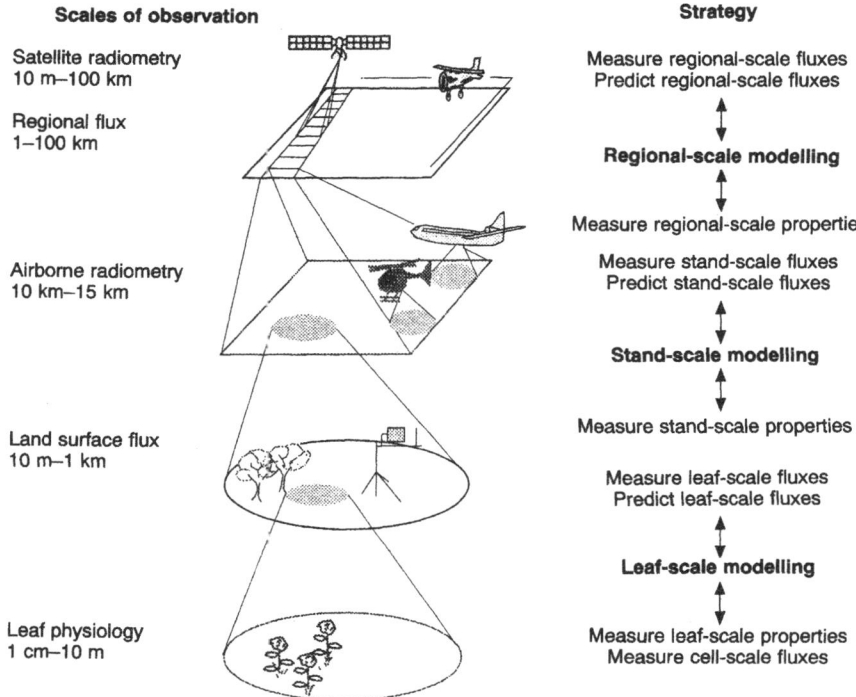

Satellite radiometry   Measure regional-scale fluxes
10 m–100 km   Predict regional-scale fluxes

Regional flux
1–100 km   **Regional-scale modelling**

                                              Measure regional-scale properties

Airborne radiometry   Measure stand-scale fluxes
10 km–15 km   Predict stand-scale fluxes

                                              Stand-scale modelling

                                              Measure stand-scale properties

Land surface flux   Measure leaf-scale fluxes
10 m–1 km   Predict leaf-scale fluxes

                                              Leaf-scale modelling

Leaf physiology   Measure leaf-scale properties
1 cm–10 m   Measure cell-scale fluxes

Fig. 2. Scales of organisation in a research project, corresponding to scales of organisation in nature (organism, ecosystem, landscape, region). Reproduced from Jarvis (1995) with permission.

disciplines have different tools and vocabularies. The questions that can usefully be posed at each level are different. For example, at the higher levels the question might be 'what is the rate of change in the forest area?' and 'how long does it take for abandoned farm to revert to forest?', whereas at the physiologist's level the questions might be 'will the stomatal behaviour change when the plants are growing at an elevated CO_2?' and 'will turnover of carbon in the soil be accelerated by the higher temperatures?' The benefits of recognising the hierarchy are as follows. (i) By formally declaring the hierarchy, it is possible to provide context and relatedness to what would otherwise be many disparate studies, and for the entire community of scientists to have fruitful discussion about the overall problem, which involves changes and sensitivities at *all* scales. (ii) Scaling-up procedures can be checked by reference to work at the next highest level of organisation; for example, models of plant photosynthesis and respiration are usually calibrated using biochemical and physiological data and used in the forward mode to make predictions about the annual totals of carbon assimilated (Grace *et al.*, 1995; Lloyd *et al.*, 1995). Such models can be spot-checked by flying appropriate CO_2 and water flux sensors over the landscape for selected days. (iii) The perceived natural hierarchy, as defined in the diagram, is subjected to scrutiny by the entire community of scientists, and some parts may be revealed to be redundant, whilst new parts may be identified as relevant and deserving study. This is important as the final predictions about the combined effect of land-use change and climate change on the forested region as a whole are likely to be based on a model that has a similar structure to this diagram.

Are such procedures necessary and desirable? As we have seen, they are contrary to the traditions established in physics and chemistry, and they fly in the face of established science because they ask scientists to look beyond the bounds of their discipline and to interact outside these limits. It will be an adventure, and it is still too early to say what the outcome will be.

References

Allen, T.F.H. & Star, T.B. (1982). *Hierarchy: Perspectives for Ecological Complexity.* Illinois: University of Chicago Press.

Collins, N.M. & Wood, T.G. (1984). Termites and atmospheric gas production. *Science*, 224, 84–85.

Ehleringer, J.R. & Field, C.B. (eds.) (1993). *Scaling Physiological Processes: Leaf to Globe.* New York: Academic Press.

Grace, J., Lloyd, J., McIntyre, J. *et al.* (1995). Carbon dioxide uptake by an undisturbed tropical rain forest in South-West Amazonia. *Science*, 270, 778–780.

Jarvis, P.G. (1995). Scaling processes and problems. *Plant, Cell and Environment*, 18, 1079–1089.

Lakes, R. (1993). Materials with structural hierarchy. *Nature*, 361 511–515.

Levin, S. (1992). The problem of pattern and scale in ecology. *Ecology*, 73, 1943–1976.

Lloyd, J., Grace, J., Miranda, A.C. *et al.* (1995). A simple calibrated model of Amazon rainforest productivity based on leaf biochemical properties. *Plant, Cell and Environment*, 18, 1129–1145.

Luan, J. (1995). Simulation of forest ecosystem dynamics, with respect to the problem of hierarchy. PhD thesis, University of Edinburgh.

Luan, J., Grace, J. & Muetzeldeldt, R. (1996). A hierarchical model of forest growth and succession. *Ecological Modelling*, in press.

McNaughton, K.G. & Jarvis, P.G. (1991). Effects of spatial scale on stomatal control of transpiration. *Agricultural and Forest Meteorology*, 54, 279–302.

Norman, J.M. (1993). Scaling processes between leaf and canopy levels. In *Scaling Physiological Processes: Leaf to Globe*, ed. J.R. Ehleringer & C.B. Field, pp. 41–76. New York: Academic Press.

Oechel, W.C., Hastings, S.T., Voulitis, G., Jenkins, M., Riechers, G. & Grulke, N. (1993). Recent changes of arctic tundra ecosystems from a net carbon sink to a source. *Nature*, 361, 520–523.

Passioura, J.B. (1981). The interaction between the physiology and the breeding of wheat. In *Wheat Science – Today and Tomorrow*, ed. L.T. Evans & W.J. Peacock. Cambridge: Cambridge University Press.

Rowe, J.S. (1992). The integration of ecological studies. *Functional Ecology*, 6, 115–118.

Waddington, C.H. (1977). *Tools for Thought*. St Albans, UK: Paladin.

Zimmerman, P.R., Greenberg, J.P., Wandiga, S.O. & Crutzen, P.J. (1982). Termites: a potentially large source of atmospheric methane, carbon dioxide and molecular hydrogen. *Science*, 218, 563–565.

G.R. SQUIRE and G.J. GIBSON

Scaling-up and scaling-down: matching research with requirements in land management and policy

Introduction

Knowledge of land and rural industries is needed by a wide range of people and organisations, from those who work the land to the administrations that determine the broad policies for use of land. Government and industry support research institutes, universities, colleges and private laboratories to produce such knowledge. This paper explores the match between the knowledge that farmers and administrations need and the knowledge gained from research. Features of rural systems, the questions asked regarding them and the bodies that supply the answers are briefly described. The methodology of research is then scrutinised to suggest how research, implemented ostensibly to solve problems, retreats from the scale and complexity of the problem in order to get results of publishable precision. Consequently, the answers provided by the research are, by themselves, often unable to solve the original problem. It is argued that reversing reductionism through a 'system model' is a valuable means of combining knowledge from a range of scales so as to elucidate processes operating at the original scale and degree of complexity. The system model must, however, incorporate the uncertainties in prediction that arise from intrinsic variability in biological and economic systems, from noise in the physical environment and from incomplete knowledge of the scientific concepts underlying the model. It is concluded that research would be better matched with requirements if (i) policy-makers and scientists together set the scale and complexity of any enquiry; (ii) some measurements were made as near as possible to the scale and complexity of the original problem; (iii) steps were taken to fill persistent gaps in knowledge; and (iv) much more attention was paid to discovering the scaling laws governing the main processes in biological and economic systems (see Marshall, Crawford and Porter, this volume).

Systems, questions and research enterprises

Systems

Rural industries, such as agriculture, forestry, waste disposal and tourism, are complex systems. They comprise physical resources, biological organisms, work, profit, people and societies. They use the energy of solar radiation to convert CO_2, water and nutrients to complex biological substances through the biochemistry of plants and animals. Many factors conspire to determine the organisms that can survive or be cultivated. Environmental factors interact with socio-economic factors such as the local availability of raw materials, the nature of local and worldwide markets for produce, urban pressures, and the policies and regulations imposed by government (Fig. 1). The range of products, in turn, determines the nature of the farms and estates and the vitality of the rural economy.

These rural systems, as all biological systems, show *heterogeneity* and *coupling*. Heterogeneity is evident from the variability between individual plants in a field, from the patchwork of fields and rough land in a catchment, and from the different climatic and economic regions of a country. Coupling is evident from mutual interactions between soil and plants, hosts and parasites and, on the regional or global scale, vegetation and the atmosphere. Economic systems exhibit coupling through the interactions of producers and consumers, mediated by the laws of supply and demand. Heterogeneity and coupling increase the complexity of a system and can endow it with behaviour that changes with scale (Marshall *et al.*, this volume). Nevertheless, heterogeneity and coupling are often ignored when questions are asked about rural systems, and ignored still when research sets about trying to answer them.

Questions

Questions are asked at two broad temporal and spatial scales: that of the farmer and land manager, and that of national and local policy. At the scale of management, questions are mainly about the short-term use of fields, hillsides, farms and plantations: they concern processes in the upper part of Fig. 1. They seek answers regarding what to grow, how to grow it and the profit it will bring. At this scale, the physical environment, the biological organisms, the economics of the enterprise and the social conditions can all influence the answers. Changes in land use are made at this scale, even though the stimulus for the change might come from economic or environmental influences at regional, national and global scales.

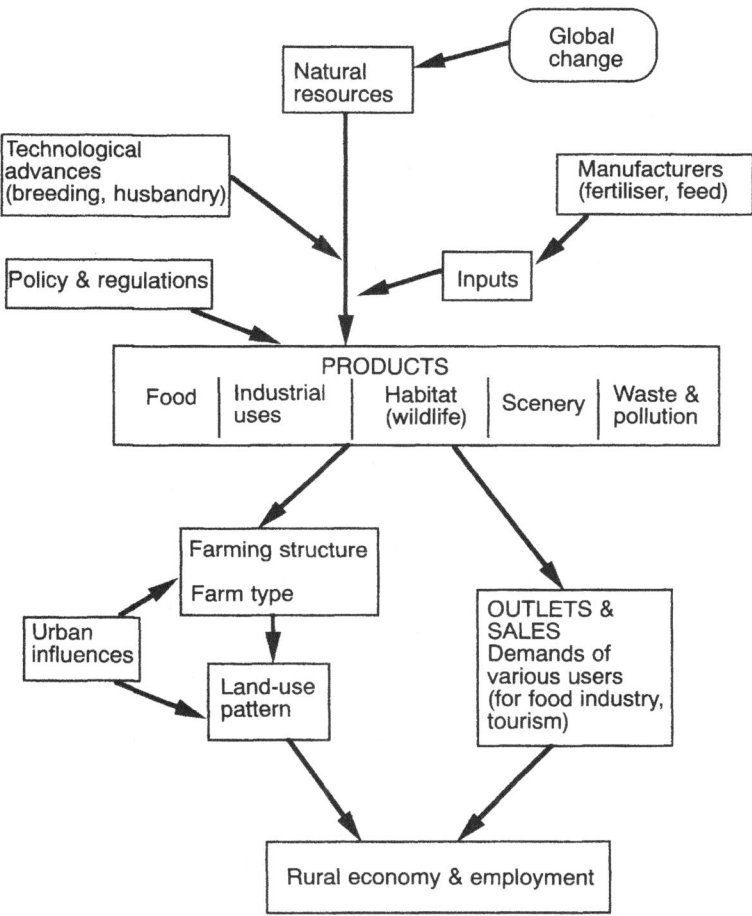

Fig. 1. Representation of the main components of a rural system.

At the scale of policy, questions mainly concern issues of economics and societies – feeding the population, the structure of farms, inputs in relation to markets and sales, employment, and, secondary, 'knock-on' effects to industry and the wider environment. The domain of these questions is the lower part of Fig. 1. The answers, nevertheless, require some biological and physical information. This information may not be important when food is abundant and the choice of crops is determined by subsidies and regulations. It is important, however, when subsistence agriculture dominates the economy, when agricultural products are intrinsically very profitable (as in many tropical plantation economies), when agriculture may harm individuals or the quality of their lives or when a disaster makes the food chain uncertain or unsafe (e.g. nuclear fall-out from Chernobyl).

At the scale of management, the answers come from the managers' own experience, often with advice from specialists. At the scale of policy, administrations might ask questions through specialist advisers, but the answers ultimately come from economists, scientists and technologists. At both scales, the source of much of the information is usually a research organisation.

Research programmes and organisations

The work of a typical research organisation with a remit for land use covers the flows of resources through primary and secondary production to products such as food, industrial material, habitat and scenery. Studies of the economy and the people may be included within such enterprises (Fig. 2). The programmes of the Scottish Office Agriculture Environment and Fisheries Department (SOAEFD) and land-based research in Canada are each integrated in this way, as are the research and development enterprises serving fruit and beverage crops in some tropical countries. In other places, biological and environmental science, economics and social science are managed and funded separately. Irrespective, however, of which organisation funds which subject in Fig. 2, many answers require that quantities defined in physical units, such as tons per hectare or nutrient concentration, are compared with, or translated into, an economic or social unit, such as the ranking of a habitat for animals, financial profit or level of employment. The conversion of these 'currencies' is a major difficulty that has not been resolved.

The research programme in Fig. 2 is classified in terms of the object of study. Research can also be categorised by the immediacy of its impact on practical problems. Many sponsors encourage research into fundamental questions of natural philosophy, or long-term studies that may eventually yield practical benefits but which have no immediate applications, as well as research on problems requiring answers in the short term. It is nevertheless often difficult to classify a research project as being 'fundamental' or 'applied', since solving applied problems often requires a measure of fundamental research. However, most public sponsors of research and development in the UK now do so using terms such as 'basic', 'strategic', 'applied' and 'experimental development'.

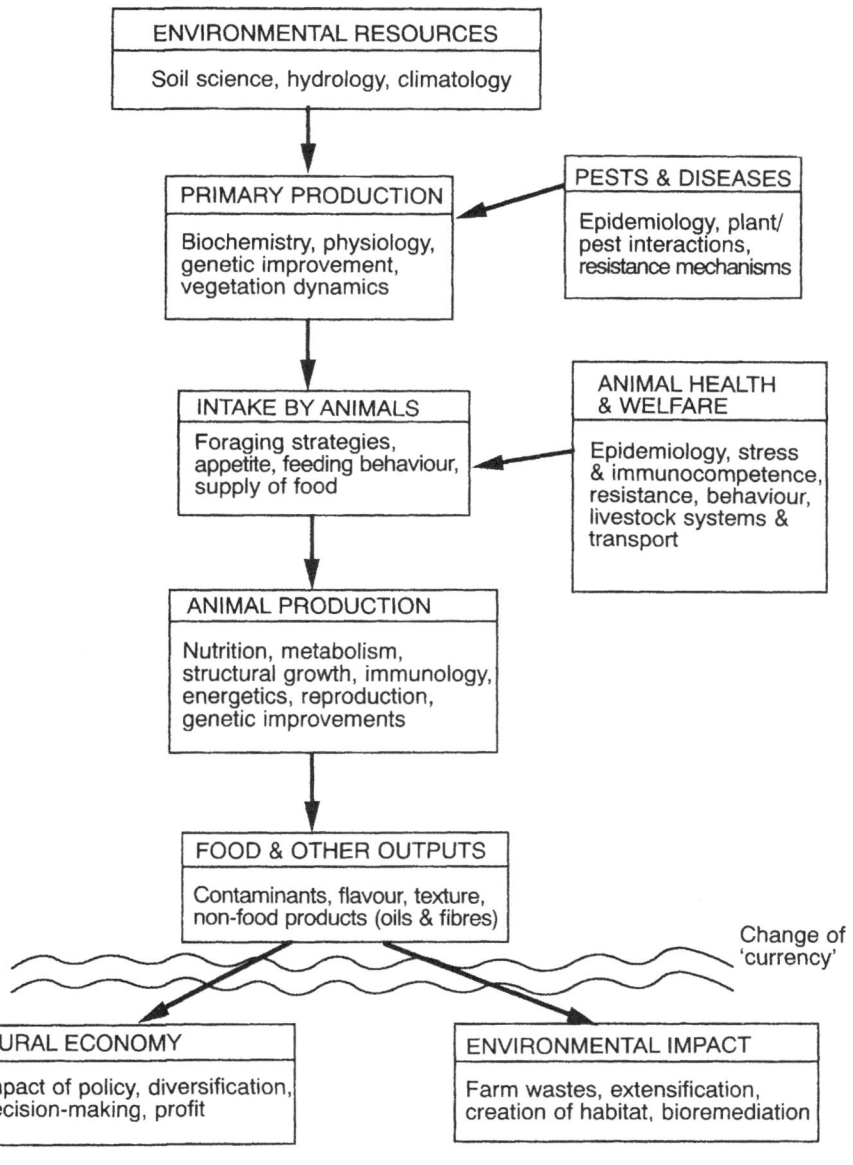

Fig. 2. Topics in a mainly biological research programme on organisms, land and land usage.

Scale of measurement or of interest, however, is largely independent of the subject categories in Fig. 2 and of a basic/applied classification. For instance, moving down Fig. 2 need not imply increasing scale, at least until the concept of land use is reached, when study is usually limited to the scale of the plant stand or the herd, and above. Neither should moving down the figure imply that the research becomes more applied. Some research on soil or molecular biology is directly applied, while research on the structure of complex economic and ecological systems is fundamental (Kay, 1991).

A research programme, and the knowledge it yields, can be envisaged, therefore, as lying in a three-dimensional 'space' parameterised by subject, scale and applicability (Fig. 3a). No point of this space is redundant since, in theory, there is no intrinsic limit to the knowledge that science can produce in those subjects that it can legitimately address (Medawar, 1984). However, at any time, there will be volumes of the space that cannot be penetrated because of some impediment in methodology at those combinations of subject, scale and applicability.

A similar approach can be employed to classify policy questions by embedding them in a three-dimensional space with axes representing the notions of subject, scale and precision (Fig. 3b). In this case subject would identify the bio-physical and socio-economic processes with which the question deals as well as those that have implications for its answer; the scale would represent that level at which processes should be considered, while precision would be that required to provide an answer sufficiently definitive to guide management or policy.

Examples of land-use problems

It is often the experience in rural systems that many questions of husbandry and policy, when embedded in the space of Fig. 3b, correspond to combinations of subject and scale that are not being addressed in any research project (i.e. the corresponding region is empty in Fig. 3a). This is particularly true when the stated question is economic or social, while its answer requires the use of bio-physical knowledge that may not be easily 'convertible' into the socio-economic currency. Several examples of current concern serve as illustration.

Set-aside is a major and abrupt change of land use in the UK, where it now forms the third largest area of arable land after wheat and barley. Land previously used for crops is used either temporarily or permanently to support another type of non-food vegetation in order to reduce perceived food surpluses. There can be several objectives and constraints in the use of set-aside. For instance, it might be required that

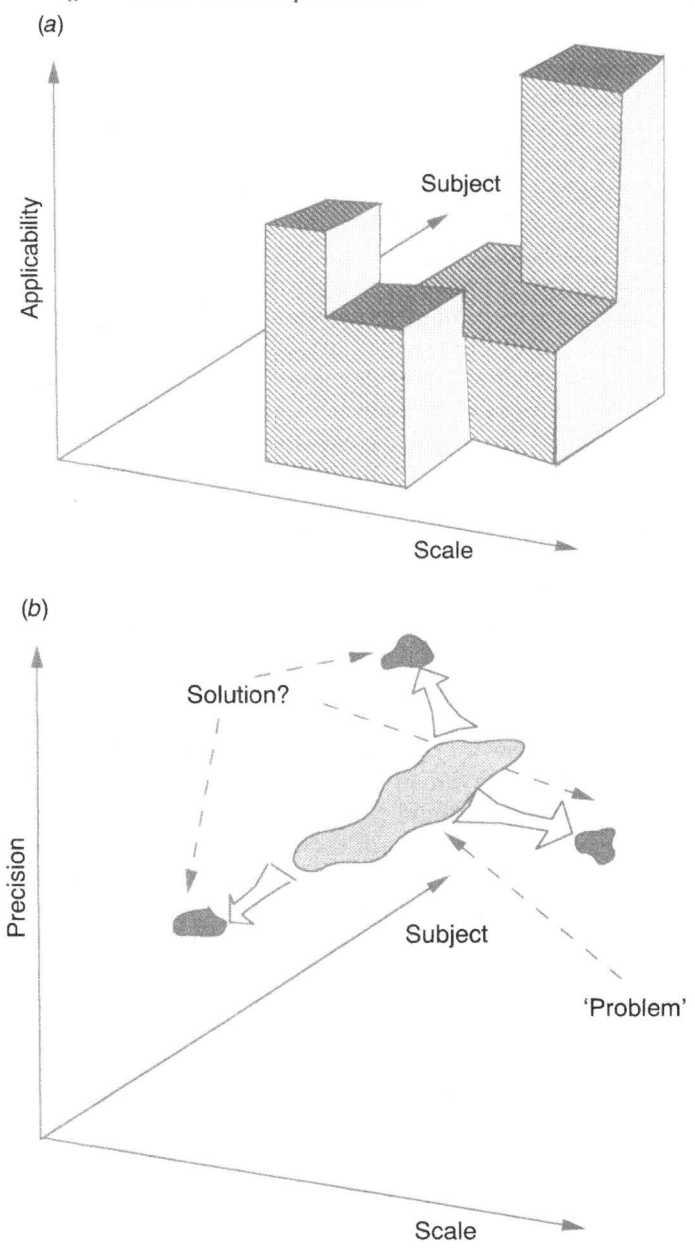

Fig. 3. Schematic representation of a research programme (*a*) and a biological system (*b*) in three-dimensional space.

the new vegetation should not lead to greater losses of nitrate or water into neighbouring water courses, should not increase the risk to neighbouring fields of colonisation by weeds, pests or pathogens, and should not adversely affect the performance of arable crops grown on the land at a future time. In some areas, the aim may be to provide a haven for scarce wildlife. To achieve these objectives, choices must be made at several scales. First, the administration must set conditions and incentives for transferring arable fields into set-aside. Second, farmers must decide whether incentives are adequate and conditions not too restrictive, and then which fields to convert. Finally, a range of farming operations must be selected, which affect variously the structure of the soil before the vegetation cover develops, the botanical composition of the cover, the timing and frequency of cutting so as to reduce seeding or promote certain assemblages, and the subsequent treatment of the soil after set-aside, for example to reduce weed growth.

Conversion to set-aside is implemented at the scale of the machine on the field, but its effects are seen at both smaller and larger scales. The movements of nitrogen and water out of the set-aside field are determined by processes at the scale of the pore and crack in the soil and depend on soil physical structure and the distribution of organic matter, plant structures and soil microbes. The influences of weed seeds, pathogens and vertebrates (such as rabbits, for example, which thrive in set-aside) operate at many scales, from the individual plant outcompeted, invaded or eaten, to the movement of populations over many kilometres. These organism-scale and regional-scale processes involve complex heterogeneity and couplings, and decisions regarding the configuration of set-aside fields in the landscape should ideally be made at a scale much larger than most individual farms and estates!

Set-aside, therefore, presents many questions whose answers demand knowledge at scales from the soil pore, through the field and catchment, to the country. The organisations that fund research into set-aside, mainly the Ministry of Agriculture, Fisheries and Food (MAFF) and SOAEFD, needed information and advice quickly and opted to commission research organisations to implement an extensive set of trials at different sites. These trials are mostly at the scale of the field and farming operation. They are top-down, imposing various treatments and observing the results on a range of quantities such as nitrate in the field drains, composition of the vegetation and yield after set-aside (Clarke, 1992). This approach is arguably the most economical and pragmatic option for getting some knowledge quickly. When finished, the trials will undoubtedly provide general indications of best practice, yet they are necessarily repetitive and inevitably are giving some very variable, unexpected and inexplicable results. They might also say little about processes occurring at scales above the field.

The reason for choosing empirical trialling was probably the paucity of knowledge about processes in the vegetation and soil occurring at the scale of the plant community and rhizosphere. For this reason, much other research into land and vegetation consists of field trials that hardly refer to the great amount of scientific knowledge at the scales of the soil pore, cell, leaf and organ. Examples include research on practices to reduce leaching of nitrate from arable land, and on restoration and regeneration of diversity in managed land. Research on critical loads of atmospheric pollutants – carried originally to answer questions about water quality – cannot, for similar reasons, be applied directly to answer questions about pollutant loading on vegetation.

Other parts of Fig 3*a* can be identified that are also scarcely inhabited, yet very relevant to matters of land usage. One example concerns the way individual farmers and managers make decisions on the basis of the information they have. Decisions on land use are made for many reasons other than financial gain and may appear quite illogical when judged only by scientific or economic criteria (Wynne, 1992). The following argument considers how gaps in knowledge occur within such crucial areas.

Scaling-up and scaling-down in research methodology

The biological and economic systems where study might provide answers to questions of land use are complex and have much uncertainty associated with them. Invariably, the problem is too difficult to solve to the required degree of precision at the existing scale and complexity. The many variables needed to specify the system's state, and the uncertainty implicit in its behaviour, mean that prohibitive amounts of data would be required if reproducible results were to be reached from experimentation.

Faced with this problem, most scientists identify those facets of the system that fall within their own areas of interest and restrict their study to them. Frequently, this restriction in scope, and in the dimension of the system being studied, is accompanied by a change in scale from that of the original phenomenon which concerns the farmer or policy maker to one that is more convenient for the scientist. The scale is sometimes moved up, at other times down. For example, if the phenomenon relates to the stand, field or ecosystem, an experimenter might impose some form of environmental control that reduces the scale – as when a plant or piece of turf becomes the subject of attention rather than the stand or hillside. In doing this, much of the heterogeneity and coupling at the original scale is lost. Or if the phenomenon is in the detailed structure of the soil or genome, an experimenter might move up in scale (to estimate parameters governing flows of water and gas in bulked soil, for example), thereby losing some of the fine-scale heterogeneity and coupling. Examples of this reductionist approach are common throughout science (Grace, van Gardingen & Luan, this volume). Food microbiologists seek to understand the growth characteristics of bacteria in food by measuring bacterial growth in laboratory media (Buchanan, Stahl & Whiting, 1989). These media may reproduce the environment of the food in terms of factors such as pH, temperature and salt content. However, they are not capable of representing the effects of texture, or the heterogeneity that exists in many foodstuffs. In this instance, a loss of realism that results from reductionism can be measured to some extent by comparing the prediction with opportunist observations of the real system. However, when considering questions of land-use policy, such observations are unlikely to be available. Indeed, the purpose of the science is to anticipate the nature of such observations.

The reductionist approach is illustrated in Fig. 3b. The subspace occupied by the original question – the 'problem' of Fig. 3b – is projected to one or more regions of limited scope and scale for which it is feasible to achieve some specified precision. This latter degree of precision is often that which is sufficient to allow demonstration of statistical significance through experimentation and to result in reproducible, publishable science. From this step, however, the original phenomenon and the scale-dependent methods used to investigate its various projections often become confused. The methods themselves might become the phenomenon in the mind of the experimenter. ('Illusion' and 'measurement' have the same root!) Consequently, the boundaries of the investigations, in the space of Fig. 3b, are drawn ever tighter and farther away from the original phenomenon.

In order for the science to impact on policy, then, it is necessary to re-constitute the behaviour of the real system from its several facets that have been studied through the reductionist approach – reversing the process illustrated in Fig. 3b – while being able to understand the re-constituted system to the required degree of precision.

Reversing reductionism

System modelling is one approach to tackling this problem, whereby the behaviour of a system is synthesised from a knowledge of component processes by representing these processes and their interactions mathematically. The approach has become attractive to scientists. High-quality graphics allow results to be summarised and presented as colourful user-friendly computer packages that can be marketed as decision-support systems.

However, this *post-hoc* reversal is not always sensible, particularly if the precision at which the scientists have worked has been based on criteria other than the need to answer the question as posed. As an illustration, consider the following hypothetical example, based on real systems currently being studied. Consider a population of parasites that exits in two locations: on a grass sward and within a host species that grazes the sward. Further assume that the dynamics of the parasite population in each location (in the absence of any migration between them) are described by a simple logistic equation and that transmission between locations is controlled by two rate parameters: m_s, which characterises the rate of transmission between sward and host, and m_h, which characterises the rate at which the parasite is excreted from the host to the sward. For the hypothetical example, it is assumed that m_s is proportional to a time-varying climatic factor C. Figure 4a is a schematic representation of this system. Suppose that the motivation for studying such a system – the 'policy question' – is to determine from observations of the climate, C, those times at which the parasite population within the host (H) exceeds some critical value, perhaps to aid pharmaceutical intervention.

Following a reductionist approach, a scientist might focus his attention on the relationship between m_s and C – the 'scientific question' – developing a model for this one aspect of the system. Suppose that experiments and modelling demonstrate that a relation of the form $m_s = kC$ accounts for 90% of the variance in the observed values of m_s. While the elicitation of such a strong correspondence may appear to be a successful outcome of an experimental programme, when viewed in the context of the whole system the picture changes. Figure 4b depicts the temporal variation of H for two realisations of the model system of Fig. 4a, with the uncertainty in the scientist's model represented as a stochastic variation in the parasite uptake rate. On day t, this uptake rate is calculated as $m_s(t) = k(1 + \varepsilon_t)C_t$, where ε_t is generated from a normal distribution of mean 0 and variance 0.1. For simplicity, $k = 1$, and C_t is generated from a moving average process. The simulations use two different sequences $\{\varepsilon_t\}$ but are otherwise identical. When the two trajectories in Fig. 4b are compared, it is apparent that, while they exhibit qualitative similarity, they provide contradictory answers to the policy question, with marked differences in the position of peaks

(a)

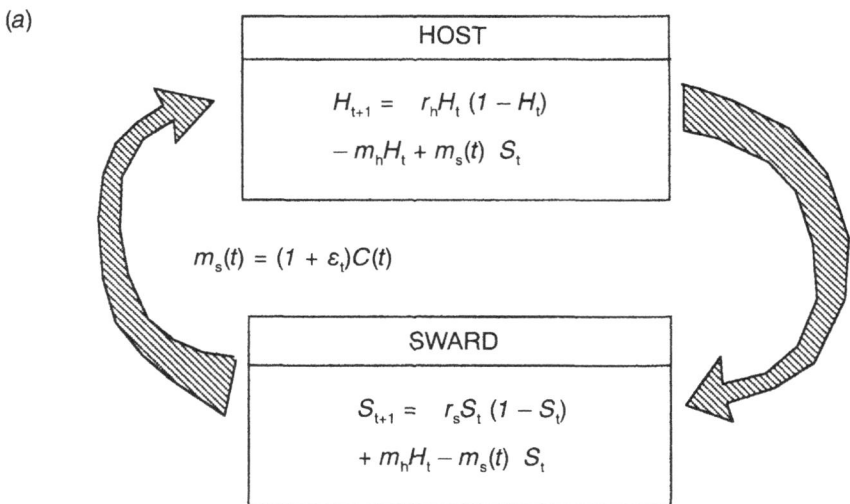

(b)

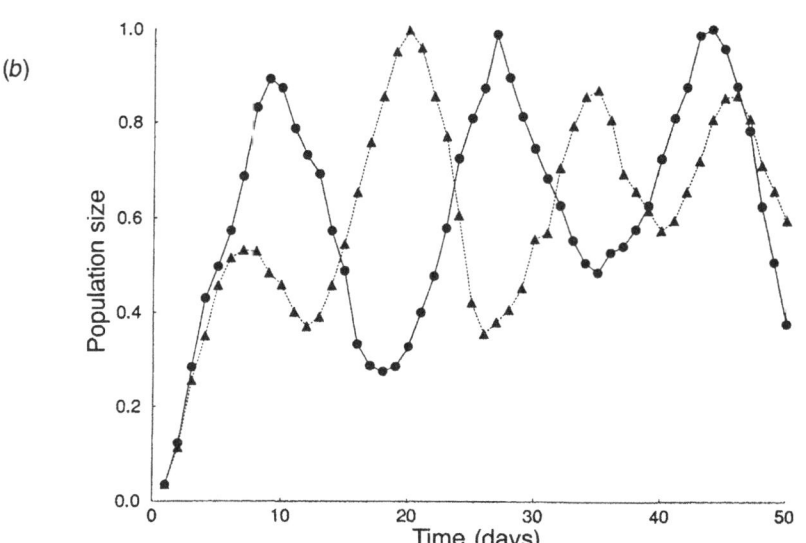

Fig. 4. (a) Schematic diagram of hypothetical host/parasite sward system and (b) the parasite population dynamics within the host for different stochastic realisations of the model: 1st realisation (●) and 2nd realisation (▲).

predicted by the respective models. Therefore, a model that might appear highly effective in the context of one particular aspect of a system may be inadequate when viewed as a component of the larger system, preventing effective reversal of the reductionist approach to enable questions at the policy level to be answered.

A real example of how uncertainty in submodels can seriously bias the predictions of a system model was discussed by Elston and Glasbey (1991). They considered the 1980 Agriculture and Food Research Council (AFRC) energy model, a decision-aid giving deterministic predictions of animal weight gain as a function of feed intake. It showed that, once the uncertainty in the estimated parameters of submodels was taken into account, the deterministic predictions of expected weight gain for a given diet could be biased by up to 20%. Furthermore, when outcomes for individual animals were compared with the prediction, biases of up to 150% were observed.

While these examples do not deal with policy questions at the national level, the points they illustrate are equally valid, if not more so, on that scale. There are several examples in the published literature, and among current research projects, where the systems approach has been applied, through interdisciplinary initiatives, in an attempt to answer policy questions. Examples include the Scottish Climate Change Initiative (1996), the Scientific Committee on Problems of the Environment (SCOPE) initiative in Australia (Freeman, 1992), and the ongoing North Eastern Land Use Project (NELUP), which is nearing completion at Newcastle University, UK. While the full impact of such research has yet to be measured, there has been a tendency for models developed within such initiatives to provide a framework for further research, rather than provide answers for policy-makers, a view expressed by Goodall (1983), who said of the SCOPE initiative:

> It is not suggested that the model is by any means perfect, that it incorporates the whole of relevant existing knowledge, or that it can be applied in practice without modification. It does, however, provide a valuable framework for further advance; and it is hoped that a full account of experience in the project will be of value to others who may be contemplating similar large-scale interdisciplinary modelling tasks.

This suggests that the main impacts of at least some initiatives may be within the world of science. The following section considers the question of how such large-scale interdisciplinary initiatives involving system modelling should be tackled in order to maximise their chances of breaking out of the world of science and influencing government, policy and society.

Matching policy with science through system modelling

There is no doubt that a system approach is of value in transferring the knowledge gained from basic research to the domain of the policy-maker. Less clear is how the approach should be applied in order to maximise impact on policy. This final part of the paper discusses some issues of general relevance that have arisen from recent research (Gibson *et al.*, 1994) on the feasibility of applying the systems approach to understanding the Scottish rural economy. This work considers how system models can best be used to answer questions of policy and to specify the type of strategic research that is likely to have most impact on policy.

Anticipation of questions

It is not feasible to define a system model of a complex entity, such as the rural economy of a region, that can provide answers to all the questions regarding it that managers and policy-makers might ask. This would require that a myriad of biological, economic and social processes be represented within a single model, with levels of precision that are unrealistic in the light of current scientific knowledge, existing data or the inherent unpredictability of the real world. In practice, it is better to design system models to suit particular questions, which can demand very different treatment. So before embarking on a multidisciplinary system-modelling initiative, the policy objectives should be stated in terms of questions that the model should be capable of answering, together with an indication of the precision required for answers to be useful.

Having identified the aims of a multidisciplinary initiative, it is necessary to translate these into requirements for basic science, so that existing relevant research and current shortfalls in scientific knowledge can be identified. A means of making this connection is to use a system approach to specify those processes that impinge on the policy issues, and the interactions and coupling that exist between the processes. The conceptual model produced by such an exercise can go far in determining the nature of component projects within a co-ordinated initiative, as well as the collaborative links between them, and how existing knowledge can be best exploited. By first designing the 'big picture' in this way, the scope of processes studied under the reductionist approach has a better chance of matching that of the real system. Successful reversal of reductionism, however, requires more than simply the design of an overall system, or framework, within which to embed the basic research. It also requires a thorough analysis of the behaviour that the system can be expected to exhibit. Only when this has been done can appropriate research objectives for the component projects be specified. Indeed, such an analysis may reveal that the scientific goals implied by the policy goals are unrealistic, requiring some relaxation of the policy-makers' demands.

Taking account of uncertainty

What should be considered within such an analysis before detailed scientific work is undertaken? One issue that must be examined is the effect of uncertainty on the behaviour of the system. The question of uncertainty within system models is one that has seldom been given the attention it deserves, and rarely at a sufficiently early stage in the course of the research projects. In a recent report to the AFRC (Gibson *et al.*, 1993), Biomathematics and Statistics Scotland surveyed the role of uncertainty in modelling within the Scottish Agricultural Research Institutes, identifying four main sources of uncertainty:

- intrinsic stochasticity – which arises in systems whose behaviour is governed by random events, as in the dynamics of small populations
- uncertainty in extrinsic factors, such as climatic variables in deterministic crop models
- uncertainty in model parameters estimated from data sets
- uncertainty in scientific knowledge and the appropriate form of the model.

The importance of the first three was illustrated earlier in this paper. The last of the sources of uncertainty has recently been considered by Draper (1995), who formulated a Bayesian framework for dealing with uncertainty in models. He carried out two case studies (forecasting world oil prices and the probability of failure of a subsystem within the space shuttle) that demonstrated the unrealistic confidence which may be placed in predictions if model uncertainty is ignored.

Through a careful analysis of the sensitivity of the conceptual system model to these sources of uncertainty, a number of questions that have direct bearing on the success of the initiative can be answered. It should be possible to state the levels of certainty with which subsystems should be representable, in order for the whole system model to be capable of answering questions of management and policy. The analysis may warn against the use of inappropriate methodologies and indicate the need to develop new methodologies, or carry out further experimental work.

Such analysis enables the research to be given priorities and provides a guide for the timetable and allocation of effort within an initiative.

The details of such an analysis need not be quantitative, though standard approaches such as sensitivity analysis undoubtedly have a role to play. Indeed, it may not be possible to quantify certain aspects of the system, and the analysis may involve a degree of qualitative judgement. Consider the hypothetical host/parasite system discussed previously, in which the sensitivity of the system behaviour to moderate stochastic uncertainty in the uptake rate was demonstrated. The choice of statistical distributions for this uncertainty must be made by the system analyst, perhaps based on historical information or the opinions of experts. It is vital that the sources of uncertainty are not ignored simply because they cannot be easily quantified. It would be informative to apply this kind of analysis to more complex models of real host/parasite systems, such as that developed by Roberts & Grenfell (1991; 1992) for nematodes in ruminant populations. Knowledge of the effect of stochastic uncertainty in the uptake and excretion of the parasite and the immune response of animals could provide valuable guidance to future experimental programmes and indicate the context in which the model could be used as a reliable decision-aid.

Conclusions

The following conclusions are derived from our experience, particularly as a result of recent work within the SOAEFD-funded feasibility study of a system approach.

Much greater interaction is needed between those who formulate policy and those who plan and carry out the research. The initial analysis of any system should indicate the scale and scope of the research to be commissioned to answer the problem. If a problem is so complex that an answer with acceptable precision is unlikely to be found, then this conclusion should be communicated to policy-makers.

While it is accepted that the scope and scale of a problem might have to be adjusted in order to render it tractable, some link should always be kept, by making some measurements at or around the scale and complexity of the original phenomenon. In particular, models should be tested at a range of scales including, as near as possible, that of the phenomenon.

Steps should be taken to fill persistent gaps in knowledge at any scale or level of complexity that have been found to hinder progress with a series of problems. In relation to land and vegetation, for instance, gaps have arisen at the scale of the rhizosphere and of the plant community, and in the subject of why farmers and managers make the choices they do.

The gap between bio-physical and socio-economic research is an extreme case, caused mainly by the respective researchers tendencies to decrease the scale in bio-physical research and increase it in socio-economic approaches. Ways must be found to close the gap, while retaining scientific method and rigour.

Research might concentrate more than it does on identifying the scaling laws governing the process being examined (Marshall *et al.*, this volume), and on defining and using conservative factors and processes that are not strongly influenced by scale. (The base temperature and temperature response of development in plants and some invertebrates is an example of a conservative trait.)

Finally, it is our opinion that a system approach should not in any way restrict science. Rather it should free scientific thought from the very tight boundaries that it often imposes on itself, at scales far from those at which most biological processes operate on land and in vegetation.

Acknowledgements

This work was funded by the Scottish Office Agriculture Environment and Fisheries Department.

References

Buchanan, R.L., Stahl, H.G. & Whiting, R.C. (1989). Effects and interactions of temperature, pH, atmosphere, sodium chloride and sodium nitrite on the growth of *Listeria monocytogenes. Journal of Food Protection*, 52, 844–851.

Clarke, J. (ed.) (1992). *Set-aside*. Farnham, UK British Crop Protection Council (Monograph 50).

Climate Change Initiative (1996). Climate change – from impact to interaction. *Agricultural and Forest Methodology*, 79, in press.

Draper, D. (1995). Assessment and propagation of model uncertainty. *Journal of the Royal Statistical Society series B*, 57, 45–98.

Elston, D.A. & Glasbey, C.A. (1991). Variability within systems models: a case study. *Agricultural Systems*, 37, 309–318.

Freeman, G. (1992). The SCOPE regional model. *Agricultural Systems and Information Technology*, 4, 2, 7–8.

Gibson, G.J., Glasbey, C.A., Kempton, R.A. & Elston, D.A. (1993). *Modelling in the SARIs: The Role of Uncertainty*. Report to Agricultural and Food Research Council. Edinburgh: Scottish Office Agriculture and Fisheries Department.

Gibson, G.J., Crawford, J., Sibbald, A., Aspinall, R. & Doyle, C. (1994). *Systems Modelling and the Rural Economy: a Study into the Feasibility of Using Mathematical Models as a Decision Aid to Policy Makers*. Edinburgh: Office Agriculture and Fisheries Department.

Goodall, D.W. (1983). Introduction. In *Pastoral and Social Problems in a Semi-arid Environment – A Simulation Model*, ed. T.G. Freeman & P.R. Benyon, p. 4. Melbourne: CSIRO/UNESCO.

Kay, J.J. (1991). A nonequilibrium thermodynamic framework for discussing ecosystem integrity. *Environmental Management*, 15, 483–495.

Medawar, P.B. (1984). *The Limits of Science*. Oxford: Oxford University Press.

Roberts, M.G. & Grenfell, B.T. (1991). The population dynamics of nematode infections of ruminants: periodic perturbations as a model for management. *IMA Journal of Mathematics Applied in Medicine and Biology*, 8, 83–93.

Roberts, M.G. & Grenfell, B.T. (1992). The population dynamics of nematode infections of ruminants: the effect of seasonality in the free-living stages. *IMA Journal of Mathematics Applied in Medicine and Biology*, 9, 29–41.

Wynne, B. (1992). Misunderstood misunderstandings: social identities and public uptake of science. *Public Understanding of Science*, 1, 281–304.

P.M. ATKINSON

Scale and spatial dependence

Introduction

To understand and monitor the environment at the global scale, it is necessary to measure physical properties of the environment over the entire surface of the Earth. Since the 1960s, meteorological satellite sensors with moderate and coarse spatial resolutions have provided remotely sensed images of reflected radiation for the entire surface of the Earth. To use such images it is necessary to measure the property of interest (such as vegetation amount) at the ground and relate these ground measurements to co-located pixels of the imagery, for example via a regression relation. Then the relation may be applied to the remaining pixels to estimate the property of interest over the extent of the image. A problem is that often the measurements made at the ground are much smaller than the image pixels. Then, one must scale-up the measurements made at the ground so that they represent the same area as each of the image pixels. This paper describes the geostatistical operation of regularisation, which is useful in understanding the process of scaling-up, and gives an example of scaling-up involving remotely sensed imagery.

Recent analysis of remotely sensed imagery has raised many questions relating to scale (for example, Woodcock & Strahler, 1987; Clark, 1990; Lam & Quottrochi, 1992; Atkinson, 1993a; Bian & Walsh, 1993; McGwire, Friedl & Estes, 1993). The objective here is to unravel such issues and present a clear exposition on (i) the meaning of scale; (ii) the effects of changing scale; and (iii) the potential strategies for tackling the problems associated with comparing measurements made at one scale with those made at another.

The importance of scale to the study of continental and global forms and processes was heightened when the first astronauts returned photographs of the Earth. Such images prompted people to think about the limited extent and fragility of the environment, and the long-term effect that human interaction with the environment may have at the global scale. To understand and monitor what is happening to the environment at the global scale it is necessary to measure physical properties such as carbon flux, surface albedo and sea surface temperature over the entire globe.

The first satellite-borne sensors (such as the National Oceanic and Atmospheric Administration (NOAA) series of satellites, starting in 1960 with the Television and Infrared Observation Satellite (TIROS), and including the Improved TIROS Observation Satellite (ITOS), the NOAA series (NOAA 1 to NOAA 13) and the TIROS-N series) were intended primarily for observing the *atmosphere* and the *oceans* with relatively coarse spatial resolutions (here the term spatial resolution is used to mean the area of the patch of ground represented by each pixel in the remotely sensed image). For example, the Advanced Very High Resolution Radiometer (AVHRR) on-board the NOAA 6 to NOAA 13 series has a spatial resolution of 1.1 km (by 1.1 km).

In 1972, the first of the Landsat series of satellites was launched carrying a Multispectral Scanning System (MSS), which had a spatial resolution of 79 m. In 1982, the Landsat 4 satellite was launched carrying a Thematic Mapper (TM) sensor with a spatial resolution of 30 m, and in 1986 the Système Probatoire d'Observation de la Terre (SPOT) satellite was launched with a High Resolution Visible (HRV) MSS with a spatial resolution of 20 m. These sensors were intended primarily for *land* applications such as land cover monitoring (for example, Bauer *et al.*, 1979), and geological, geomorphological and vegetation mapping at *local* scales (although complex mosaics were produced at national scales (for example, Merson, 1983)).

The Landsat and SPOT satellite sensors are generally unsuitable for global-scale applications, mainly because of the very large amount of data necessary to provide complete coverage. However, in the mid-1980s, several researchers realised that the earlier meteorological satellite sensors are appropriate for mapping and monitoring properties, particularly those of vegetation, at the global scale (Justice *et al.*, 1985; 1991; Tucker *et al.*, 1985; Townshend *et al.*, 1991), and they undertook analyses to demonstrate the advantages (Justice, Townshend & Markham, 1987; Townshend & Justice, 1988).

Many properties in remotely sensed scenes, including atmospheric properties, vary appreciably over short times, so that often one cannot be sure what the remotely sensed data represent from the imagery alone. It is necessary to go into the field and measure the property of interest (for example, vegetation amount) at several places within ground resolution elements or GREs (the patch of ground that a pixel represents) (Dozier & Strahler, 1983). These observations may then be related to the corresponding pixels of the image (Fig. 1), a regression relation established and this relation extended to the remainder of the image to estimate the property at the unobserved locations at the ground (Curran & Williamson, 1985).

Problems associated with scale arise because the measurements made of the property at the ground may be made for an area that is very small in relation to the GRE that each remotely sensed image pixel represents. In many cases, the observations may be so small in relation to the image pixels that they may be approximated as points. The task then is to scale-up the measurements at the ground so that they match the size of the image pixels. For local and regional scale applications using Earth resources satellites, this is problematic and requires careful sampling design. For global applications, this is near impossible and requires (in addition to sampling design) international collaboration, extensive field-based measurement and considerable expense. Even then it is difficult to be sure that the problems associated with scaling-up have been adequately tackled.

In this paper, the problems associated with scaling-up are examined using a set of techniques for spatial analysis known as geostatistics (Journel & Huijbregts, 1978; Isaaks & Srivastava, 1989; Cressie, 1991). Geostatistics is based on and accounts for spatial dependence: the likelihood that observations close in space are more alike than those further

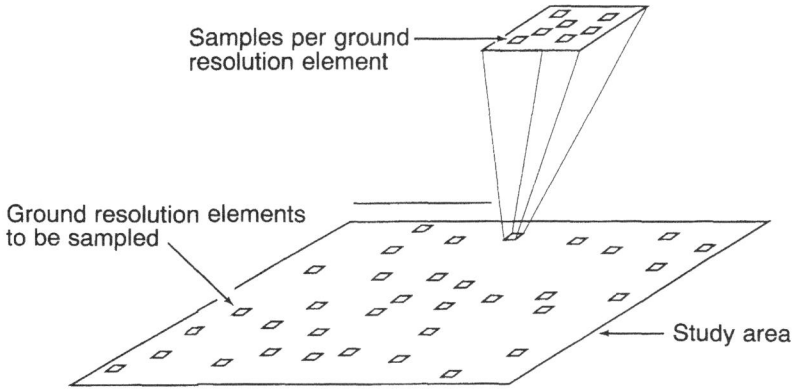

Fig. 1. Diagrammatic representation of two-level sampling of properties at the ground for relation with remotely sensed imagery. The ground resolution element refers to the area on the ground represented by a single image pixel (after Curran & Williamson (1985). *International Journal of Remote Sensing*, Taylor and Francis).

apart. Using a geostatistical framework, it is possible to design efficient sampling strategies that allow scaling-up.

Scale, measurement and spatial variation

Scale

The term scale can mean many things depending on what is described (Woodcock & Strahler, 1987; Lam & Quottrochi, 1992). Strictly, scale refers to the ratio of the size of a representation of an object to its actual size. In cartography, for example, a map has a scale that defines how large or small it is in relation to the real world. The convention is that a small-scale map has a relatively small size ratio (for example, 1:100 000) and a large-scale map has a relatively large size ratio (for example, 1:10 000) (Foody & Curran, 1994). This is often a source of confusion as 1:10 000 is only a large scale in relation to 1:100 000. Compared to 1:1000 and even 1000:1 it is a small scale. Further, to many people the ratio 1:100 000 seems to imply a larger scale (difference in size) than the ratio 1:10 000.

More generally, the term scale is used in place of the word size. We are familiar with phrases such as scale of operation, scale of analysis and scale of investigation. In these instances, the term scale is used simply to mean size, and it is implicit that we mean 'in relation to something else' (where that something else may be, for example, the everyday human scale). This contradicts the cartographic scale above. For example, a map that covers the globe has a small cartographic scale, but an investigation that covers the globe is large scale. This more general use of the word scale is adopted throughout since it avoids the ambiguities associated with ratios and fractions when describing scales of variation, forms and processes, and scales of measurement.

It is possible to distinguish scale in terms of the forms of underlying phenomena (and the processes that create them) and the sampling framework that is used to measure them. Thus, a tropical forest might extend across a large region and the processes that act upon it might extend over the same scale. Rainfall that covers the entire forest is a large-scale process. A different scale relates to measurement. If it is possible to measure only a small part of the forest, the analysis would be small scale. Then, variation (forms and processes) at larger scales would not be detected. If, alternatively, it is possible to sample the entire forest, the study would be large scale, and the larger scale variation may be detected. The interaction between the underlying forms and processes, and the sampling frame determines the nature and scale of the observed phenomena.

In the following two sections these two aspects of scale are considered. The scale of measurement is considered first because the scales of variation that are detectable depend upon it.

The scale of measurement

The scale of measurement depends on the sampling frame or sampling strategy. The sampling frame can itself be divided into the spatial or geometrical characteristics of each individual *observation* and the spatial coverage and spatial extent of the *sample*.

The support is the size, geometry and orientation of the space on which an observation is defined. The size of support is similar to the spatial resolution of remotely sensed imagery as used above to refer to the size of individual pixels. Therefore, the terms size of support and spatial resolution will be used interchangeably throughout.

The support is an important concept because it is inextricably linked to the scale of measurement (Moellering & Tobler, 1972; Thornes, 1973; Openshaw, 1977). The effect of the support on the scales of natural variation that are detected has been demonstrated most notably in remote sensing (for example, Clark, 1990; Sèze & Rossow, 1991). To understand how the support affects the scale of measurement and the scale of underlying spatial variation that is detectable from such measurement, one must examine how the support interacts with the underlying spatial variation to produce a single value.

In practice, values are derived through measurement over a support v, a finite element of space that has a specific size, geometry and orientation. Then, an observation $z_v(\mathbf{x}_0)$ can be treated as a realisation of the random variable $Z_v(\mathbf{x}_0)$, the spatial mean or integral of $Z(\mathbf{x})$ over v centred on $\mathbf{x}_0$. (Terms are defined in the glossary on p. 56.) Formally,

$$Z_v(\mathbf{x}_0) = \frac{1}{v} \int_{v(\mathbf{x}_0)} Z(\mathbf{y}) \, d\mathbf{y} \qquad (1)$$

where $Z(\mathbf{y})$ is the property Z defined on a punctual or point support. This means that the underlying spatial variation is averaged over the support to produce a single mean value. Note that all measurements of natural phenomena are made through integration in this way; strictly there is no such thing as a point or support of zero size.

The support may also vary in shape and orientation. For example, in remote sensing the support has a centre weighting function shaped like a bell such that underlying spatial variation at the centre of the support receives more weight than that towards the edges. This function is the result of the point spread function (PSF) of the sensor, which is caused, among other things, by atmospheric effects on reflected radiation between the Earth and the sensor. Where the support has an anisotropic shape, orientation may be important.

The support is only one aspect of the spatial sampling framework or sampling strategy. Three more aspects are the sampling scheme, sampling intensity and sample size. The sampling scheme refers to the spatial distribution of the sample observations. Examples are the random, stratified random, centric systematic and systematic (including square grid and equilateral triangular) sampling schemes. The sampling intensity refers to the number of observations in a unit area and the sample size is the number of observations in total. Together, the sampling scheme, sampling intensity and sample size define the spatial coverage and spatial extent of measurement; the spatial coverage refers specifically to the set of distances between pairs of observations in a given direction and the spatial extent refers to the largest distance between pairs of observations in a given direction.

Two aspects of the scale of the sampling framework have been identified. The first is the support and the second is the spatial coverage and spatial extent of the sample. The importance of these aspects of the scale of measurement will become clear when spatial variation is examined through spatial dependence.

Scales of variation and spatial dependence

The scale of variation or the scale of a particular form or process is simply its size. This is easier to conceptualise for objects than for continuous variation, but the principles are the same. In the simplest case, we may consider an image model in which discs (value 1) are distributed randomly over a background (value 0), with the restriction that no discs overlap (for example, Strahler, Woodcock & Smith, 1986). The scale of variation is analogous to the size (in one dimension) of the discs, with the largest scale of variation equal to the disc diameters. For most natural phenomena, however, spatial variation exists at a range of scales (Mandelbrot, 1982).

Spatial dependence is useful in understanding scale because (i) it simplifies our view of spatial variation; (ii) it identifies the scale of the underlying variation, forms or processes; and (iii) it provides a link between spatial variation and the sampling frame (and, in particular, the support).

Most environmental phenomena distributed spatially over the surface of the Earth are spatially dependent, at least at some scale. For example, spatial dependence at a variety of scales has been detected in geological properties (Journel & Huijbregts, 1978), soil properties (Webster, 1985), ecological properties (Rossi *et al.*, 1992) and, most recently, in remotely sensed images (Yoder *et al.*, 1987; Ramstein & Raffey, 1989; Wald, 1989; Webster, Curran & Munden, 1989; Cohen, Spies & Bradshaw, 1990; Bian & Walsh, 1993; Gohin & Langlois, 1993; McGwire *et al.*, 1993; Atkinson & Curran, 1995). Spatial dependence is intuitively necessary; if properties were spatially independent at all scales, all realisations or values at all places would be equal and there would be no form or structure.

We may represent spatial dependence with the variogram, a function that relates semivariance to lag. The semivariance is defined as half the expected squared difference between the random functions (RF) $Z(x+h)$ at a particular vector distance and direction of separation or lag h (Matheron, 1965). The variogram $\gamma(h)$ is then given by

$$\gamma(\mathbf{h}) = 1/2 \, \mathrm{E}[\{Z(\mathbf{x}) - Z(\mathbf{x} + \mathbf{h})\}^2] \qquad (2)$$

The experimental (or measured) variogram may be estimated for supports v of size $|\,v\,|$ by

$$\hat{\gamma}_v(\mathbf{h}) = 1/2 \, P(\mathbf{h}) \sum_{(i,j)\,\in D(\mathbf{h})}^{P(\mathbf{h})} \{z_v(\mathbf{x}_i) - z_v(\mathbf{x}_j)\}^2 \qquad (3)$$

where $D(\mathbf{h})$ is defined as $\{(i, j) \mid \mathbf{x}_i - \mathbf{x}_j = \mathbf{h}\}$ (ensuring that the lag distance between $\mathbf{x}_i$ and $\mathbf{x}_j$ is $\mathbf{h}$), and $P(\mathbf{h})$ is the cardinality of $D(\mathbf{h})$.

For most geostatistical procedures, a mathematical model must be fitted to the experimental variogram. One must check that the model is conditional negative semidefinite (CNSD) to ensure that all possible linear combinations of the RF described cannot be negative. For most applications, the model is selected from several that are known to be CNSD or 'authorised' for the dimensions of the space over which the RF is defined. The more common models are presented by Webster & Oliver (1990).

In most cases, the model fitted to an experimental variogram approaches and intercepts the ordinate at some positive finite value known as the nugget variance, c_0. The nugget variance results from measurement error and from extrapolating what may be an inappropriate model to the ordinate where there are no data. It has been suggested that where the observations are adjacent or overlapping (as with remotely sensed imagery) the nugget variance may be used as an estimate of measurement error (Curran & Dungan, 1989; Atkinson, 1993a).

The variogram identifies the scale of the measured spatial variation through its range and shape. The variogram of the discs in the above image model has a range that is equal to the diameter of the discs (Fig. 2a). If the distribution of the diameter sizes is made size dependent so that there are many small discs and few large ones, the shape of the variogram will reflect this (Fig. 2b). To understand how the underlying disc model affects the shape of the resulting variogram consider that each disc has its own variogram with a given range and that each of these variograms is superimposed to produce one average variogram for the entire area. Then the shape of the variogram reflects the distribution of ranges and consequently the distribution of disc sizes. This model extends to the description of processes and spatial variation in general.

Interestingly, the variogram in Fig. 2b describes a fractal surface (Mandelbrot, 1982). We can tell this (without knowing the underlying disc model) because the mathematical function that describes the variogram is the fractional Brownian model given by

$$\gamma(h) = 1/2\,h^{\theta} \tag{4}$$

where θ ranges from greater than 0 to less than 2. In Fig. 2b it is equal to 0.5. The fractal dimension D may be computed from this variogram model as

$$D = 2 - \theta/2 \tag{5}$$

The importance of fractals to issues of scale cannot be understated. While there is some controversy over how they should be computed (Klinkenberg & Goodchild, 1992) and their utility as yet seems to be limited to measuring surface roughness (Brown, 1987; Rees, 1992) and to simulation (Lovejoy & Schertzer, 1986), no other mathematical

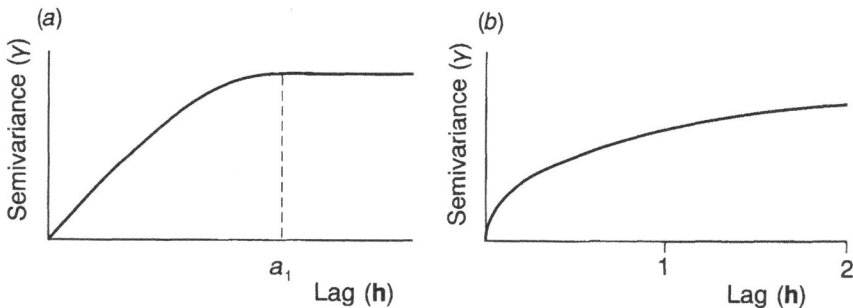

Fig. 2. Variogram of disc model with (a) discs of the same diameter and (b) discs of varying diameters.

model has provided as realistic a description of the underlying spatial variation and forms found in nature (Barnsley, 1989).

The experimental variogram of natural phenomena is itself defined for a support of given positive (that is, non-zero) size. If it were measured on a different size of support, it would have a different shape and would imply a different set of scales of variation. Therefore, it is important to know the support for which it is defined. Further, it would be very useful if the effect of the support on the variogram (and, therefore, on measurable scales of variation) could be modelled. The effect of the scale of measurement on the scales of variation that are detectable from the variogram is considered next.

Scaling and regularisation

In this section, the emphasis is on *changing* the scale of measurement, and how different scales of measurement affect the scales of natural phenomena that we observe. Here, scale of measurement refers on the one hand to the support or spatial resolution and on the other to spatial coverage and spatial extent. These two aspects must be considered together if we are to tackle the problems associated with scaling-up.

Changing the support and regularising the variogram

The effect of imposing the support of the sampling frame on the underlying phenomenon can be modelled through spatial dependence. Moreover, the effect of changing the size of support can also be modelled. This is important because scale is most relevant when comparing data obtained at one scale with those obtained at another.

Equation (6) (Journel & Huijbregts, 1978) describes the relation between the punctual semivariance and the regularised (defined on a support of positive size) semivariance at a given lag

$$\gamma_v(\mathbf{h}) = \bar{\gamma}(v, v_{\mathbf{h}}) - \bar{\gamma}(v, v) \tag{6}$$

Here $\bar{\gamma}(v, v_{\mathbf{h}})$ is the integral of the punctual semivariance between two supports of size $|v|$ whose centroids are separated by $\mathbf{h}$, given formally by

$$\bar{\gamma}(v, v_{\mathbf{h}}) = \frac{1}{v^2} \int\limits_v \int\limits_{v(\mathbf{h})} \gamma(\mathbf{y}, \mathbf{y}') \, d\mathbf{y} \, d\mathbf{y}' \tag{7}$$

where **y** describes an observation of size $| v |$ and **y**$'$ describes indepen-
dently another observation of equal size and shape at a lag **h** away.
The quantity $\bar{\gamma}(v, v)$ is the average punctual semivariance within an
observation of size $| v |$ and is written formally as

$$\bar{\gamma}(v,v) = \frac{1}{v^2} \int\limits_{v} \int\limits_{v} \gamma(\mathbf{y},\mathbf{y}') \, \mathrm{d}\mathbf{y} \, \mathrm{d}\mathbf{y}' \qquad (8)$$

where **y** and **y**$'$ now cover the same pixel independently.

Equation (6) provides a means by which to assess the effect of size
of support (and shape and orientation of support) on the nature and
scale of measured spatial variation (Clark, 1977; Jupp, Strahler &
Woodcock, 1988; 1989; Atkinson, 1993a). Equation (6) implies that the
only variation detectable from the sample values (made on a support v)
is that described by the term on the left-hand side of Equation (6). So
what happens to the variation described by the second term on the
right-hand side of Equation (6) (the within-block variance)? The answer
is that it is completely obscured from analysis by integration over the
support, as described in Equation (1).

From Equation (6), the effect of regularisation over the support is to
remove small-scale variation in favour of large-scale variation occurring
within the spatial extent of the sampling frame. If much of the variation
is small scale in relation to the support (for example, small discs in the
above model), then much variation will be removed. If, however, all
the discs are large in relation to the support, then only a small amount
of variation (corresponding to a smoothing of the disc edges) will be
removed.

Figure 3 shows two sets of variograms that have been computed from
airborne MSS data in wavebands in the red (0.63–0.69 μm) (Fig. 3a)
and near-infrared (0.76–0.90 μm) (Fig. 3b) wavelengths, and at two
spatial resolutions of 1.5 m and 2 m (Atkinson, 1993a). The variograms
have been fitted with the exponential model given by

$$\gamma(h) = 1 - \exp(-h/r) \qquad \text{for } 0 < h \qquad (9)$$
$$\gamma(0) = 0$$

where r is the distance parameter of the model such that an effective
range a' may be computed as $a' = 3r$.

The effect of coarsening the spatial resolution is to remove short-
range variation from the variograms. The variograms representing a
spatial resolution of 2 m then describe the longer-range variation that
remains. Regularisation has important implications for remotely sensed

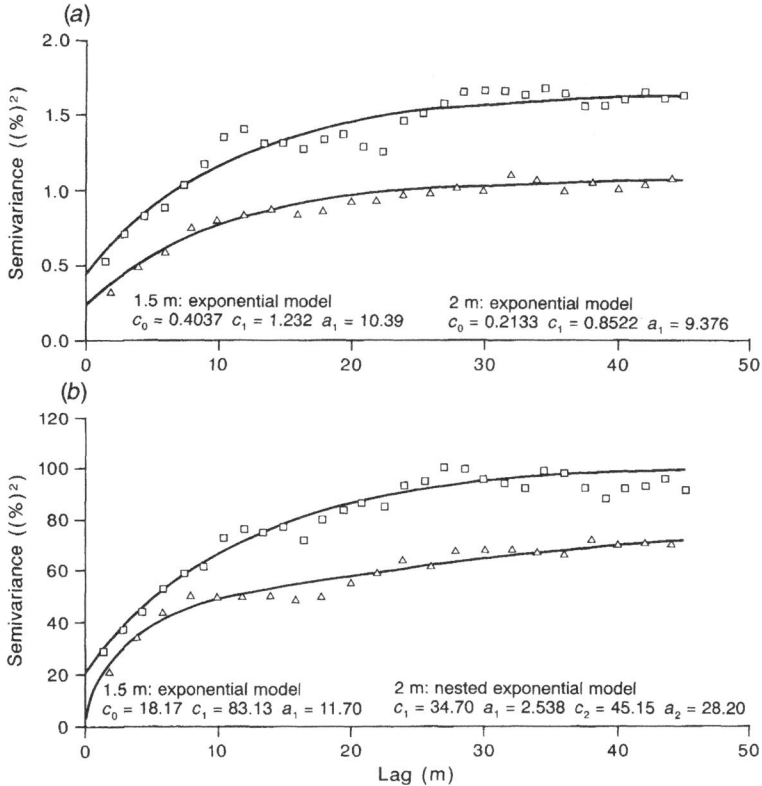

Fig. 3. Experimental variograms of airborne MSS imagery at spatial resolutions of 1.5 m (aircraft at 600 m, □) and 2 m (aircraft at 800 m, △) in the red (*a*) and near-infrared (*b*) wavebands. (After Atkinson (1993a). *International Journal of Remote Sensing*, Taylor and Francis.)

images at global scales (for example, AVHRR images) and their comparison with point measurements made at the ground.

Given Equation (6), it should be possible to regularise the variogram (that is, increase the size of support on which it is defined) to new larger sizes of support without measuring on that new support. The support must be approximated discretely to allow the integration of the semivariance between cells. The method provides estimates of the regularised semivariance at several discrete lags to which a mathematical model may then be fitted.

In certain circumstances, for example for core samples in a soil survey, the experimental support may be so small in relation to the new support that Equation (6) may be used directly. In others, for example where the spatial resolution of remotely sensed imagery is the coarsest possible for the particular sampling intensity, it may be necessary to, first, estimate the punctual or point variogram by deregularisation. This may be achieved by choosing a suitable model for the punctual variogram and determining its coefficients by applying and reapplying Equation (6) until the fit between the theoretically regularised variogram and the experimentally regularised (or measured) variogram is acceptable.

The importance of Equation (6) is not simply for understanding the effect of the support and regularising the variogram. Summary statistics, such as the dispersion, variance $D^2(v, V)$ and *a priori* variance $D^2(v, \infty)$, and simulations may be produced from the regularised variogram for the new support without actually measuring on that support (Zhang, Warrick & Myers, 1990).

Equation (6) provides a clear link between the scale of the underlying spatial variation evident in the variogram and the scale of measurement evident in the support.

Changing the spatial coverage

The second aspect of the scale of measurement that is important when changing scale is the spatial coverage and spatial extent of the sampling frame. Problems arise with Equations (3) and (6) in practice because as the spatial resolution coarsens so the area covered tends to increase. Over a small area such as an agricultural field, a spatial resolution of 2 m or 3 m may be appropriate, whereas at the global scale, a spatial resolution of 1 km or more may be appropriate. The problem is that if the region of analysis is changed then the nature of the property being studied may change and new forms of spatial variation, particularly at larger scales, may be introduced that were not previously detectable.

Of particular significance are the problems associated with crossing edges and borders. For example, within an agricultural field, spatial variation in spring-barley may be continuous and spatially dependent. When the analysis is extended to the next field it may have to deal not only with spring-barley, but also with sugar-beet, and also the spatial variation between the two fields. If the spatial resolution coarsens as the spatial extent increases, the spatial variation between fields will soon take over from that within the fields (Equation (6)). Increasing the spatial extent further and coarsening the spatial resolution, watersheds will eventually be crossed and the spatial variation between fields will be smoothed out in favour of that between regions (of, for example, mountain and lowland agriculture). Finally, as the global scale is approached, variation between sea and land takes over, and on the land surface variation between places tens, hundreds and thousands of kilometres apart is detected. Whatever happened to the field of spring-barley?

A further important issue with regard to changing cover type is the difference between categorical and continuous spatial variation. Generally, in remote sensing, the spatial variation encountered is continuous. Often this variation is made categorical by classifying the data (for example, into land cover). Environmental variables tend also to be continuous (except where they are absent and the presence or absence of a particular property could be encoded as categorical data). The relevance of these data types to the present discussion is that generally regularisation tends to make categorical variation continuous through mixing at subpixel scales (Jasinski, 1990). However, sometimes as the scale (extent) of analysis increases, the introduction of new forms of spatial variation may mean that categorical data are a more appropriate representation (for example, when crossing field boundaries).

In summary, the action of changing the sampling frame, for example from that suitable for measuring phenomena in a field of spring-barley to that suitable for measuring phenomena over the entire globe, involves two changes to the scale of measurement. The first is that the spatial resolution coarsens, and this means that less small-scale variation is detectable. The second is that the spatial coverage and spatial extent of the analysis increase, and this means that new forms of spatial variation, particularly those at larger scales, are detectable.

Scaling-up

Scaling-up is the process of transforming the sampling frame from a small to a large scale. Scaling-up is necessary when measurements made on a sampling frame at one scale are to be related to those made on another sampling frame at a larger scale. For example, in remote sensing the objective is often to scale-up measurements made at the ground to match the sampling frame of the remotely sensed imagery.

To scale-up successfully, both the support and the spatial extent of the small-scale sampling frame must be increased to match those of the large-scale sampling frame. The second of these may be achieved readily by ensuring that the coverage of the small-scale sample is equal to that of the large-scale sample. The first is more problematic. In remote sensing, image pixels with spatial resolutions in excess of 1 km (for example, AVHRR imagery) may need to be related to measurements made at the ground with a much smaller support, sometimes less than 1 m in size. Such ground data have a support that is so small in relation to the image pixels that they may be approximated as points.

In certain circumstances, the problems of scaling-up can partly be circumvented by using remotely sensed imagery with intermediate and fine spatial resolutions to bridge the gap between the properties at the ground and the remotely sensed imagery with a coarse spatial resolution. In the following examples, such data are assumed not to be available. Further, the scaling relation is assumed to be linear. The important issue of non-linear scaling is beyond the scope of the present chapter.

Hierarchical sampling

It is possible to compare directly point measurements made at the ground with image pixels. However, the simple correlation between the two variables will be expected to be low (and, therefore, the accuracy of predictions made from the imagery will be low also). This strategy is not recommended and some attempt should be made to scale-up the point measurements made at the ground.

Assuming that the support of the variable at the ground is 1 m and the spatial resolution of the imagery is 1.1 km, it is clear that complete coverage of the GRE is near impossible (requiring some 1 210 000 observations). Therefore, to estimate the mean value for each of several GREs one must sample *within* the GREs (Fig. 1). The question for sampling to obtain regularisation within the support is 'what sampling scheme and what sampling intensity should be used?'. The question of what support to choose might also be considered, although this is beyond the scope of the present discussion.

Both theoretical and empirical studies have shown that systematic sampling schemes are more efficient than more random alternatives (Yates, 1948; Quenouille, 1949; Whittle, 1954; Matérn, 1960; Burgess, Webster & McBratney, 1981). This is because most environmental variables are spatially dependent so that observations close together tend to be alike. Random schemes inevitably have some observations close together so that information is duplicated. Systematic schemes ensure that the observations are as far apart and as statistically independent as possible, maximising information. The only drawbacks of systematic sampling are that it may not be possible to estimate the precision with which the mean value is estimated (Dunn & Harrison, 1993) and that there is some risk of bias caused by periodicity (Finney, 1950). Generally, the equilateral triangular grid is the most efficient scheme. The square grid is almost as precise and is more convenient to implement both in the field and on the computer. Here, it is assumed that the square grid will be used to sample within the GREs.

The remaining question is 'what sampling intensity should be used to estimate the mean value within each GRE given that the sampling scheme is a square grid?' The answer depends on the spatial dependence in the variable, and in particular on the scales at which the underlying forms or processes operate. Kriging is an optimal technique for linear unbiased estimation (Journel & Huijbregts, 1978). It is optimal because it weights the sample observations according to how statistically dependent they are on the estimate. Observations that are close spatially to the estimate are likely to be similar in value and, therefore, receive more weight. Observations that are far away receive little or no weight. The exact weights are determined by referring to the variogram.

Kriging not only estimates optimally, but it also estimates the minimum estimation or kriging variance as a by-product of the kriging process. The kriging variance depends only on the variogram and the configuration of sample observations in relation to the estimate. Therefore, if the variogram is known, the sampling intensity necessary to achieve a specified error can be determined prior to the actual field survey.

The above method is illustrated now with data from a survey of a field of spring-barley at Broom's Barn Experimental Station, which lies 5 km west of Bury St Edmunds in Suffolk, UK. Three properties, the normalised difference vegetation index (NDVI), the percentage cover of crop and the dry biomass of the crop, were measured on 2 May 1987 along a 100 m transect of 100 observations spaced 1 m apart. For further details of the study, see Atkinson (1991). The objective was to estimate the mean of the three measured properties within GREs of 79 m (for comparison with Landsat MSS image pixels).

Experimental variograms were estimated using Equation (3) for each property (Fig. 4). Mathematical models were then fitted to the experimental values of semivariance by weighted least squares approximation. The model providing the best fit (in the least squares sense) to the variogram of NDVI is exponential, while the models for the variograms of percentage cover and dry biomass are both circular:

$$\gamma(h) = c_0 + c_1 \{1 - (2/\pi)\cos^{-1}(h/a) + (2h/\pi a)\sqrt{(1 - (h^2/a^2))}\} \tag{10}$$

$$\text{for } 0 < h \le a$$

$$\gamma(h) = c_0 + c_1 \qquad\qquad\qquad\qquad \text{for } h > a$$

$$\gamma(0) = 0$$

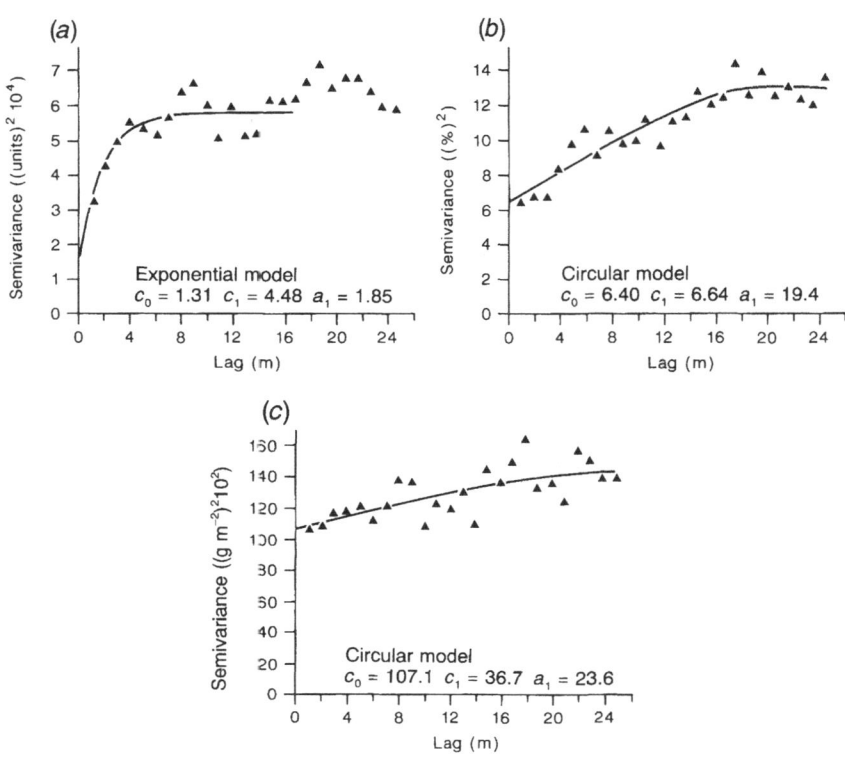

Fig. 4. Experimental variograms of (*a*) NDVI, (*b*) percentage cover of crop and (*c*) dry biomass (after Atkinson (1991). *International Journal of Remote Sensing*, Taylor and Francis).

where a is the range, c_0 is the nugget variance and c_1 is the spatially dependent or structured component. Note that the models were fitted to different lags in each case to improve the fit, and that the variogram for the NDVI is not directly comparable to the others as it was computed from data with a support of 1.5 m.

Each variogram was input in turn to the kriging equations and the kriging errors computed for a range of sample sizes for both a square grid and, for comparison, a random sampling scheme with the same sample size (Fig. 5). From the graphs it is possible to determine the sample size necessary to achieve a desired precision in advance of the actual survey. Thus, if a standard error of 0.5% was considered acceptable for estimating the mean NDVI, a sample size of approximately 20 observations would be necessary to achieve it (given that the sampling

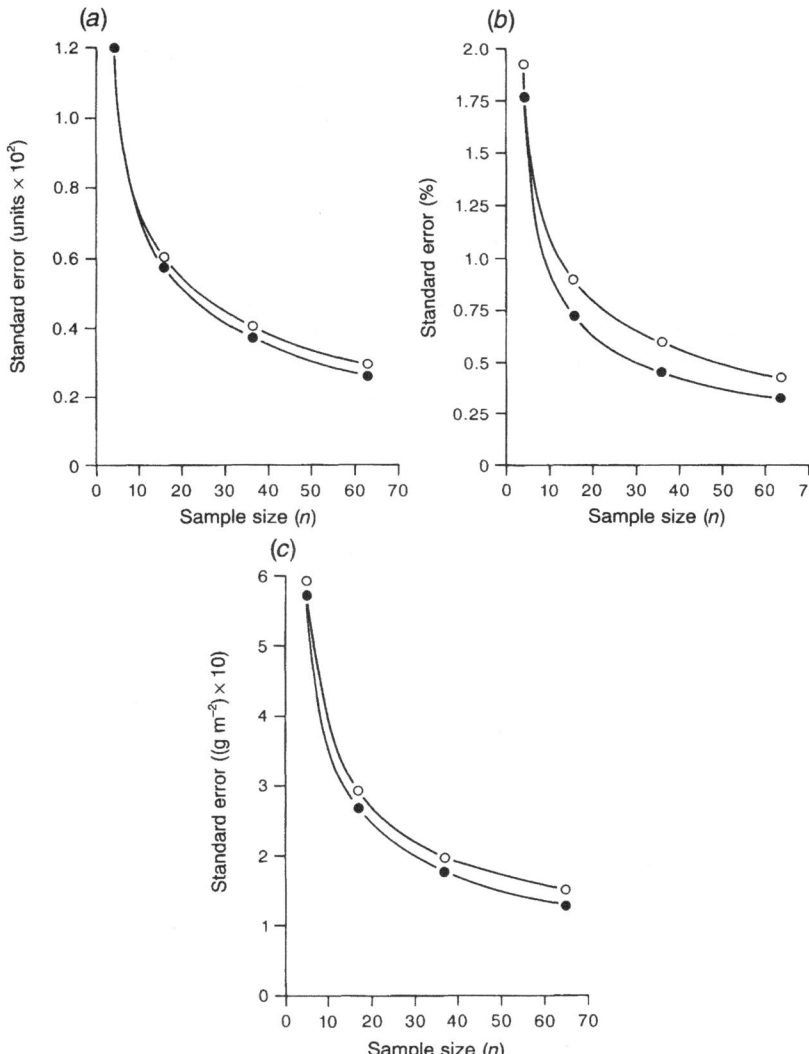

Fig. 5. Kriging error (●) and classical standard eror (○) plotted against sample size for (*a*) NDVI, (*b*) percentage cover of crop and (*c*) dry biomass (after Atkinson (1991). *International Journal of Remote Sensing*, Taylor and Francis).

scheme is a square grid). Alternatively, if the resources available for survey are limited then the graphs can be used to determine the precision attainable with a maximum affordable sample size. If the precision achievable is not acceptable the survey can be abandoned before further effort is wasted in the field.

In practice, there are several problems associated with the above strategy. Of these, the most important is that the support of each pixel is centre weighted because of the PSF of the sensor. Consequently, the spatial extent of each pixel exceeds that of the nominal GRE. To account for this 'over-regularisation' a correspondingly larger area on the ground could be sampled. To account for the centre weighting of the support two alternatives can be considered. In the first, the sample observations are weighted to reflect their position in relation to the PSF. In the second, the observations are positioned to reflect the shape of the PSF; that is, more observations are placed at the centre of the GRE and less towards the edges.

Another problem associated with the practical implementation of the above strategy is that often it is not known in advance where the GREs will occur at the ground. To counter this effect, an area on the ground that is twice that of the GRE could be sampled, although in practice selecting the nearest neighbour is likely to give sufficient accuracy. In addition, it may not be known where the ground observations are located on the imagery. This is especially the case for imagery with a coarse spatial resolution where the large scale of measurement means that the imagery is difficult to interpret in terms of features at the ground. These issues compound the problems related to scaling-up.

The second prerequisite to scaling-up is that the spatial coverage and spatial extent of the ground sample match that of the remotely sensed imagery. The reason for this is that the full dynamic range of variation in both properties must be measured so that the bivariate distribution function (BDF) between both variables is adequately characterised. Strictly, this does not necessitate an even spatial coverage. If the full range of spatial variation in both variables is encountered in a small region then it is sufficient that only that region be sampled. However, experience has shown that variation tends to increase as the scale of analysis increases, and in some cases there is no identifiable limit to this variation. It may be desirable to undertake analyses to determine the form of spatial variation so that the sampling strategy may be optimised (Oliver & Webster, 1986).

In global terms, the above recommendation implies that it is not sufficient to measure in one region or country alone: measurements at the ground should be evenly (that is, systematically) spaced to cover the entire surface of the Earth. In practice, the BDF may not be stationary from place to place but rather may take a different form depending on the location of the measurements at the ground. For example, the relation between the biomass and the NDVI derived from AVHRR pixels may be different for tropical forests from that for temperate zones. If such non-stationarity is expected, then it may be preferable to model each region or zone independently. Within regions of stationarity, an even spatial coverage should be provided with an extent matching that of the imagery. To do this necessitates sampling (for example, Dancy, Webster & Abel, 1986).

Interpolation and smoothing

In many cases, it may not be possible to design the sampling strategy according to the spatial variation in the property of interest and the sampling frame of the imagery as above. This may be because the sample has been undertaken already, because it may be undertaken by some other organisation or because there is insufficient authority or resources to design a suitable sampling strategy. If the sample covers the region evenly in the traditional sense, for example with a random sampling scheme, then the problems associated with scaling-up may far exceed those above where an optimal hierarchical sampling approach is designed. There are potential solutions to these problems, but they have not been widely implemented and must be seen as experimental at the present time.

Assume that there is a random spatial sample of the property at the ground defined on a support of 1 m and it is desired to relate these data to AVHRR image pixels. The 'point' measurements must be scaled-up so that they represent the GRE of the image pixels. The problem is that there is not a spatial sample within each GRE, but merely a single observation. It is not possible to average the variation to estimate the mean value, as in Equation (1). Therefore, alternatives must be used. The two alternatives considered here are block kriging (Burgess & Webster, 1980) and conditional simulation (Journel & Alabert, 1989; Rubin & Journel, 1991; Atkinson, 1993b; Dungan, Peterson & Curran, 1994). For reasons that should become clear, conditional simulation is the preferable alternative when the estimates are to be compared with another variable.

Block kriging is the form of kriging used to estimate over a support or domain larger than that of the observations. It can be used, for example, to interpolate to the 1.1 km support of the AVHRR pixels from a sparse sample of point observations at the ground. Its success depends on the form and scale of spatial dependence in the property of interest represented with the variogram. If observations close to the estimate are related to it statistically, block kriging may provide a useful estimate. This may seem the ideal solution since kriging automatically accounts for the size of support of the estimate. However, a problem is that kriging involves smoothing, where the term smoothing is used to mean regularisation independent of that which occurs naturally through integration over the support. Smoothing occurs because information from neighbouring observations is used in the averaging process, effectively extending the support out to these data.

Smoothing is not a problem in itself, and the estimate over each AVHRR GRE will be the best possible from the available data. However, problems occur when the estimated values are related to a second variable such as the remotely sensed imagery. As a result of smoothing, the univariate distribution function (UDF) of the ground variable and its BDF with the remotely sensed variable will be altered. The UDF will have a reduced variance in relation to the true UDF defined on a support of 1.1 km and the BDF will reflect this change. One possible solution to this problem involves regularising the variogram defined for 1 m supports (using Equation (6)) to estimate the dispersion variance of the 1.1 km supports. Given the differences in the dispersion variances (for 1 m and 1.1 km supports) it should be possible to correct the UDF and BDF for smoothing.

A second potential technique for interpolation to the 1.1 km supports is conditional simulation (Morrisey, 1991; Morrisey & Greene, 1993). The aim of conditional simulation is not to estimate optimally but to interpolate while retaining the second-order moments of the UDF (Deutsch & Journel, 1992).

There are many methods of conditional simulation. The following procedure is useful in understanding the effect of conditional simulation on the BDF. Kriging is used, first, to estimate optimally and, second, to estimate the amount of variation lost through smoothing. These two components make up the conditionally simulated values. This means that the dispersion variance and variogram of the original distribution are maintained, and that the data are not smoothed. The problem is that the variation added back in is orthogonal (unrelated) to the kriged values. As a result, the BDF between the conditionally simulated values and the remotely sensed variable may be altered. The extent to which such modifications to the BDF affect regression relations derived from the BDF has yet to be determined.

Conclusions

This chapter was presented in three sections, the first dealing with the meaning of scale, the second examining the effect of changing the scale (of measurement) and the third proposing some geostatistical solutions to the problems associated with scaling-up. The major points of this paper can be summarised.

1. The concept of scale in the present context is best understood in terms of the *size* of phenomena.
2. The two scales of importance are (i) the scale of spatial variation, forms or processes, and (ii) the scale of measurement.
3. The scale of measurement may be further divided into the support (which defines the limit to small-scale variation that may be detected), and the spatial coverage and spatial extent (which defines the limit to large-scale variation that may be detected).
4. Spatial dependence (represented with the variogram) is a guide to the scale of the underlying spatial variation, forms and processes as viewed through a particular sampling frame. The shape and range of the variogram indicates the set of scales of variation, forms or processes within the region studied.
5. As a result of regularisation over the support of the observations, small-scale variation may not be detectable from the experimental variogram. The variogram and regularisation are useful in understanding and comparing between measurements made on different supports.
6. As the spatial coverage and spatial extent of the survey increase, new forms of variation may be introduced and this may lead to non-stationarity and other problems.
7. It is inappropriate to compare point measurements of properties at the ground directly with image pixels with GREs of size 1 km or more.
8. Scaling-up requires regularisation of point measurements so that small-scale variation is removed in favour of large-scale variation (which is correlated with the remotely sensed imagery). To achieve this a sampling strategy should be designed so that natural averaging within GREs is possible. If kriging is used to estimate the mean value within each GRE, the estimation variance may be predicted in advance of the survey.

9.   In addition to natural averaging, even spatial coverage should be attained to ensure a full characterisation of the BDF.

10. Problems arise when the sampling strategy at the ground does not allow regularisation within the support. A sparse, even coverage (for example, a simple random sampling scheme) requires careful processing to achieve unbiased relations.

Acknowledgements

The author thanks the Natural Environment Research Council for its CASE award while at the University of Sheffield and Rothamsted Experimental Station, and the Leverhulme Trust for its research award while at the University of Bristol.

Glossary

$Z_v(\mathbf{x}_0)$	random variable, the spatial mean or integral of $Z(\mathbf{x})$ over v centred on $\mathbf{x}_0$
$z_v(\mathbf{x}_0)$	realisation of the random variable $Z_v(\mathbf{x}_0)$
$Z(\mathbf{y})$	property Z defined on a punctual or point support
$\gamma(\mathbf{h})$	variogram defined on a punctual or point support
$\hat{\gamma}_v(\mathbf{h})$	experimental (and regularised variogram)
D	fractal dimension
$\bar{\gamma}(v, v_\mathbf{h})$	integral punctual semivariance between two supports of size $\mid v \mid$ whose centroids are separated by $\mathbf{h}$
$\bar{\gamma}(v, v)$	integral punctual semivariance within an observation of size $\mid v \mid$
$D^2(v, V)$	dispersion variance
$D^2(v, \infty)$	*a priori* variance
a	range of the variogram model
c_0	nugget variance of the variogram model
c_1	spatially dependent or structured component of the variogram model

References

Atkinson, P.M. (1991). Optimal ground-based sampling for remote sensing investigations: estimating the regional mean. *International Journal of Remote Sensing*, 12, 559–567.

Atkinson, P.M. (1993a). The effect of spatial resolution on the experimental variogram of airborne MSS imagery. *International Journal of Remote Sensing*, 14, 1005–1011.

Atkinson, P.M. (1993b). Simulating Landsat TM images with the variogram. In *Towards Operational Applications*, ed. K. Hilton, pp. 25–32. Nottingham: Remote Sensing Society.

Atkinson, P.M. & Curran, P.J. (1995). Defining an optimal size of support for remote sensing investigations. *IEEE Transactions on Geoscience and Remote Sensing*, 33, 1–9.

Barnsley, M.F. (1989). *Fractals Everywhere*. San Francisco: Freeman.

Bauer, M.E., Cipra, J.E., Anuta, P.E. & Etheridge, J.B. (1979). Identification and area estimation of agricultural crops by computer classification of Landsat MSS data. *Remote Sensing of Environment*, 8, 77–92.

Bian, L. & Walsh, S.J. (1993). Scale dependencies of vegetation and topography in a mountainous environment in Montana. *Professional Geographer*, 45, 1–11.

Brown, S.A. (1987). A note on the description of surface roughness using fractal dimension. *Geophysical Research Letters*, 14, 1095–1098.

Burgess, T.M. & Webster, R. (1980). Optimal interpolation and isarithmic mapping of soil properties II. Block kriging. *Journal of Soil Science*, 31, 333–341.

Burgess, T.M., Webster, R. & McBratney, A.B. (1981). Optimal interpolation and isarithmic mapping of soil properties IV. Sampling strategy. *Journal of Soil Science*, 32, 643–659.

Clark, C.D. (1990). Remote sensing scales related to the frequency of natural variation: an example from paleo-ice-flow in Canada. *IEEE Transactions on Geoscience and Remote Sensing*, 28, 503–515.

Clark, I. (1977). Regularization of a semi-variogram. *Computers and Geosciences*, 3, 341–346.

Cohen, W.B., Spies, T.A. & Bradshaw, G.A. (1990). Semivariograms of digital imagery for analysis of conifer canopy structure. *Remote Sensing of Environment*, 34, 167–178.

Cressie, N.A.C. (1991). *Statistics for spatial data*. New York: Wiley.

Curran, P.J. & Dungan, J.L. (1989). Estimation of signal to noise: a new procedure applied to AVIRIS data. *IEEE Transactions on Geoscience and Remote Sensing*, 27, 620–628.

Curran, P.J. & Williamson, H.D. (1985). The accuracy of ground data used in remote sensing investigations. *International Journal of Remote Sensing*, 6, 1637–1651.

Dancy, K.J., Webster, R. & Abel, N.O.J. (1986). Estimating and mapping grass cover and biomass from low-level photographic sampling. *International Journal of Remote Sensing*, 7, 1679–1704.

Deutsch, C.V. & Journel, A.G. (1992). *GSLIB Geostatistical Software Library*. Oxford: Oxford University Press.

Dozier, J. & Strahler, A.H. (1983). Ground investigations in support of remote sensing. In *Manual of Remote Sensing*, 2nd edn, ed. R.N. Colwell, pp. 959–986. Falls Church, VA: American Society of Photogrammetry.

Dungan, J.L., Peterson, D.L. & Curran, P.J. (1994). Alternative approaches for mapping vegetation quantities using ground and image data. In *Environmental Information Management and Analysis: Ecosystem to Global Scales*, ed. W. Michener, S. Stafford & J. Brunt, pp. 237–261. London: Taylor and Francis.

Dunn, R. & Harrison, A.R. (1993). Two dimensional systematic sampling of land use. *Journal of the Royal Statistical Society, Series C: Applied Statistics*, 42, 585–601.

Finney, D.J. (1950). An example of periodic variation in forest sampling. *Forestry*, 23, 96–111.

Foody, G.M. & Curran, P.J. (1994). Scale and environmental remote sensing. In *Environmental Remote Sensing from Regional to Global Scales*, ed. G.M. Foody & P.J. Curran, pp. 223–232. Chichester: Wiley.

Gohin, F. & Langlois, G. (1993). Using geostatistics to merge in situ measurements and remotely sensed observations of sea surface temperature. *International Journal of Remote Sensing*, 14, 9–19.

Isaaks, E.H. & Srivastava, R.M. (1989). *Applied Geostatistics*. Oxford: Oxford University Press.

Jasinski, M.F. (1990). Sensitivity of the normalized difference vegetation index to sub-pixel canopy cover, soil albedo and pixel scale. *Remote Sensing of Environment*, 32, 169–187.

Journel, A.G. & Alabert, F. (1989). Non-Gaussian data expansion in the Earth sciences. *Terra Nova*, 1, 123–134.

Journel, A.G. & Huijbregts, C.J. (1978). *Mining Geostatistics*. London: Academic Press.

Jupp, D.L.B., Strahler, A.H. & Woodcock, C.E. (1988). Autocorrelation and regularization in digital images I. Basic theory. *IEEE Transactions on Geoscience and Remote Sensing*, 26, 463–473.

Jupp, D.L.B., Strahler, A.H. & Woodcock, C.E. (1989). Autocorrelation and regularization in digital images II. Simple image models. *IEEE Transactions on Geoscience and Remote Sensing*, 27, 247–258.

Justice, C.O., Townshend, J.R.G., Holben, B.N. & Tucker, C.J. (1985). Analysis of the phenology of global vegetation using meteorological satellite data. *International Journal of Remote Sensing*, 6, 1271–1318.

Justice, C.O., Townshend, J.R.G. & Markham, B.L. (1987). MODIS spatial resolution study. *International Journal of Remote Sensing*, 8, 1119–1121.

Justice, C.O., Dugdale, G., Townshend, J.R.G., Narracott, A.S. & Kumar, M. (1991). Synergism between NOAA-AVHRR and Meteosat data for studying vegetation development in semi-arid West Africa. *International Journal of Remote Sensing*, 12, 1349–1368.

Klinkenberg, B. & Goodchild, M.F. (1992). The fractal properties of topography: a comparison of methods. *Earth Surface Processes and Landforms*, 17, 217–234.

Lam, N.S.-N. & Quottrochi, D.A. (1992). On the issues of scale, resolution and fractal analysis in the mapping sciences. *Professional Geographer*, 44, 88–98.

Lovejoy, S. & Schertzer, D. (1986). Scale invariance, symmetries, fractals, and stochastic simulations of atmospheric phenomena. *Bulletin of the American Meteorological Society*, 67, 21–32.

Mandelbrot, B. (1982). *The Fractal Geometry of Nature*. New York: Freeman.

Matérn, B. (1960). Spatial variation. *Meddelanden Från Statens Skogsforskningsinstitut*, 49, 1–144.

Matheron, G. (1965). *Les Variables Régionalisées et Leur Estimation*. Paris: Masson.

McGwire, K., Friedl, M. & Estes, J.E. (1993). Spatial structure, sampling design and scale in remotely sensed imagery of a California savanna woodland. *International Journal of Remote Sensing*, 14, 2137–2164.

Merson, R.H. (1983). A composite Landsat image of the United Kingdom. *International Journal of Remote Sensing*, 4, 521–527.

Moellering, H. & Tobler, W. (1972). Geographical variances. *Geographical Analysis*, 4, 34–50.

Morrisey, M.L. (1991). Using sparse raingages to test satellite-based rainfall algorithms. *Journal of Geophysical Research*, 96, 18 561–18 571.

Morrisey, M.L. & Greene, J.S. (1993). Comparison of two satellite-based rainfall algorithms using Pacific Atoll raingage data. *Journal of the American Meteorological Society*, 32, 411–425.

Oliver, M.A. & Webster, R. (1986). Combining nested and linear sampling for determining the scale and form of spatial variation of regionalized variables. *Geographical Analysis*, 18, 227–242.

Openshaw, S. (1977). The modifiable areal unit problem. *Concepts and Techniques in Modern Geography, Catmog 38*. Norwich, UK: Geo-Abstracts.

Quenouille, M.H. (1949). Problems in plane sampling. *Annals of Mathematical Statistics*, 20, 355–375.

Ramstein, G. & Raffey, M. (1989). Analysis of the structure of radiometric remotely sensed images. *International Journal of Remote Sensing*, 10, 1049–1074.

Rees, W.G. (1992). Measurement of the fractal dimension of ice-sheet surfaces using Landsat data. *International Journal of Remote Sensing*, 13, 663–671.

Rossi, R.E., Mulla, D.J., Journel, A.G. & Franz, E.H. (1992).

Geostatistical tools for modeling and interpreting ecological spatial dependence. *Ecological Monographs*, 62, 277–314.

Rubin, Y. & Journel, A.G. (1991). Simulation of non-Gaussian space random functions for modelling transport in ground water. *Water Resources Research*, 27, 1711–1721.

Sèze, G. & Rossow, W.B. (1991). Effects of satellite data resolution on measuring the space/time variations of surfaces and clouds. *International Journal of Remote Sensing*, 12, 921–952.

Strahler, A.H., Woodcock, C.E. & Smith, J.A. (1986). On the nature of models in remote sensing. *Remote Sensing of Environment*, 20, 121–139.

Thornes, J.B. (1973). Markov chains and slope series: the scale problem. *Geographical Analysis*, 5, 322–328.

Townshend, J.R.G. & Justice, C.O. (1988). Selecting the spatial resolution of satellite sensors required for global monitoring of land transformations. *International Journal of Remote Sensing*, 9, 187–236.

Townshend, J., Justice, C., Lei, W., Gurney, C. & McManus, J. (1991). Global land cover classification by remote sensing: present capabilities and future possibilities. *Remote Sensing of Environment*, 35, 243–255.

Tucker, C.J., Vanpraet, C.L., Sharman, M.J. & Van Ittersum, G. (1985). Total herbaceious biomass production in the Senegalese Sahel: 1980–1984. *Remote Sensing of Environment*, 17, 233–249.

Wald, L. (1989). Some examples of the use of structure functions in the analysis of satellite images of the ocean. *Photogrammetric Engineering and Remote Sensing*, 55, 1487–1490.

Webster, R. (1985). Quantitative spatial analysis of soil in the field. *Advances in Soil Science*, 3, 1–70.

Webster, R., Curran, P.J. & Munden, J.W. (1989). Spatial correlation in reflected radiation from the ground and its implications for sampling and mapping by ground-based radiometry. *Remote Sensing of Environment*, 29, 67–78.

Webster, R. & Oliver, M.A. (1990). *Statistical Methods for Soil and Land Resources Survey*. Oxford: Oxford University Press.

Whittle, P. (1954). On stationary processes in the plane. *Biometrika*, 41, 434–449.

Woodcock, C.E. & Strahler, A.H. (1987). The factor of scale in remote sensing. *Remote Sensing of Environment*, 21, 311–322.

Yates, F. (1948). Systematic sampling. *Philosophical Transactions of the Royal Society*, Series A, 241, 345–377.

Yoder, J.A., McClain, C.R., Blanton, J.O. & Oey, L.-Y. (1987). Spatial scales in CZCS-chlorophyll imagery of the southeastern U.S. continental shelf. *Limnology and Oceanography*, 32, 929–941.

Zhang, R., Warrick, A.W. & Myers, D.E. (1990). Variance as a function of sample support size. *Mathematical Geology*, 22, 107–122.

C.E. WOODCOCK, J.B. COLLINS
and D.L.B. JUPP

Scaling remote sensing models

Introduction

The idea that remote sensing can be useful for parameterising environmental process models is becoming well established (see for example Mooney & Field, 1989; Avissar, 1992; Hall *et al.*, 1992). The synoptic coverage of satellite sensor images is well-suited to provide spatial data for input to environmental process models. Remote sensing images are particularly useful for capturing the heterogeneity in landscapes, which is critical for distributed-parameter models. Figure 1 illustrates the way remote sensing is typically used in this context. In this figure, remote sensing images are used in conjunction with a remote sensing model to produce spatially distributed landscape parameters that serve as inputs to environmental process models. 'Remote sensing models' in this case could refer to any of a wide range of methods for extracting information from an image. One example of a remote sensing model might be a regression between leaf area index (LAI) and the normalised difference vegetation index (NDVI) (Running *et al.*, 1989; Band *et al.*, 1991). In this example, LAI might subsequently be an input parameter required for an environmental process model concerned with transpiration, photosynthesis or net primary production. Another example might be a surface energy balance model, where the required derivatives from remote sensing might be surface albedo. In this case, the remote sensing model may involve land cover classification, with different albedos assigned for each land cover type.

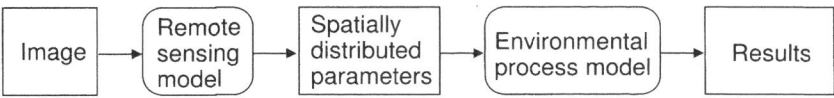

Fig. 1. A schematic drawing of a simple conceptual model for the use of remote sensing in conjunction with environmental process models. Remotely sensed images provide spatially distributed parameters for input to environmental process models via a remote sensing model.

Interest has grown in recent years in studying environmental processes over larger areas (Ehleringer & Field, 1993). Many environmental process models were developed for use at the scale of leaves or plants, meaning they required inputs at spatial resolutions, using a linear scale, of the order of 10^{-1} to 10^1 m, and typically could be applied to areas of the order of 10^3 to 10^5 m. As interest has moved to regional scales, where inputs are required of the order of 10^2 to 10^4 m for areas of the order of 10^5 to 10^7 m, the use of remote sensing has become essential to apply these models for larger areas. However, there remain significant questions regarding the best ways to use satellite remote sensing in the scaling process. There are two competing requirements in using remote sensing to scale environmental process models. The first is that the satellite sensor image must have fine enough spatial resolution to capture the pertinent heterogeneity in the landscape. In essence, this requirement is that enough detail exists in the image data to allow differentiation of varying states of the desired phenomena. Second, the satellite sensor data must be of sufficient extent to cover the area of interest. In general, as detail (spatial resolution) increases in an image, the extent or coverage decreases. One key to the scaling issue is the appropriate selection of satellite sensor imagery, given this trade-off. In the past. there have been very few choices regarding satellite sensor imagery. At the scale of fields or forest stands, imagery from either the SPOT or Landsat satellite sensors has been the obvious choice. For regional scales, data from the NOAA AVHRR series of satellites have been essentially the only choice. However, in the future, there will be a much wider selection of data available from a wide array of sensors. The question of which kind of satellite sensor data to use and the appropriate scaling method will become more relevant.

This chapter presents a discussion of the role of remote sensing in scaling environmental process models. The approach presented is based on an understanding of the effect of the spatial properties of images on the scaling approach selected. The methods for analysing spatial properties of images are applied to an image of a forested region of the Sierra Nevada Mountains in California. The intent is to provide a framework for considering alternative scaling strategies, including the selection of appropriate imagery.

In the discussion of the role of remote sensing in scaling, it is necessary to discuss landscapes and their organisation to help form a link between landscapes and remotely sensed images. A nested-hierarchical model for landscapes is used, following the treatment by Woodcock & Harward (1992). In this hierarchical model of landscapes, each level in the hierarchy corresponds to a different scale. In a forested landscape, for example, the most fundamental element might be individual leaves, whose approximate sizes define the finest scale in the model. The next level in the hierarchy might be trees, followed by groups of trees of similar species or size. These patches, or stands, would then form a third scale in the landscape. All patches of the same kind would combine to form forest classes, which would be a fourth level in the hierarchy. These different forest types might then combine to form a general class of *forest*, which exists with other classes at this level, such as *grassland*, *water* or *urban land*. In this model, each successive level in the hierarchy is more general and is formed by combining elements from the level below.

The reason for using such a model of the landscape is that particular investigations are concerned with phenomena that vary at particular scales. For example, in studies of insect defoliation, the variation at the scale of individual trees might be most important. For studies of hillslope processes, it is the variation at the field or stand scale that is most relevant. The remotely sensed imagery to be used must be suitable for the scale of the investigation. For example, to study acid rain damage to forest stands over a large region, it would make little sense to use imagery with a resolution fine enough to resolve individual leaves. In this situation, the volume of data would prove overwhelming. Conversely, images with pixels so large that each pixel includes many forest stands would not be useful, as the effects in each stand could not be resolved.

The effects of different landscape scales in images

A key to the scaling process using remote sensing is an understanding of the magnitude of the effects in images resulting from processes acting at different scales in the landscape. Using the nested-hierarchical model, this approach corresponds to understanding how much of the variance in an image results from the different levels in the hierarchy. Using the vegetation example above, this would correspond to assessing the magnitude of the effects owing to individual leaves, trees, stands of tree, types of tree, and so on.

We seek a hierarchical partitioning of a digital image corresponding to the hierarchical partitioning of vegetation mentioned above. In all cases, the image pixel is necessarily the smallest data element and corresponds to the lowest level of the hierarchy. Which of the levels of the vegetation hierarchy correspond to the next level of the data hierarchy depends on the resolution of the imagery. For fine resolution imagery collected from airborne scanners, individual pixels could be aggregated into trees, which then form the next level of the hierarchy. When using SPOT HRV or Landsat TM (Thematic Mapper) data, the next level may be fields or forest stands, since individual trees would probably not be resolvable. In any case, any landscape elements that are identifiable in a digital image could be aggregated to the next higher level of the vegetation hierarchy. For example, forest stands in a SPOT HRV scene could be aggregated to vegetation classes. In the following discussion, we consider digital images as a four-level hierarchy in which pixels aggregate to regions corresponding to forest stands, then to vegetation classes, then to the entire image.

A nested-hierarchical model of spatial data is provided by Moellering & Tobler (1972) and is elaborated by Collins, Woodcock & Jupp (1995). The model begins with the concept of a regionalised variable, which assigns some value $Z(\mathbf{s})$ to a vector location $\mathbf{s}$ within a spatial domain set D (Cressie, 1993, p. 52).

$$\{Z(\mathbf{s}) : \mathbf{s} \in D\}$$

D corresponds to the entire area represented by the image. The hierarchical model describes the image as being composed of a number of vegetation classes, defined as disjoint subsets of the domain set D and denoted D_i. Each class D_i is in turn composed of a number of regions, denoted D_{ij}. Regions are in turn composed of pixels, denoted D_{ijk}. So each entity at each level of the hierarchy is associated with some area within the image. The mean value of the regionalised variable $Z(\mathbf{s})$ over some area T is given by

$$\mu(T) = \int_T Z(\mathbf{s}) \frac{d\mathbf{s}}{A_T} \tag{1}$$

where A_T is the area of the domain set T. Then the mean of the entire image is $\mu(D)$, the mean of a class on the first level of the hierarchy is $\mu(D_i)$, and so on.

It should be noted that the discussion above is a simplification of the true nature of remotely sensed data. The observed value for a pixel is not exactly a simple integration of the variable $Z(\mathbf{s})$ over some well-defined area on the ground. In reality, the sensor point spread function must be considered, implying that the observations are not truly disjoint and that values of $Z(\mathbf{s})$ are not weighted equally in all areas of the scene. The formulation above amounts to assuming a square-wave response function for the sensor. Under this assumption, we may define our observed pixel values as

$$x_{ijk} = \mu(D_{ijk}) \tag{2}$$

Using the above statistics, it is possible to create a new set of images that are derivatives of the original image and that contain only the effects associated with an individual scale. Values for these new images at each pixel can be calculated for the entire image (I), and for scales of classes (C), regions (R), and pixels (P), respectively, as

$$I = \mu(D) \tag{3}$$
$$C_i = \mu(D_i) - \mu(D) \tag{4}$$
$$R_{ij} = \mu(D_{ij}) - \mu(D_i) \tag{5}$$
$$P_{ijk} = x_{ijk} - \mu(D_{ij}) \tag{6}$$

I is the image effect, C_i is the effect associated with class i, R_{ij} is the effect associated with region j of class i, and P_{ijk} is the residual or pixel effect associated with pixel k of region j of class i.

Adding the above four equations indicates that an observed pixel value is equal to the sum of the effects of all levels of the hierarchy:

$$x_{ijk} = I + C_i + R_{ij} + P_{ijk} \tag{7}$$

Entities on lower levels of the hierarchy necessarily are smaller than objects on any higher level. According to Moellering & Tobler (1972), this ordering of levels by areal size can be taken as a surrogate for scale or resolution. Data at different levels of the hierarchy thus correspond to different geographical scales.

A useful measure of the spatial structure of images is the semivariogram, which measures semivariance as a function of distance (Curran, 1988; Woodcock, Strahler & Jupp, 1988a,b; Atkinson, this volume). Assuming stationarity within the image, the semivariance ($\gamma(h)$) is

$$\gamma(h) = \frac{1}{2} E[Z(s) - Z(s + \mathbf{h})]^2 \tag{8}$$

where $\mathbf{h}$ is a vector lag. In this presentation, only distance (h) is considered, and all variograms are assumed isotropic. This equation can be used to calculate a semivariogram from any image – including the original image, I, and the new images related to the effects associated with

(a)

(b)

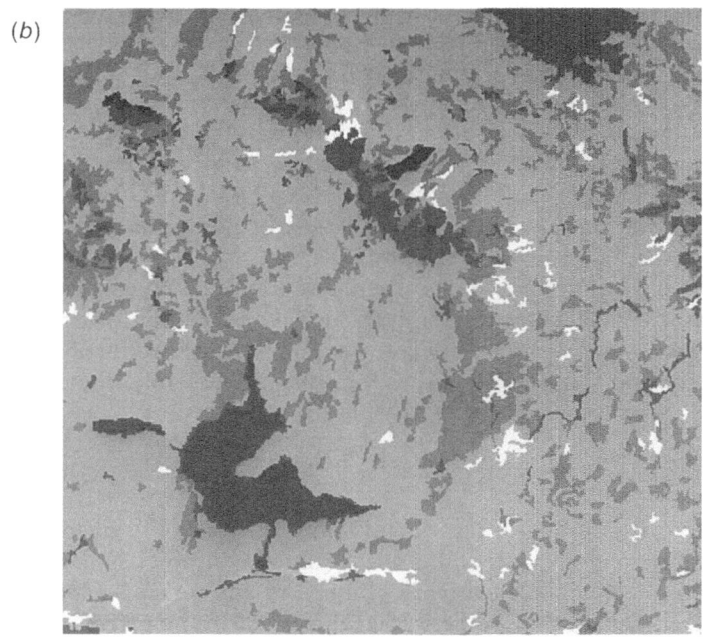

(c)

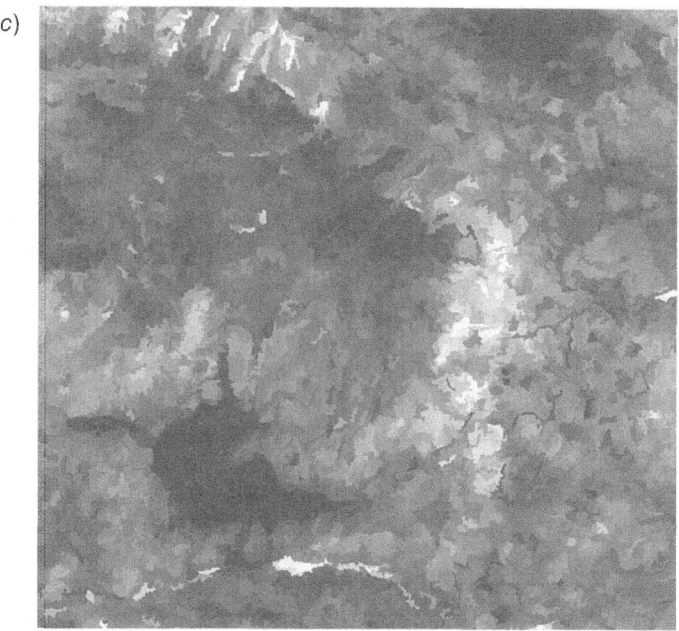

Fig. 2. NDVI-derived data from Landsat TM images for a portion of the Plumas National Forest in California. (*a*) The original image. The subimage covers 1.5 km on a side. (*b*) The mean NDVI for the class of each pixel for the same area as in (*a*). (*c*) The mean NDVI for the region of each pixel.

classes C, regions R and pixels P to create separate semivariograms for each: γ_I, γ_C, γ_R and γ_P.

Note that all these formulations are for a single image, which is what is used below in this chapter. The generalisation to the multivariate case to allow use with multispectral imagery is direct.

The Plumas National Forest data set

A data set from the Plumas National Forest in the Sierra Nevada Mountains in California is used to illustrate the above ideas (Fig. 2). This image was used as part of a project to map and inventory forest vegetation using satellite remote sensing and a forest canopy reflectance model (Woodcock *et al.*, 1994). It is used in this chapter because of the availability of data on region and class membership for each pixel in the image. The original image (I) is NDVI derived from the Landsat TM, which has a 30 m spatial resolution (Fig. 2*a*). The classes in this image correspond to the general vegetation types of conifer forest, hardwood forest, brush, grass, water and meadow. Figure 2*b* shows the mean NDVI for the class of each pixel, or $\mu(D_i)$. The regions in the image correspond to vegetation stands. For example, within the conifer class, individual stands are differentiated on the basis of species composition, size and crown cover of trees. Following the nested-hierarchical model, the conifer class is simply the union of all the conifer stands. Similarly, a vegetation stand, or region, is simply all the pixels inside its boundary. Figure 2*c* shows the mean NDVI for the region of each pixel, or $\mu(D_{ij})$. The methods used to create the regions and classes in this image are given in detail in Woodcock *et al.* (1994).

The class image defined above, *C*, is simply Fig. 2*b* minus the global mean for the image, $\mu(D)$. The region image, *R*, is simply Fig. 2*c* minus Fig. 2*b*, and the pixel image, *P*, is Fig. 2*a* minus Fig. 2*c*. Figure 3 shows a transect of pixel values from the same scan line for each of these four images: *I*, *C*, *R* and *P*. In this figure, the *y*-axis is NDVI and the *x*-axis is position in the image. *I* and *P* both contain pixel-scale variation, so brightness fluctuates from one pixel to the next. This effect is apparent in the high frequency variation in the image. *R* and *C* have less high frequency variation, as the value for each pixel is a mean calculated from a larger area. Since regions are smaller than classes,

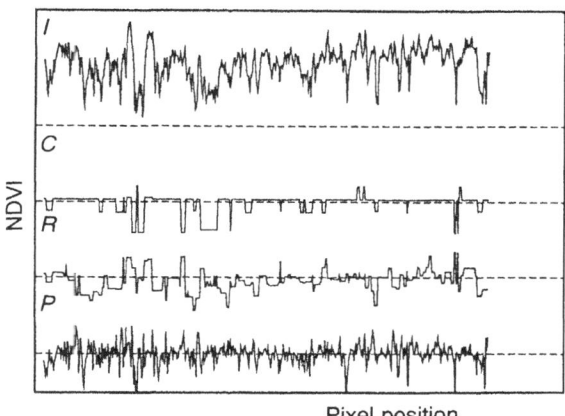

Fig. 3. Transects along the same scan line for the original image *I*, the class image *C*, region image *R* and the pixel image *P*.

there is higher frequency variation in the region image than the class image.

Figure 4 shows the semivariograms for each of the four images. Interpretation of these semivariograms is helped by understanding a few concepts about variograms (Curran, 1988). First, semivariance tends to rise to the level of the global variance for the image, which is called the *sill*. The distance to the sill is called the *range of influence* and is related to the size of objects in the image. Prior work in remote sensing using variograms has focused on the relationship between landscape parameters and variograms (Jupp & Strahler, 1988, 1989; Woodcock *et al.*, 1988a,b; Cohen & Spies, 1991), kriging and sample design (Curran & Williamson, 1988; Webster, Curran & Munden, 1989; Atkinson, 1991; Atkinson, Curran & Webster, 1992; Atkinson, this volume), or selecting spatial resolutions (Atkinson & Curran, 1995).

The semivariogram for the original image ($\gamma(h)$), shows the highest overall variance and can be interpreted as containing the effects of all scales. Notice that it rises quickly as a function of lag and then continues to increase slowly throughout the remainder of the graph. The fact that it never reaches its sill, or levels off, is significant as it indicates that there are objects larger in size than the 1500 m range of this graph.

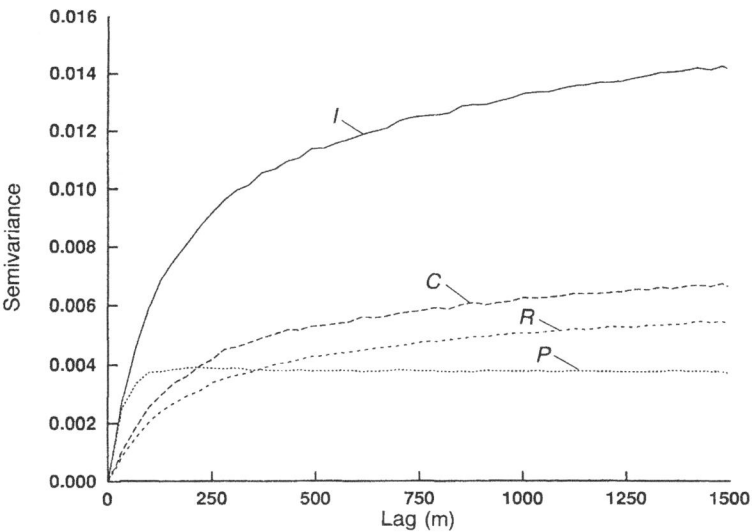

Fig. 4. Semivariograms for *I*, *C*, *R* and *P*.

This interpretation is supported by Fig. 2a, which shows many features exceed this size, such as the lake in the lower left corner.

The semivariograms for P, R and C illustrate that these scales in the landscape affect semivariance differently. P contains only effects within regions and, therefore, it rises sharply to a well-defined sill at a lag of approximately 100 m. Beyond that lag, P exhibits random spatial structure. The semivariograms for R and C rise more slowly at short lags than either I or P. The rate of rise for these images is a function of two factors. The first factor is the frequency as a function of lag with which pixel pairs fall across boundaries in the images (Jupp et al., 1988; 1989). At short distances, these frequencies are low, and they increase with lag. For this effect, the size of the objects in the image is the critical factor. Based on this factor alone, the semivariogram for C would be expected to rise more slowly than for R, as classes are larger than regions (Fig. 2b,c). However, note in Fig. 4 that the semivariance for C rises faster than that for R.

A second factor that influences the rate of increase in the semivariance with lag is the magnitude of the differences across boundaries. As the magnitudes of these differences increase, so does the rate of rise of the semivariance. Based on this factor alone, the semivariance for C would be expected to rise faster than that for R. Remember that C is calculated at each pixel by subtracting the mean for the entire image ($\mu(D)$), from the mean for the class of that pixel, ($\mu(D_i)$). R results from the subtraction of the mean for the class, ($\mu(D_i)$), from the mean for the region, ($\mu(D_{ij})$). Thus, the magnitudes of differences are less in R than in C.

Finally, to compare accurately the rates of increase of semivariance for different images, the comparison needs to be done relative to the sill for each image. Figure 5 shows all four semivariograms, each scaled for its own sill as follows:

$$\gamma_s(h) = \frac{\gamma(h)}{\sigma^2}$$

where $\gamma_s(h)$ refers to a scaled semivariance, and σ^2 is the global variance for the image. Therefore, a value of one for a scaled semivariance indicates that the sill has been reached. This figure illustrates the differences between the pixel scale and the region and class scales. Processes acting at the pixel scale have a very short range of influence and do not contribute to spatial structure in the image beyond lags of about 100 m. But the semivariance of processes acting at larger scales increases more slowly, with that for classes rising slightly faster than that for regions.

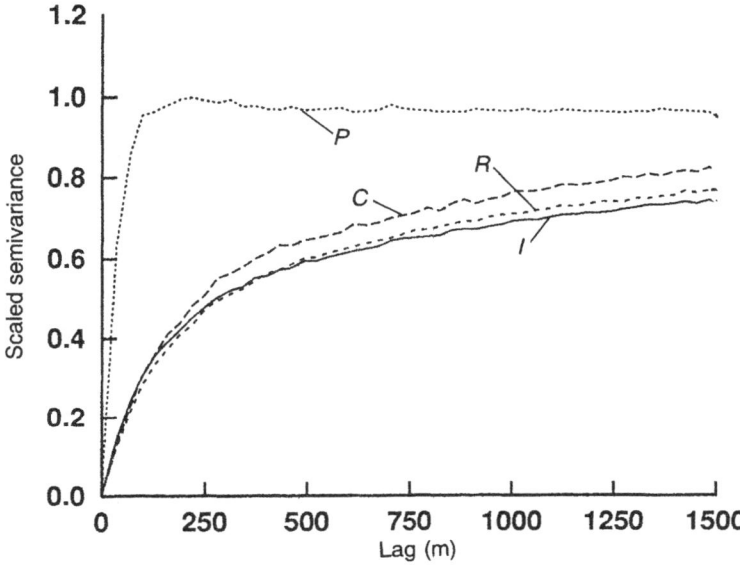

Fig. 5. Scaled semivariograms for *I*, *C*, *R* and *P*.

Alternative approaches to scaling using remote sensing

Central to the use of remote sensing in scaling environmental process models is the concept of spatial resolution. The tradeoff mentioned above between detail and extent in satellite imagery requires close attention. The ideal spatial resolution is the largest pixel size that allows recovery of the desired landscape parameters. For landscape parameters to be recovered, the image must still 'carry' the signal related to those parameters. From the perspective of the treatment above, this becomes the magnitude of the variance remaining in an image related to the landscape scale of interest. Therefore, the variograms presented in the previous section provide an indication of the utility of various resolutions for studying different scales of information in a landscape.

From the perspective of evaluating the utility of spatial resolutions, the key data from the variograms are the distances between the semivariance and the sill. This distance is most readily observed in the scaled semivariograms in Figure 5. The distance from the semivariance to the sill is an indication of the proportion of the original variance remaining as a function of lag. When broken down by different landscape scales, this provides a useful indication of the expected strength of the 'signal' in a satellite sensor image of a particular resolution for each landscape scale.

Given this perspective, it is interesting to consider alternative scaling strategies involving remote sensing. The first is referred to as the GIS (Geographic Information Systems) approach and is the most commonly used. Figure 6 is a schematic representation of this scaling approach and shows the use of local imagery to produce local-scale results, which then have to be aggregated across large areas to yield regional results. In this approach, it would be assumed that there is sufficient detail in the local imagery to recover the desired landscape parameters. The most likely problem encountered with this approach would be insufficient areal coverage, as the use of remote sensing images with a high degree of detail implies limited coverage. To overcome this limitation, many images would have to be processed to yield regional results.

A second scaling approach uses regional-scale satellite sensor images, which when processed yield regional-scale derivatives for input to environmental process models (Fig. 7). Note that this figure includes use of the local-scale remote sensing model with the regional-scale images. The variogram analysis above provides a way to evaluate the validity of this approach. Specifically, the variograms yield information about the information content of particular landscape scales at different resolutions. For example, in the vegetation mapping scheme used for the Plumas National Forest, the estimation of tree size is done using the Li–Strahler model at the level of individual stands (or regions) (Li & Strahler, 1985). However, the data required are intra-region variances – so variance at the pixel scale is essential for this application. The conclusion from the variograms presented in Figs. 4 and 5 is that spatial resolutions above approximately 100 m would not be useful, as the required pixel-scale variance would have been removed from the image. However, if land cover classes are of interest, there is still considerable

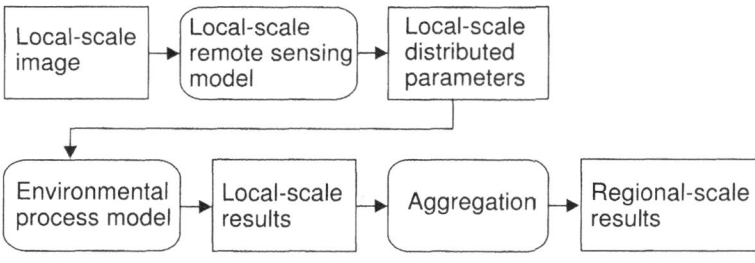

Fig. 6. Schematic drawing of the GIS scaling strategy. Regional-scale results are achieved by aggregating local-scale results.

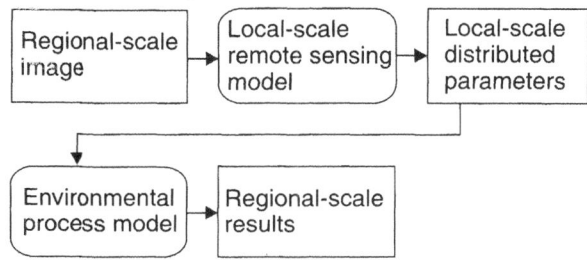

Fig. 7. Schematic drawing of the second scaling strategy, which involves using an image with coarser resolution. A problem that is not addressed is that the nature of the remote sensing model changes when coarse-resolution imagery is used.

variance remaining related to that landscape scale at considerably coarser spatial resolutions.

A third approach combines elements of both the prior two approaches and emphasises the importance of the remote sensing model. Most remote sensing models have been developed and applied at local scales, where they are the most easily calibrated and validated. The example at the beginning of this chapter involving estimation of LAI from NDVI helps illustrate this point. It is much easier to collect field measurements of LAI for calibration and validation purposes for 30 m Landsat TM pixels than for 4 km NOAA AVHRR GAC pixels. The third approach, as illustrated in Fig. 8, involves calibrating the regional-scale remote sensing model through an approach much like the GIS approach discussed above. Through the use of local-scale imagery, local-scale results can be aggregated to the size of pixels in regional-scale imagery. These data can then be used to calibrate the remote sensing model to be used in conjunction with the regional-scale imagery. The regional-scale remote sensing model could then be used in the processing of regional-scale imagery. This approach is differentiated from the second to highlight problems that can occur when remote sensing models calibrated at the local scale are applied to regional-scale satellite sensor images. Particularly if the relationship between the image and the landscape parameters is non-linear, this approach is necessary. This third approach shares the same concerns as the second with respect to detail and the variance remaining in images related to the landscape scale of interest.

To illustrate the relationship between the scaling alternatives discussed here and the variograms for various scales of variation in images, let us return to the problem discussed above of estimating LAI over large areas from remotely sensed images. For this discussion, we

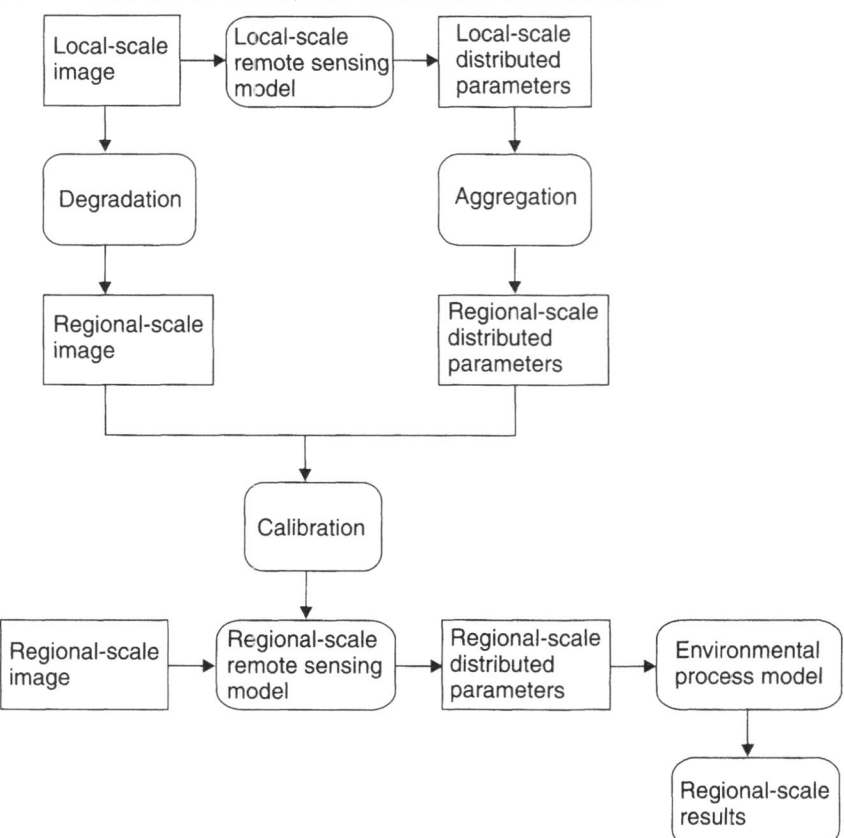

Fig. 8. Schematic drawing of the third scaling strategy. Local-scale data are used to produce local-scale ditributed parameters, which are aggregated to the regional scale. Simultaneously, the local-scale image is degraded to a regional scale and is used with the regional-scale distributed parameters to calibrate a regional-scale remote sensing model, overcoming the major problem with the method presented in Fig. 7. The regional-scale remote sensing model is then used to process regional-scale images.

assume that LAI is a function of both NDVI and vegetation class (Baret & Guyot, 1991). Therefore, there would be a different relationship between LAI and NDVI for each vegetation class.

To use the GIS approach to scaling, local-scale imagery with small enough pixels to detect the NDVI variation within classes would be required. This implies sufficient signal at the pixel or region scales in the images, which is evaluated as the difference between the sill and the observed variograms for the pixel or region scales. Figure 5 suggests that virtually no variation at scales finer than regions could be detected at resolutions much higher than 100 m. However, some variation at the scale of regions is still present at resolutions as high as 250 m. To determine LAI for input to environmental process models at regional scales, such relatively fine-scale imagery would have to be used, and the process would have to be repeated over many local-scale images.

The second scaling alternative implies imagery with large pixels, in which case there is not likely to be sufficient detail in the pixel or region scales to allow applying separate functions for estimating LAI for each vegetation class. For the Plumas image, this situation is illustrated by the lack of pixel- and region-scale variance at distances beyond about 500 m. Lacking sufficient signal at the pixel and region scales, some simplified relationship between NDVI and LAI would have to be used. The problem is that it is unclear what function to use to relate NDVI to LAI, as more than one vegetation class may exist in any single pixel. The temptation is to use the relationships developed at local scales using fine-resolution imagery. The problem is that there is no way to validate this approach. The strength of this approach is the ease with which LAI estimates over large areas can be produced. The drawback is the nature of the function relating LAI to NDVI, and missing the pixel- and region-scale variation in LAI.

The third scaling approach addresses this problem of the function to be used with coarse-resolution imagery through calibration using local-scale imagery. For this approach, the functions relating LAI to NDVI initially developed at the local scale would be applied at the local scale. The LAI estimates would then be aggregated to the pixel size of the coarse-scale images and empirically related to observed NDVI values from the coarse-resolution images. This empirically derived function could then be applied over large areas to produce regional estimates of LAI. This approach requires more work than the second option but has a more solid foundation. It also shares the limitation of the second approach with respect to finding pixel- and stand-scale variation in LAI.

Conclusions

Remote sensing contributes to the scaling of environmental process models by providing the landscape parameters necessary as inputs to the models. When attempting to apply these environmental process models over large areas, there are several approaches available. The selection of an appropriate approach is linked to the relationship between the spatial resolution of the satellite sensor imagery and the scale of the landscape parameters of interest. Isolating the effects associated with different landscape scales in variograms helps in the evaluation of scaling strategies. The critical factor is the magnitude of the image variance related to the landscape scale of interest that remains in images as a function of resolution.

References

Atkinson, P.M. (1991). Optimal ground-based sampling for remote sensing investigations: estimating the regional mean. *International Journal of Remote Sensing*, 12, 559–567.

Atkinson, P.M. & Curran, P.J. (1995). Defining an optimal size of support for remote sensing investigations. *IEEE Transactions on Geoscience and Remote Sensing*, 31, 768–776.

Atkinson, P.M., Curran, P.J. & Webster, R. (1992). Cokriging with ground-based radiometry. *Remote Sensing of Environment*, 41, 45–60.

Avissar, R. (1992). Conceptual aspects of a statistical-dynamic approach to represent landscape subgrid-scale heterogeneities in atmospheric models. *Journal of Geophysical Research* 97, 2729–2742.

Band, L.E., Peterson, D.L., Running, S.W. *et al.* (1991). Forest ecosystem processes at the watershed scale: basis for distributed simulation. *Ecological Modelling*, 56, 171–196.

Baret, A.F. & Guyot, A.G. (1991). Potentials and limits of vegetation indices for LAI and APAR assessment. *Remote Sensing of Environment*, 35, 161–173.

Cohen, W.B. & Spies, T.A. (1991). Semi-variograms of digital imagery for analysis of conifer canopy structure. *Remote Sensing of Environment*, 34, 167–1787.

Collins, J.B., Woodcock, C.E. & Jupp, D.L.B. (1995). Spatial dependence and nested hierarchical scene models. In *1995 ACSM/ASPRS Annual Convention and Exposition Technical Papers*, 535–544.

Curran, P.J. (1988). The semivariogram in remote sensing: an introduction. *Remote Sensing of Environment*, 24, 493–507.

Curran, P.J. & Williamson, H.D. (1988). Selecting a spatial resolution for estimation of per-field green leaf area index. *International Journal of Remote Sensing*, 9, 1243–1250.

Ehleringer, J.R. & Field, C.B. (1993). *Scaling Physiological Processes: Leaf to Globe*. San Diego, CA: Academic Press.

Hall, F.G., Huemmrich, K.F., Goetz, S.J., Sellers, P.J. & Nickeson, J.E. (1992). Satellite remote sensing of surface energy balance: successes, failures, and unresolved issues in FIFE. *Journal of Geophysical Research* 97, 19 061–19 089.

Jupp, D.L.B., Strahler, A.H. & Woodcock, C.E. (1988). Autocorrelation and regularization in digital images I: basic theory. *IEEE Transactions on Geoscience and Remote Sensing*, 26, 463–473.

Jupp, D.L.B., Strahler, A.H. & Woodcock, C.E. (1989). Autocorrelation and regularization in digital images II: simple image models. *IEEE Transactions on Geoscience and Remote Sensing*, 27, 247–258.

Li, X. & Strahler, A.H. (1985). Geometric-optical modelling of a conifer forest canopy. *IEEE Transactions on Geoscience and Remote Sensing*, 23, 705–721.

Moellering, H. & Tobler, W. (1972). Geographical variances. *Geographical Analysis*, 4, 35–50.

Mooney, H.A. & Field, C.B. (1989). Photosynthesis and plant productivity – scaling to the biosphere. In *Photosynthesis*, ed. W.R. Briggs, pp. 19–44. New York: Alan R. Liss.

Running, S.W., Nemani, R.R., Peterson, D.L. *et al.* (1989). Mapping regional forest evapotranspiration and photosynthesis by coupling satellite data with ecosystem simulation. *Ecology*, 70, 1090–1101.

Webster, R., Curran, P.J. & Munden, J.W. (1989). Spatial correlation in reflected radiation from the ground and its implications for sampling and mapping by ground-based radiometry. *Remote Sensing of Environment*, 29, 67–78.

Woodcock, C.E. & Harward, V.J. (1992). Nested-hierarchical scene models and image segmentation. *International Journal of Remote Sensing*, 13, 3167–3187.

Woodcock, C.E., Strahler, A.H. & Jupp, D.L.B. (1988a). The use of variograms in remote sensing: I. Scene models and simulated images. *Remote Sensing of Environment*, 25, 323–348.

Woodcock, C.E., Strahler, A.H. & Jupp, D.L.B. (1988b). The use of variograms in remote sensing: II. Real digital images. *Remote Sensing of Environment*, 25, 349–379.

Woodcock, C.E., Collins, J., Gopal, S. *et al.* (1994). Mapping forest vegetation using Landsat TM imagery and a canopy reflectance model. *Remote Sensing of Environment*, 50, 240–254.

B. KRUIJT, S. ONGERI and P.G. JARVIS

Scaling of PAR absorption, photosynthesis and transpiration from leaves to canopy

Introduction

Prediction of carbon fixation and water loss of vegetation in a changing climate requires a model that is sensitive to those environmental variables that are likely to change. In addition, the number of structural, optical and physiological parameters of the vegetation used by the model should be kept to a minimum to allow application at regional and global scales. Such parameters may in practice vary spatially within the vegetation and vary slowly in time. This paper evaluates approaches to minimise the requirement for spatially varying parameters.

Physiological and optical parameters are frequently known at the scale of individual leaves, shoots or plants, but not at the vegetation scale, because they are determined from measurements made in a cuvette or chamber. Values of such parameters in the literature are mostly given at leaf scale for this reason, but also because, for a given plant species, leaves are considered basic elements of a vegetation canopy.

If only leaf-scale parameters are known, carbon and water fluxes for a vegetation canopy as a whole can only be predicted by '*bottom-up*' scaling, calculating the fluxes for each leaf separately and then summing them over all leaves in the canopy (Jarvis, 1993). Such an approach requires detailed knowledge of the variation of parameters throughout the canopy, depending on position and age, as well as on knowledge of the distribution of foliage and of environmental variables in the canopy. These heavy demands on model input often make 'bottom-up' scaling impractical for regional- and global-scale applications. As an alternative, one can measure canopy fluxes directly above the canopy, using eddy co-variance techniques, and determine bulk canopy parameters from these fluxes, in what can be classified as a '*top-down*' approach. The bulk parameters then characterise a 'big leaf' model, in which the whole canopy is treated as a single unit with one set of properties. 'Big leaf' models are more appropriate than 'bottom-up' models for application at regional and global scales, but there are not enough measured surface flux records available for most landscapes to allow application of a 'top-down' approach. In contrast, scattered leaf-scale information is frequently available for the species in different vegetations.

A useful approach, therefore, is to develop a 'big leaf' model using leaf-scale parameters through 'bottom-up' scaling (see Berry *et al.*, this volume). A 'bottom-up' model can be used to predict canopy fluxes for part of a data set, and these fluxes can then be used to define bulk canopy parameters. Alternatively, mathematical rules can be defined to relate canopy-scale parameters to leaf-scale parameters.

In general, parameters at canopy scale are not identical to parameters at leaf scale, and they cannot be treated as simple averages or totals of the parameters of all leaves in the canopy. Two phenomena are largely responsible for this. First, the leaves in a canopy are not independent but interact with each other and with the surrounding atmosphere, modifying their own and their neighbour's environment. For example, the radiative absorptivity of a canopy as a whole may differ from the absorptivities of the individual leaves, because leaves shade each other. Second, several processes that are defined by leaf-scale parameters respond non-linearly to environmental variables within a canopy. Photosynthesis is a typical example of such a process, because its dependence on photosynthetically active radiation (PAR) saturates at high PAR flux density, whereas PAR varies strongly with depth into a canopy. Therefore, in averaging leaf-scale parameters within a canopy, the parameters must be weighted by the actual leaf-scale flux they are predicting for a given set of environmental conditions (Sellers *et al.*, 1992; McNaughton, 1994).

The aim of this paper is to investigate these two particular problems in scaling-up through comparisons of predictions made with a spatially explicit, three-dimensional canopy radiation and physiology model, MAESTRO (Wang & Jarvis, 1990a), and with two simplified models. First, the role of interactions amongst leaves in PAR absorption by canopies and the importance of horizontal heterogeneity in the spatial distribution of foliage will be addressed. Simulations made with two contrasting 'bottom-up' models will be compared, one assuming horizontal heterogeneity (SAIL, Verhoef, 1984) and the other treating grouping of foliage into separate tree crowns explicitly (MAESTRO). Second, the integration of non-linear responses of photosynthesis and transpiration will be addressed for the special case of *full acclimation* of leaf physiology to the local PAR environment in the canopy. As a test of this assumption, simulations will be compared made with MAESTRO, a detailed 'bottom-up' model, and with a 'bulk physiology' model, which is similar to a 'big leaf' model except that the bulk PAR absorption is still modelled by MAESTRO. The specific objective is to evaluate the hypothesis put forward by Sellers *et al.* (1992) that canopy fluxes can be simulated as a 'big leaf' based on the physiology of the topmost leaves and the bulk canopy PAR absorptivity, provided that full acclimation occurs.

The three models, varying in their degree of spatial detail, are described in the next two sections and are schematically depicted in Fig. 1. MAESTRO computes the distribution of radiation absorption in an individual tree crown, taking into account shading and scattering by surrounding trees and by leaves within the crown itself. Physiological processes are then simulated for a large number of grid points within the tree canopy. SAIL calculates radiation absorption and reflection in a more sophisticated way, but only for the bulk of the foliage, which it assumes to be horizontally homogeneous in distribution. The third class of model considered here represents the use of scaled-up canopy parameters, in a 'big leaf' or 'bulk physiology' model for which bulk canopy radiation absorptance is given as measured or externally simulated input.

This paper considers questions relating to scaling-up in *space*. There are other interesting issues relating to scaling-up in *time*, such as the non-linear effects of minute-scale dynamic responses of stomata and photosynthesis to sun flecks. Such responses are not represented in any of the models considered here. The notation used in the analysis is listed at the end of this chapter.

Interactions and PAR absorption

SAIL (Verhoef, 1984) is a model of the interactions between electro-magnetic radiation and vegetation surfaces; it was originally designed to predict the bi-directional reflectance distribution function (BRDF) of vegetation canopies. It is an example of a so-called turbid medium approach in which individual canopy elements are treated as small absorbing and scattering particles with defined optical properties and orientations. The canopy is treated as horizontally uniform and consequently the radiation field is a function of the vertical co-ordinate only. The canopy is considered as a single layer over a Lambertian reflecting soil background. These assumptions lead to the expectation that such a model would best represent dense, horizontal, uniform canopies composed of relatively small leaves.

SAIL is an extension of the earlier SUITS model (Suits, 1972) and as such is based on the Kubelka–Munk solution to the radiative transfer

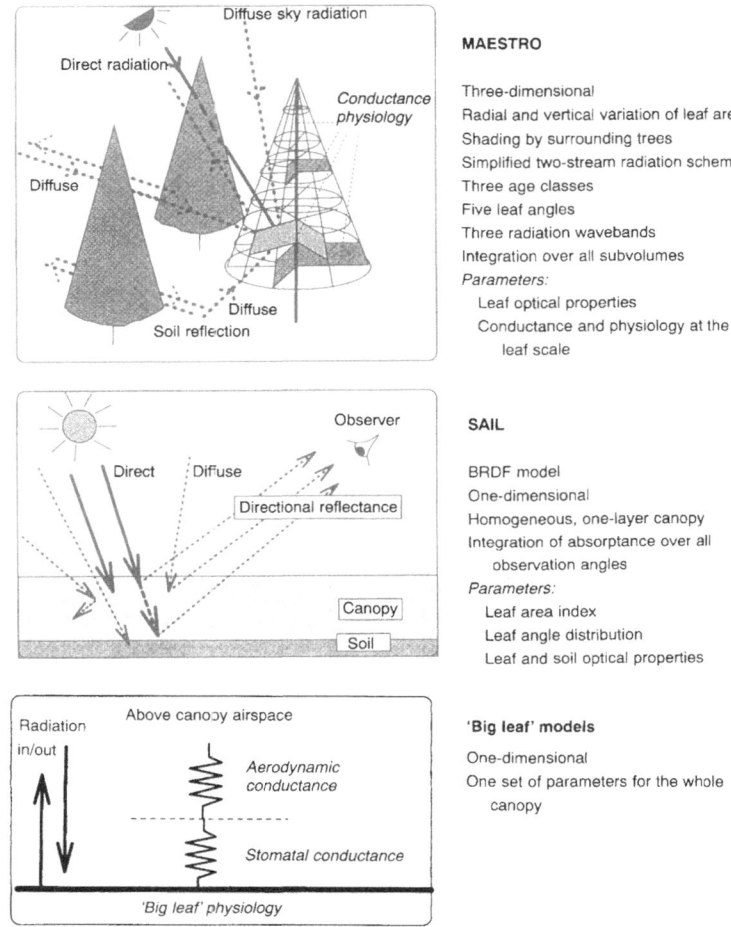

Fig. 1. Three different levels of canopy model complexity: three-dimensional, one-layer and 'big leaf'.

equation of the canopy. Diffuse light within the canopy is described in terms of up- and down-welling fluxes perpendicular to the plane of the medium. The direct beam flux is assumed to exist only in the downward direction, so that a direct flux incident on vegetation or soil surfaces is reflected as diffuse light (Lambertian reflectance). The canopy reflectance is found from the radiance in the viewing direction by solving a set of differential equations containing terms for each flux, and the various absorption and scattering coefficients for the canopy, which in turn depend upon leaf area index (LAI), leaf angle distribution and leaf optical properties.

The model predicts the spectral reflectance, for a given sun–observer geometry (expressed by the sun and observer zenith angles and the relative azimuth angle of the observer from the sun), from the following input parameters: canopy architecture (LAI and leaf angle distribution), leaf and soil optical properties, and the ratio of diffuse to direct solar radiation, as depicted in Fig. 1. A range of spectral bands can be treated in this way, but SAIL is most commonly used with PAR (400 nm to 700 nm) and NIR (near-infrared) (700 nm to 1100 nm) (e.g. Goward & Huemmrich, 1992).

The model predicts the reflected radiation, which includes components for both canopy and soil, as a fraction of the incident radiation in a particular waveband. Therefore, in order to calculate the canopy absorption the soil reflection component must be separated. The canopy absorption may then be found from:

$$I_{ac} = I_i - I_r - I_{as}$$

where I is any radiation, subscript a refers to absorbed, c refers to canopy, i refers to incident, r to reflected and s to soil.

To calculate the total radiation absorbed by the canopy in 1 hour, the model was integrated over all observer zenith and relative azimuth angles, for all positions of the sun during that time period, using a ten-point Gaussian quadrature integration method.

A comparison of predicted canopy absorption at PAR wavelengths was made with the three-dimensional canopy model MAESTRO (Wang & Jarvis, 1990a).

MAESTRO contains a broadly similar, if slightly simpler, radiative transfer model but differs markedly from SAIL in that horizontal homogeneity of the foliage distribution is not assumed. By contrast, MAESTRO contains grouping of the foliage at two scales: within a stand, grouping is into defined tree crowns while within crowns, the foliage is allowed to vary in both the vertical and horizontal directions according to empirical, continuous functions (Wang, Jarvis & Benson, 1990).

MAESTRO splits the trees into a number of subvolumes (Fig. 1) and calculates the penetration of radiation to each subvolume by considering direct and diffuse radiation separately, depending on solar zenith angle, azimuth angle and leaf inclination angle. The radiation to each subvolume is calculated as the probability of light penetration along a given light path, taking into account the leaf area densities in the surrounding trees, and in the 'target' tree. Additionally, MAESTRO evaluates penetration of scattered radiation by surrounding leaves and the soil by constructing, for each point, an 'equivalent one-dimensional canopy', where the point concerned is positioned according to its relative location along the path of PAR from canopy entry to the soil. The methods for calculating scattering are similar to those used in SAIL, except that only single scattering is considered. This is a good approximation for PAR wavelengths for which leaf absorptivity is high. The total absorbed radiation in each subvolume is finally found by integrating over all leaf inclination angles for the direct component as well as over all diffuse irradiance angles for the diffuse component. The radiation routines of MAESTRO have been tested by Wang & Jarvis (1990a).

The comparison was made using experimental data from a stand of Sitka spruce (*Picea sitchensis*) at Blairadam, Scotland, collected in June 1992 by J. Massheder, Edinburgh. These data included hourly incident PAR and fraction of beam and diffuse radiation.

Results

Figure 2a shows the variation of canopy and total absorptance during part of a day, as simulated by SAIL. It can be seen that for low LAI (< 3) canopy absorptance decreases towards noon, when the sun is highest in the sky, because there is a greater likelihood at that time of radiation penetrating the canopy to reach the soil than at the lower solar elevation angles occurring earlier or later in the day. The total radiation absorptance increases slightly towards noon, indicating that the amount of PAR being absorbed by the soil is increasing significantly. When the sun is low in the sky (early, or late in the day), PAR has a much longer pathlength through the canopy before reaching the soil and is consequently more likely to be intercepted by the vegetation; this is evident as a higher canopy absorptance. For dense canopies (LAI = 7), decrease in canopy absorptance towards noon is much less marked, since, at these large leaf areas, little solar radiation is able to reach the soil without being intercepted by the foliage, even for very small solar zenith angles.

Figure 2b shows a comparison between leaf and canopy absorptance as predicted by SAIL. Even at low leaf areas (LAI = 2), the absorptance of a canopy of reflective leaves (i.e. leaves of low absorptance) is higher than the leaf absorptance, because PAR reflected or transmitted by leaves is absorbed by other leaves in the canopy. For very highly absorbent leaves, little radiation is reflected or transmitted to be intercepted by other leaves, but there are still gaps in the canopy through which some radiation penetrates to the soil where it may be absorbed,

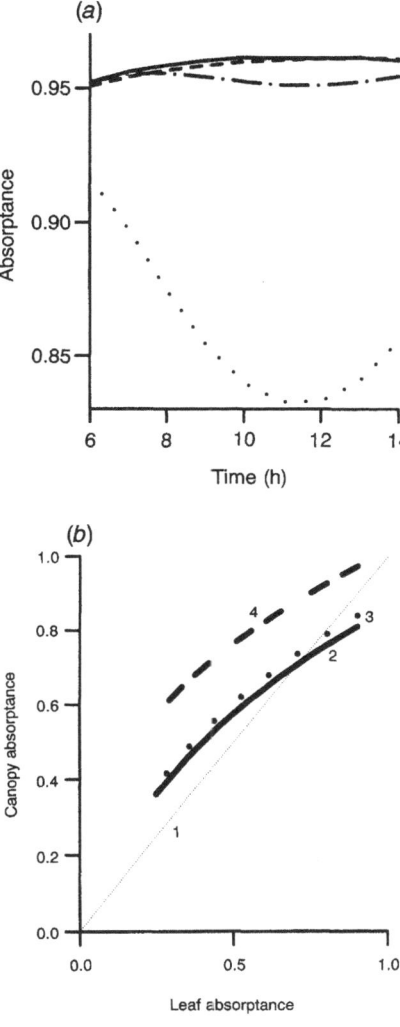

Fig. 2. (*a*) Canopy absorptance and total absorptance (canopy + soil) plotted against time of day, as predicted by the SAIL model, for two different values for LAI. LAI = 3: canopy (· · ·); total (---). LAI = 7: canopy (-·--); total (–). (*b*) Canopy absorptance plotted against leaf absorptance as predicted by the SAIL model, for two different values of LAI and two different soil reflectances (ρ_s). Leaf reflectance and transmittance were 0.1 and 0.05, respectively. 1, canopy : leaf absorptances of 1 : 1; 2, LAI = 2 and ρ_s = 10%; 3, LAI = 2 and ρ_s = 30%; 4, LAI = 7 and ρ_s = 10%.

and some of the radiation reflected by the soil escapes back upwards
through the gaps. Consequently, canopy absorptance may be less than
individual leaf absorptance, and this is more likely to occur with more
absorbent leaves and relatively sparse canopies. For dense canopies
(LAI = 7), canopy absorptance is always higher than leaf absorptance,
since the probability of a gap existing in the canopy is small. Moreover,
as leaf absorptance becomes high the canopy absorptance approaches
the value of leaf absorptance, since for dark leaves in a dense canopy
almost all light is absorbed. Higher soil reflectance results in increased
canopy absorptance, since more light is reflected from the soil back into
the canopy where it may be intercepted by leaves. All the curves show
a non-linear relationship between leaf and canopy absorptance, and
further they show variability with important canopy parameters such as
LAI. This emphasises the point that scaling-up from leaf absorptance
to canopy absorptance must be done through the use of canopy models.

Figure 3 shows a comparison between canopy absorptance predicted
by MAESTRO and SAIL for two different LAIs and two different
canopy closures. For dense, closed canopies the two models agree very
closely. However, when more gaps exist in the canopy, SAIL overestimates absorbed radiation relative to MAESTRO. For a closed
canopy with an LAI of 1, the maximum absorbed radiation is less than
for a closed canopy with an LAI of 7 because the leaf area density is
lower. For open canopies, the relative overestimation is large, indicating
that patches of bare soil significantly affect predictions of absorbed
radiation, but there is also more scatter around the 1:1 line. However,
changes in LAI are less significant for open canopies. For the open
canopies there was only 20% ground coverage (i.e. a decrease of
80%), and, as can be seen from Fig. 3, the decrease in absorbed PAR
(around 50%) is clearly not proportional to the decrease in ground
cover. These results lead us to conclude that radiation absorption
models should include terms for gaps between the vegetation in order
to represent sparse canopies more accurately.

Non-linearity in photosynthesis and transpiration

Variation of photosynthetic parameters within canopies

To integrate leaf physiological processes through a canopy, it is a prerequisite to know whether the leaf-scale parameters are constant
throughout the canopy. Numerous studies have shown that this is rarely
the case. Photosynthetic and stomatal characteristics of leaves show
phenotypic plasticity and adjust to the prevailing environment, particu-

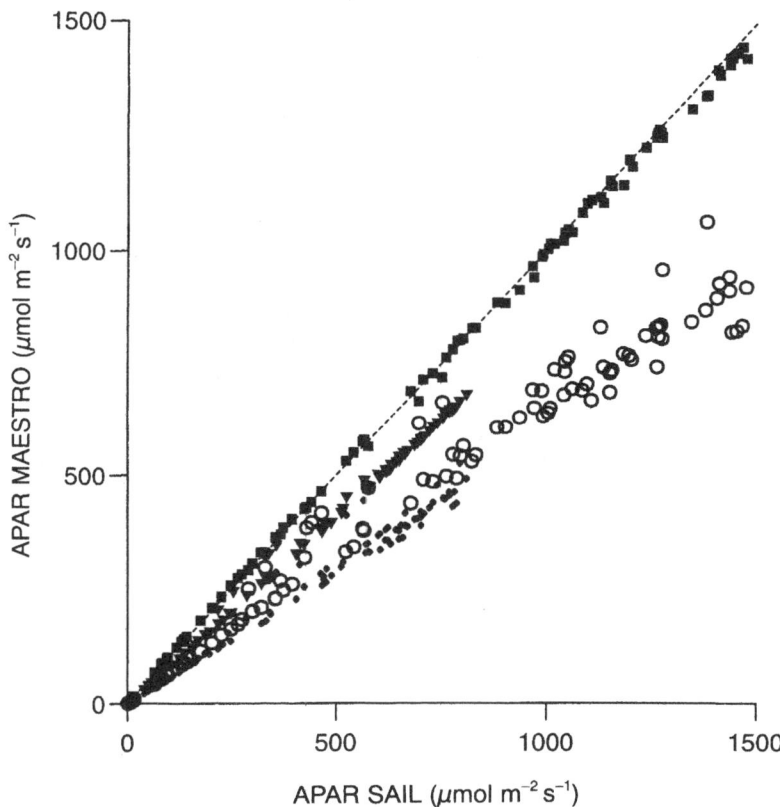

Fig. 3. Absorbed PAR (APAR) predicted by the SAIL model plotted against APAR predicted by the MAESTRO model, for two different values for LAI and for both open and closed canopies. The 'openness' of the canopy in MAESTRO was varied by assuming different radii of the tree crowns, keeping LAI the same, and thus increasing the leaf area density within crowns. The dotted line is exact agreement between the two models (1 : 1). Closed canopy: LAI = 7 (■); LAI = 1 (▼). Open canopy: LAI = 7 (○); LAI = 1 (●).

larly to temperature (e.g. Björkman, 1981a), CO_2 (Sage, Sharkey & Seemann, 1989) and radiation (Björkman, 1981b). This phenotypic plasticity may be associated with modifications during development that lead to, for example, variable leaf thickness, number of cell layers or stomatal density (Weyers, Lawson & Peng, this volume). Additionally, on shorter time scales of several days, changes occur in the amounts of photosynthetic enzymes (e.g. Björkman *et al.*, 1972) and, on much shorter time scales, in their activity (Sage *et al.*, 1989).

A number of elegant transfer experiments have demonstrated that reduction in photosynthetic capacity on transfer from high to low irradiance is associated with reduction in the amount of virtually all photosynthetic enzymes measured, including rubisco and cytochrome *f* (Evans, 1988). In other words, the photosynthetic system is a well-adjusted system that does not, in general, carry much excess capacity but which is able, over a period of five to ten days, to adjust, even in preformed leaves, either to exploit an increase in average irradiance or to re-distribute resources after a decrease in average irradiance.

The majority of transfer experiments have been done in controlled environments where the variable has been the *flux density* of radiation. By contrast, in canopies in the field, there are also large reductions in the ratio of red/far-red radiation with depth, as a result of the preferential absorption of red light by the chlorophyll above, and consequently major changes in the photo equilibrium state of phytochrome. This adds an additional component to the acclimation of photosynthesis and stomata to the local radiation environment that is generally not mimicked in experiments but which strongly affects the acclimation responses that occur during leaf development (Smith, 1986) and which are seen in canopies in the field (e.g. Lewandowska, Hart & Jarvis, 1977). Using phytochrome-deficient mutants, Smith, Samson & Fork (1993) concluded that 'reduced phytochrome levels affect some biochemical/ physiological processes, or some structural factors at the leaf level, that limit the maximum rate of photosynthesis measured under optimal conditions after acclimation to green shade'. However, in natural stands, reductions in both PAR flux density and in red/far-red ratio are largely caused by absorption by chlorophyll and are, therefore, likely to be correlated.

In forest canopies, photosynthetic capacity, maximum stomatal conductance, base respiration rate and leaf nitrogen, expressed on an area basis, all decrease with increasing depth of foliage. The evidence from transfer experiments leads one to expect that these reductions are causally related to the absorbed PAR flux density at a particular depth, integrated over about a week, and to the average red/far-red ratio of the radiation absorbed during the period of leaf development.

A number of studies in canopies (Lewandowska *et al.*, 1977; Field, 1983; Hirose & Werger, 1987; Pons *et al.*, 1989) have shown correlations between photosynthetic parameters or nitrogen concentration and radiation or depth into the canopy, but the strength of the acclimation that these correlations imply apparently varies between species and over time (Oberbauer & Strain, 1986; Neilson, quoted by Jarvis & Sandford, 1986). The field study that comes closest to simultaneous correlation of the distribution of photosynthesis, PAR absorption and nitrogen is that of Evans (1993) within a lucerne canopy. We know of no similar comprehensive study in a forest canopy.

In the following, we investigate the consequences of assuming *full acclimation to the PAR distribution* for simulating the carbon and water exchange of whole forest canopies.

Modelling photosynthesis and transpiration

First, the physiological relationships used in the present study are outlined to indicate their dependence on PAR. Photosynthesis has been modelled using the set of equations proposed by Farquhar & von Caemmerer (1982). The same equations, but with different parameters, have been used for every point in all modelled canopies. This model expresses the CO_2 assimilation rate as being either limited by the electron transport rate, J, or, if PAR saturated, by rubisco activity, which sets a maximum carboxylation rate, V_{cmax}. In both cases, CO_2 assimilation rate also depends on temperature T and internal CO_2 concentration C_i, but these relationships will not be discussed further here since these variables were kept the same for all the model comparisons. Net photosynthetic rate, A, can, therefore, be expressed as:

$$A = \min \left\{ \begin{array}{l} V_{cmax} \cdot f(T, C_i) \\ J \cdot g(T, C_i) \end{array} \right\} - R_d \qquad (1)$$

where f and g are different functions of T and C_i, and R_d is the actual 'dark' respiration rate. R_d is modelled as the product of a base respiration rate at 0 °C, R_{db}, and an exponential temperature dependence with a coefficient equivalent to a Q_{10} of about 2, where $Q_{10} = (R_d(T + 10)/R_d(T))$. J follows a PAR-response curve that can be described by (Evans & Farquhar, 1991):

$$J = \left\{ I_a + J_{max} - \sqrt{((I_a + J_{max})^2 - 4\Theta I_a J_{max})} \right\} / 2\Theta \qquad (2)$$

where I_a is PAR absorbed by the photosystems, J_{max} is the maximum electron transport rate, and Θ is a curvature factor (Leverenz *et al.*, 1990).

Since $A = g_{sp}(C_a - C_i)$, Equation (1) can be solved for A as a function of the ambient CO_2 concentration, C_a, and the stomatal conductance for photosynthesis, g_{sp}, replacing the unknown C_i. The modelling of g_s is subject to some controversy (Aphalo & Jarvis, 1993; Berry *et al.*, this volume). One of the issues is whether C_i/C_a should be conserved as a near-constant value. Here, we retain the approach of Jarvis (1976), in which stomatal conductance is expressed as:

$$g_{sp} = g_{sw}/1.6$$

$$= (g_{smax} \cdot \alpha(I_a + \beta))/(g_{smax} + \alpha(I_a + \beta)) \cdot h(T, D_v, C_a)/1.6 \qquad (3)$$

where G_{smax} and G_{sw} are the maximum and actual values of stomatal conductance for water vapour, respectively, while α and β are the initial slope and intercept of the PAR response. The function h depends on T, D_v (the water vapour saturation deficit of the ambient air) and C_a. The empirical question (3) does not automatically ensure a constant C_i/C_a, but in practice the PAR and CO_2 responses of photosynthesis and stomatal conductance are such that this ratio is kept fairly constant.

Transpiration (E) was modelled using the Penman–Monteith equation (Monteith, 1973), as expressed by Jarvis & McNaughton (1986):

$$E = c_1(1 - \Omega)g_{sw}D_v + c_2\Omega Q_a \qquad (4)$$

where c_1 and c_2 comprise temperature-dependent coefficients not dealt with further here. The factor Ω ($1 > \Omega > 0$) defines the decoupling of the transpiration process from direct dependence on D_v. The decoupling factor is a function of the ratio of g_{sw} and g_b, the aerodynamic boundary-layer conductance of the leaves, which decreases with windspeed and thus with depth into the canopy. Absorbed radiation in all wavelengths, Q_a, is commonly referred to as *available energy*.

The consequences of assuming acclimated photosynthesis

The equations for photosynthesis and transpiration are non-linear with respect to I_a. Therefore, rigorous upscaling of the parameters V_{cmax}, J_{max}, g_{smax} and R_{db} from leaf scale to canopy scale involves weighting every value with the flux predicted at each leaf and averaging this over all leaves in the canopy. Even if these parameters are constant throughout the canopy, this results in an impractical integral function for the determination of canopy-scale parameters (e.g. Sellers *et al.*, 1992; McNaughton, 1994). If, however, the parameters reflect full acclimation to the local PAR environment, as reviewed in the previous section, Sellers *et al.* (1992) show that integration becomes straightforward. Somewhat modified, their approach is reiterated here to provide the basis for a 'bulk physiology' canopy model of photosynthesis and transpiration.

Full acclimation implies that if the values of V_{cmax}, J_{max} and g_{smax} are known at any reference location in the canopy the values of these parameters in all other locations are known from:

$$B_e = B_0 \cdot \overline{I}_{ae} / \overline{I}_{a0} = B_0 \cdot p_e \qquad (5)$$

where B denotes one of the parameters, the overbars denote averages over a sufficiently long acclimation period, p is the ratio of average PAR absorbed in each canopy element (subscript e) and in a reference element (subscript 0) that can be anywhere in the canopy. The parameters are strong functions of the concentrations of photosynthetic enzymes, such as rubisco, and since a large part of the leaf nitrogen is in those enzymes, the distribution of nitrogen through the canopy will be very similar to the distribution of the parameters. Although the influence of nitrogen is not explicit in the assumption of full acclimation, its distribution is important for respiration, since it is probable that R_{db} is a function of leaf nitrogen (Ryan, 1995). It is, therefore, legitimate to assume that Equation (5) also holds for R_{db}.

If the distribution of absorbed PAR through the canopy is constant over time, if C_i/C_a is constant through similar dependence on PAR of g_{sp} and A, and if β is small, then after substitution of Equation (5) into Equations (1) to (3), photosynthesis becomes linear in p_e. For Equations (2) and (3), this requires further algebra, which will not be shown here. If we further define an analogous ratio, q_e, for the ratio of average available energy at each canopy element, Q_a, to that of the reference element, apply this to the second term in Equation (4), and assume Ω to be nearly constant through the canopy (i.e. that g_{sw} and g_b decrease similarly with canopy depth), transpiration also becomes linearly dependent on the relative absorbed radiation through the canopy.

From the foregoing, instantaneous leaf photosynthesis, conductance and transpiration in a fully acclimated canopy depend on three sets of variables: first, the values of the parameters and absorbed PAR in the reference location, second, a function of T, D_v and C_a, and, finally, p_e and q_e. In a rough or open canopy such as a forest, T, D_v and C_a do not vary much with canopy depth, whereas p_e and q_e vary strongly. Therefore, integration of photosynthesis over all leaves above a unit ground area simplifies to integrating p_e and q_e over all canopy elements up to a factor Π for PAR and Ξ for available energy, representing the radiation absorption by the canopy relative to absorption by the leaves at the reference location. These factors are then used to scale the photosynthesis in the reference location in the following way, from Equations (1), (2), (4) and (5):

$$A_c = \left(\min\left\{ \begin{array}{c} V_{cmax0} \cdot f(T, C_i) \\ J(J_{max0}, I_{a0}) \cdot g(T, C_i) \end{array} \right\} - R_{d0} \right) \cdot \sum_{e=1}^{N_e} (p_e \cdot l_e)$$ (6)

$$= A_0 \cdot \overline{I_{ac}} / \overline{I_{a0}} = A_0 \cdot \Pi$$

and

$$g_{sc} = g_{s0} \cdot \Pi$$ (7)

$$E_c = c_1(1 - \Omega)g_{sw0}D_v \cdot \Pi + c_2\Omega Q_{a0} \cdot \Xi$$ (8)

where A_0 is net photosynthetic rate and R_{d0} dark respiration, both at the reference location, subscript c refers to a bulk canopy value, l_e is the leaf area in element e, and N_e is the number of canopy elements above a unit ground area. A_c and E_c are expressed on a unit ground area basis, whereas the fluxes at the reference location are expressed on a unit leaf area basis. This is accounted for by Π and Ξ, which have units of (m^2 leaf/m^2 ground).

The approach leading to Equations (7) and (8) is in essence similar to that of Sellers *et al.* (1992), but there are some significant differences. First, Sellers *et al.* (1992) used a linear PAR dependence for photosynthesis limited by electron transport, and a quadratic equation to allow a smooth transition between rates limited by radiation and rubisco; second, stomatal conductance was modelled using the Ball–Berry approach (Ball, 1988); and third, their results were rather less flexible, because they relied on parameter values at a fixed reference, defined as the average value of leaves in the top of the canopy, and they used an exponential PAR extinction coefficient. We show that in principle any reference location can be used, as long as I_a at that point is known and the total radiation absorbed by the canopy is either measured (e.g. Russell, Jarvis & Monteith, 1989) or calculated using a radiative transfer model such as SAIL or MAESTRO.

Equations (6), (7) and (8) define a 'big leaf' model that may be applied simply using *leaf* parameters obtained at *one* canopy location. Given the assumption of full acclimation of the parameters, there is one other condition that is probably not always valid: the distribution of absorbed PAR, assumed constant, may well vary over time, because of variation in solar position or varying contributions of direct and diffuse radiation. In every canopy element, p_e will often deviate from the average by a factor $\gamma_e = (I_{ae}/I_{a0})/p_e$, and this is not always unity everywhere in the canopy, so the linearity and simple integration break down at certain times of the day. An analogous factor can be defined for q_e. Because of the importance of p_e and q_e, it is desirable to consider the values of these factors as well.

In the following, we compare predictions of photosynthesis and transpiration by a three-dimensional radiation distribution model (MAESTRO) with the integrated 'bulk physiology' approach. The foregoing treatment predicts that, in conditions of assumed full acclimation, the 'bulk physiology' model is mathematically very similar to a spatially explicit model. But it was also predicted that a variable PAR profile may introduce deviations, and the extent of such deviations may depend on canopy structure. Also, deviations of C_i/C_a and g_s/g_b from spatial constancy may cause discrepancies between a bulk model and MAESTRO.

Comparison of MAESTRO and 'bulk physiology' model

The 'bulk physiology' model used here essentially simulates fluxes from a reference leaf. Therefore, it will be referred to as 'Reference leaf model' (RLM) from now on. A comparison between MAESTRO and RLM has been made using parameters and meteorological input data from a closed stand of Sitka spruce (*Picea sitchensis*) in Scotland (Blairadam), for a week in June 1992. To investigate the sensitivity of simulated fluxes of CO_2 and water vapour to a variable PAR profile, simulations were done with MAESTRO and the RLM for different times of day, and different stand structure parameters (LAI, leaf area density, tree spacing and tree dimensions). Tree height (6 m) was kept constant, with foliage present from the ground upwards. A summary of the stand configurations used is given in Table 1.

MAESTRO was set up to calculate a moving average, with a time constant of seven days (properly initialised at the beginning of a period), for I_a in every model subvolume, as well as for I_{ac} and Q_{ac}. Equation (5) was applied to all photosynthesis, respiration and transpiration parameters. The topmost subvolume was chosen as the reference (to accord with Sellers *et al.*, 1992), and the parameters were defined for leaf surfaces normal to the sun. The parameters V_{cmax0}, J_{max0} and the reference base respiration rate, R_{db0}, were assumed to be linearly dependent on the leaf nitrogen concentration in the reference volume, according to relationships by Harley *et al.* (1992) and Field (1983, for R_{db0}). This concentration was assumed to be 4 g m^{-2}. For calculation of stomatal conductance, g_{smax0} was assumed to be 0.4 mol m^{-2} s^{-1}.

The RLM essentially consists of Equations (6), (7) and (8). The same

Table 1. *A summary of the crown and stand structural properties varied in the comparison of MAESTRO and RLM. For every combination of stand LAI and tree crown basal radius used in simulations, the tree spacing and average leaf area density ($m^2 m^{-3}$) within the crowns are given. The large crown radius of 3.9 m was chosen to approximate the properties of a horizontally homogeneous canopy*

LAI	Crown radius 0.5 m		Crown radius 1.9 m		Crown radius 3.9 m	
	Tree spacing (m)	LAD	Tree spacing (m)	LAD	Tree spacing (m)	LAD
0.5	8	10.19	–	–	–	–
1.0	2	1.27	2	0.09	4	0.08
7.0	2	8.89	2	0.62	4	0.59
	4	35.56	–	–	–	–

routines were used for photosynthesis and transpiration as in all MAES-TRO subvolumes, and the MAESTRO reference subvolume was used to represent the whole canopy. *The factors Π and Ξ were calculated by MAESTRO.* Therefore, the difference between MAESTRO and the RLM was the explicit calculation of the distribution of absorbed radiation and canopy element fluxes in MAESTRO, which enabled variation of γ_e along the PAR gradient.

Before proceeding to the results of the analysis, it is useful once more to list the assumptions on which the model parameterisations were based.

1. There is full acclimation of photosynthetic and stomatal conductance parameters to the local PAR environment
2. Acclimation to PAR is the *only* factor causing variation in the parameters through the canopy; other variables such as a dependence on needle age, complicate the analysis and are neglected here
3. The pattern of acclimation coincides with the allocation pattern of leaf nitrogen
4. Apart from radiation and windspeed, all other environmental variables are constant throughout the canopy.

Results

In Fig. 4, predictions by RLM are plotted against predictions by MAESTRO, for photosynthesis and transpiration, respectively. The

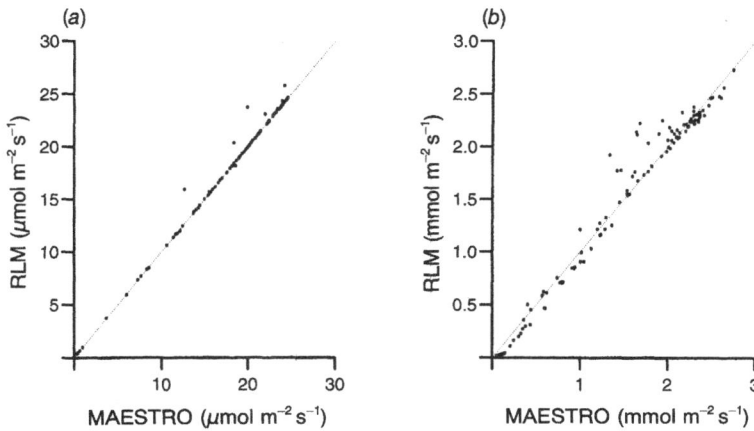

Fig. 4. Comparison of predictions of photosynthesis (CO_2 flux) (*a*) and transpiration (H_2O flux) (*b*) by RLM and MAESTRO. The simulations in this case were made for a closed Sitka spruce canopy with an LAI of 7. The spacing was 2 m and the tree crown radii were 1.9 m (see also Table 1).

plots show an almost perfect correspondence of the two models for photosynthesis, and a little more scatter for transpiration. These results show that, at least on average, the assumption of full acclimation does allow great simplification of a complex canopy model. It is interesting, though, to investigate the causes for residual scatter in the comparison, especially for transpiration.

In Figs. 5 and 6, hourly ratios of the predictions by RLM and MAESTRO are plotted against time of day, for different values of LAI and crown radius, respectively. A ratio of less than unity indicates underestimation by RLM. In general, the plots show a characteristic pattern: RLM slightly underestimates in the very early hours, during the rest of the morning transpiration is slightly overestimated, and overestimation is stronger in all cases just before sunset. Figure 5 shows a small influence of LAI on the discrepancy between the models: a large LAI appears to enhance sensitivity to time of day. Figure 6 shows little effect of different crown radii (compared for LAI = 7 only) on the discrepancy for photosynthesis or transpiration. Figure 5*b* also shows that transpiration is, on average, underestimated by RLM, by 5 to 10%.

Discussion

The essential difference between RLM and MAESTRO is the variation in the radiation distribution, and hence values of γ_e, with time in the

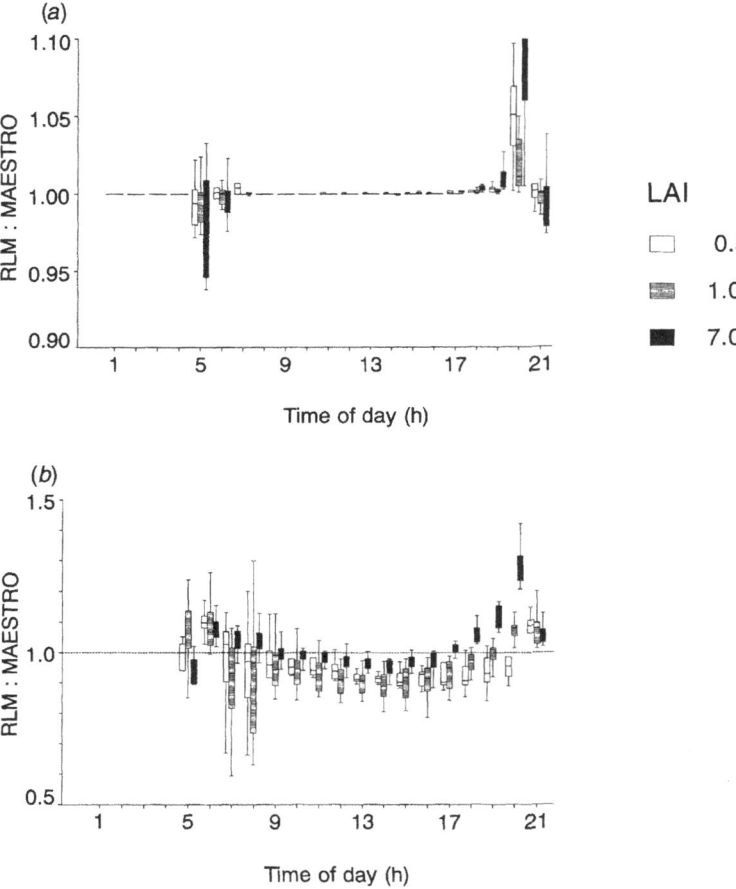

Fig. 5. The ratio of predictions of photosynthesis (CO_2 flux) (*a*) and transpiration (H_2O flux) (*b*) by RLM and by MAESTRO, plotted against time of the day, for three different LAIs, with other structural parameters varying as shown in Table 1. The results are shown as box-whisker plots, to show variability between days.

latter. In both models, rubisco-limited photosynthesis and respiration do not depend on the instantaneous value of I_a and hence do not depend on γ_e, whereas radiation-limited photosynthesis, transpiration and stomatal conductance do depend on γ_e. The performance of RLM is clearly not as good for transpiration as for photosynthesis, and this can partly be explained by a stronger overall dependence of transpiration on g_{sw} and

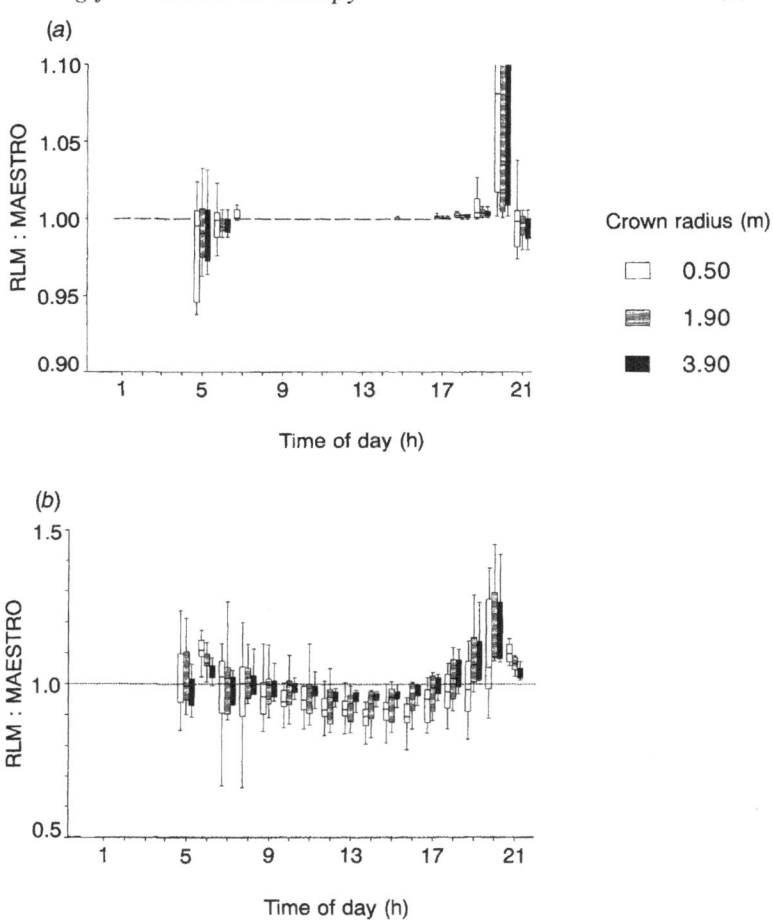

Fig. 6. The ratio of predictions of RLM and MAESTRO for photosynthesis (*a*) and transpiration (*b*) as in Fig. 5, but here the results shown are for three different crown radii, thus representing different degrees of foliage clumping, all at an LAI of 7.

hence on I_a. However, transpiration also depends quite strongly on the ratio g_{sw}/g_b, whereas the role of g_b in photosynthesis is less important. Both conductances decrease with depth into the canopy, but not always at the same rate, so Ω is not necessarily constant throughout the canopy. This undermines one of the basic assumptions of RLM for transpiration. For photosynthesis, the omission of a leaf boundary-layer conductance in the models contributed to the spatial constancy of C_i/C_a, so a re-analysis including g_b might slightly worsen the match. The spatial variation of g_b is clearly a problem in an analysis assuming full acclimation, since g_b is independent of leaf physiological processes. The observations by Berry *et al.* (this volume), showing feedback of g_b on g_s, may shed a different light on this problem.

The distribution of absorbed PAR, and hence of γ_e, is likely to deviate most strongly from the average at the low solar elevation angles that occur at dawn and dusk. At low solar elevation angles, the vertical gradient of PAR is expected to be steeper than average (i.e. I_a, and hence photosynthesis and transpiration, decreases more rapidly with depth) because the radiation reaching lower canopy levels has to travel longer distances through foliage than at higher solar elevation angles, especially in dense canopies (e.g. with LAI = 7). In this case, if the reference location for derivation of parameters throughout the canopy is placed at the top of the canopy, as was the case here, RLM will overestimate the canopy fluxes. This explains the ratios near dusk, but not those just after dawn when there is also a small underestimation by RLM.

In contrast, a reverse effect on the PAR distribution is to be expected when there is a high proportion of diffuse PAR. Diffuse radiation is omnidirectional at all times of the day and penetrates canopies more efficiently than the direct beam (Wang & Jarvis, 1990b). The PAR distribution then declines less steeply than on average, and so in this case RLM will *underestimate* canopy fluxes. Inspection of the meteorological input data used in the simulations showed that indeed most mornings were very cloudy, with a high proportion of diffuse radiation, thus providing an explanation for the low values of the ratios early in the day.

The insensitivity to crown radius, despite the very large range considered, seems surprising. After all, a strongly clumped canopy with narrow crowns and densely packed foliage within the crowns bears little similarity to a 'big leaf'. However, if we compare our results with a similar analysis by Sellers *et al.* (1992), we find that the reverse is true. Sellers *et al.* (1992) compared photosynthesis and transpiration by RLM with a *horizontally homogeneous* multilayer model. Their Fig. 9 shows larger discrepancies between the two models at low solar elevation than in the present comparison. This may be related to different diurnal variation of the spatial PAR distribution in homogeneous and clumped canopies. Closer analysis of the simulated PAR distributions in MAESTRO suggests that variation is substantial in both types of canopy, but that values of γ_e at a given time are systematically above or below unity for a more homogeneous canopy, whereas in a clumped canopy, high and low values of γ_e compensate.

This analysis leads to a very useful result that may allow substantial simplifications in the modelling of canopy fluxes. However, the basic assumption of full acclimation of physiological parameters to the local PAR environment has not yet been shown to be universally true. This study emphasises the need for further experimental work to verify these assumptions and generates a number of specific questions, including the following. Does acclimation always occur, and to what extent? Does it occur for all species in an ecosystem? Is the dependence of acclimation on p_c linear? It is also important to identify the mechanisms for the acclimation process. In particular, the role of the spectral composition of PAR, especially the red/far-red ratio, deserves more attention. Problems in the present approach that need to be further investigated include parameterisation of the stomatal and boundary-layer conductances, their possible mutual dependence and the possible dependence of g_s on photosynthesis. Incorporation of other variables influencing leaf physiology, such as leaf age, needs to be assessed as well.

Conclusions

The conclusions of this study can be summarised as follows.

1. The PAR absorbtivity by a canopy differs from that of its individual leaves as a result of reflectance and scattering of radiation by the soil and the leaves themselves.
2. Clumping (for example into crowns) reduces the radiation absorption by the canopy and needs to be taken into account by models.
3. It has been postulated for some time that most C_3 plants and canopies tend to allocate their resources in such a way as to optimise their photosynthetic response to PAR. This results in a distribution of photosynthetic parameter values that is similar (acclimated) to the local PAR gradient.
4. *Once absorbed PAR is known*, it is shown mathematically and numerically that a *hypothesis* of full acclimation to local PAR leads to a greatly simplified canopy photosynthesis and transpiration model.
5. Since simplified canopy models are of tremendous value to modelling the global climate and carbon cycle, it is important that the validity of the acclimation hypothesis is experimentally tested for several vegetation types on earth.

Acknowledgements

Olevi Kull, Institute of Ecology, Estonian Academy of Sciences, Estonia, Piers Sellers, NASA, USA and Joe Berry, Carnegie Institution of Washington, Stanford, USA contributed to the conception of this study by discussing the ideas with us. The work was supported by the NERC TIGER programme (project III.2).

Glossary

A	net photosynthesis rate
B	one of the parameters V_{cmax}, J_{max}, g_{smax} or R_{db}
C_a	ambient atmospheric CO_2 concentration
C_i	internal CO_2 concentration
c_1, c_2	temperature-dependent coefficients of transpiration
D_v	vapour pressure deficit of the atmosphere
E	transpiration rate
f, g	functions of T and C_i
g_b	leaf boundary-layer conductance
g_s	stomatal conductance, general use
g_{sp}	stomatal conductance for CO_2
g_{smax}	maximum stomatal conductance for water vapour
g_{sw}	stomatal conductance for water vapour
h	functions of T, D_v and C_a
I	any radiation, PAR
I_a	absorbed radiation, PAR absorbed by the photosystems
J	electron transport rate
l_e	leaf area in element e
N_e	total number of canopy elements e above 1 m^2 of ground
p_e	fractional average absorbed PAR (I_a)
Q_a	absorbed radiation in all wavelengths
q_e	fractional average available radiative energy (Q_a)
R_d	actual respiration rate
R_{db}	base respiration rate at 0 °C
R_{db0}	reference location base respiration
T	temperature
V_{cmax}	maximum carboxylation rate
α	initial slope of the PAR response of g_s
β	intercept of the PAR response of g_s
γ_e	relative deviation of the actual PAR profile from the average in element e

Θ	'curvature factor' of the PAR response
Π	relative canopy PAR absorption
Ξ	relative canopy absorption of available energy
Ω	decoupling coefficient of the transpiration from D_v

Subscripts

0	the reference location
a	absorbed
c	bulk canopy value
i	incident
max	maximum
r	reflected
s	by the soil
e	any canopy element

Abbreviations

BRDF	bi-directional reflectance distribution function
LAD	leaf area density
LAI	leaf area index
NIR	near-infrared radiation
PAR	photosynthetically active radiation (photon) flux
RLM	reference leaf model

References

Aphalo, P.J. & Jarvis, P.G. (1993). An analysis of Ball's empirical model of stomatal conductance. *Annals of Botany*, 72, 321–327.

Ball, J.T. (1988). An Analysis of Stomatal Conductance. PhD thesis, Stanford University, CA.

Björkman, O. (1981a). The response of photosynthesis to temperature. In *Plants and their Atmospheric Environment*, ed. J. Grace, E.D. Ford & P.G. Jarvis, pp. 273–301. Oxford: Blackwell Scientific.

Björkman, O. (1981b). Responses to different quantum flux densities. In *Encyclopedia of Plant Physiology*, Vol. 12A. *Physiological Plant Ecology I, Responses to the Physical Environment*, ed. O.L. Lange, P.S. Nobel, C.B. Osmond & H. Ziegler, pp. 57–107. Berlin: Springer Verlag.

Björkman, O., Boardman, N.K., Anderson, J.M., Thorne, S.W., Goodchild, D.J. & Pyliotis, N.A. (1972). Effect of light intensity during growth of *Atriplex patula* on the capacity of photosynthetic reactions, chloroplast components and structure. *Carnegie Institute of Washington Year Book*, 71, 115–135.

Evans, J.R. (1988). Acclimation by the thylakoid membranes to growth irradiance and the partitioning of nitrogen between soluble and thylakoid proteins. In *Ecology of Photosynthesis in Sun and Shade*, ed. J.R. Evans, S. von Caemmerer & W.W. Adams III, pp. 93–106. Australia: CSIRO.

Evans, J.R. (1993). Photosynthetic acclimation and nitrogen partitioning within a lucerne canopy. II. Stability through time and comparison with a theoretical optimum. *Australian Journal of Plant Physiology*, 20, 69–82.

Evans, J.R. & Farquhar, G.D. (1991). Modelling canopy photosynthesis from the biochemistry of the C_3 chloroplast. In *Modelling Crop Photosynthesis – from Biochemistry to Canopy (Crop Science Society of America, Publication No. 19)*. ed. K.J. Boote & R.S. Loomis, pp. 1–16. Madison, WI: Crop Science Society of America.

Farquhar, G.D. & von Caemmerer, S. (1982). Modelling of photosynthetic response to environmental conditions. In *Encyclopaedia of Plant Physiology, Vol. 12-B. Water Relations and Carbon Assimilation*, ed. O.L. Lange, P.S. Nobel, C.B. Osmond & H. Ziegler, pp. 550–587. Berlin: Springer Verlag.

Field, C. (1983). Allocating of leaf nitrogen for the maximization of carbon gain: leaf age as a control on the allocation program. *Oecologia*, 56, 341–347.

Goward, S.N. & Huemmrich, K.F. (1992). Vegetation canopy PAR absorptance and the normalised difference vegetation index: an assessment using the SAIL model. *Remote Sensing of Environment*, 39, 119–140.

Harley, P.C., Thomas, R.B., Reynolds, J.F. & Strain, B.R. (1992). Modelling photosynthesis of cotton grown in elevated CO_2. *Plant, Cell and Environment*, 15, 271–282.

Hirose, T. & Werger, M.J.A. (1987). Nitrogen use efficiency in instantaneous and daily photosynthesis of leaves in the canopy of a *Solidago altissima* stand. *Physiologia Plantarum*, 70, 215–222.

Jarvis, P.G. (1976). The interpretation of the variations in leaf water potential and stomatal conductance found in canopies in the field. *Philosophical Transactions of the Royal Society of London*, Series B, 273, 593–610.

Jarvis, P.G. (1993). Prospects for bottom-up models. In *Scaling Physiological Processes: Leaf to Globe*, ed. J. Ehleringer & C. Field, pp. 115–126. New York: Academic Press.

Jarvis. P.G. & McNaughton, K.G. (1986). Stomatal control of transpiration: scaling up from leaf to region. *Advances in Ecological Research*, 15, 1–49.

Jarvis, P.G. & Sandford, A.P. (1986). Temperate forests. In *Topics in Photosynthesis*, Vol. 7. *Photosynthesis in Contrasting Environments*. ed. N.R. Baker & S.P. Long, pp. 199–236. Amsterdam: Elsevier.

Leverenz, J.W., Falk, S., Pilström, C.-M. & Samuelsson, G. (1990).

The effects of photoinhibition on the photosynthetic light-response curve of green plant cells (*Chlamydomonas reinhardtii*). *Planta*, 182, 161–168.

Lewandowska, M., Hart, J.W. & Jarvis, P.G. (1977). Photosynthetic electron transport in shoots of Sitka spruce from different levels in a forest canopy. *Physiologia Plantarum*, 41, 124–128.

McNaughton, K.G. (1994). Effective stomatal and boundary-layer resistances of heterogeneous surfaces. *Plant, Cell and Environment*, 17, 243–262.

Monteith, J.L. (1973). *Principles of Environmental Physics*. London: Arnold.

Oberbauer, S.F. & Strain, B.R. (1986). Effects of canopy position and irradiance on the leaf physiology and morphology of *Pentaclethra macroloba* (Mimosaseae). *American Journal of Botany*, 73, 409–416.

Pons, T.L., Schieving, F., Hirose, T. & Werger, M.J.A. (1989). Optimization of leaf nitrogen allocation for canopy photosynthesis in *Lysimachia vulgaris*. In *Causes and Consequences of Variation in Growth Rate and Productivity of Higher Plants*. ed. H. Lambers, M.L. Cambridge, H. Konings & T.L. Pons, pp. 175–186. The Hague: SPB Academic Publishing.

Russell, G., Jarvis, P.G. & Monteith, J.L. (1989). Absorption of radiation by canopies and stand growth. In *Plant Canopies: Their Growth, Form and Function*, ed. G. Russell, B. Marshall & P.G. Jarvis, pp. 21–39. Cambridge: Cambridge University Press.

Ryan, M.G. (1995). Foliar maintenance respiration of subalpine and boreal trees and shrubs. *Plant, Cell and Environment*, 18, 765–772.

Sage, R.F., Sharkey, T.D. & Seemann, J.R. (1989). Acclimation of photosynthesis to elevated CO_2 in five C_3 species. *Plant Physiology* 89, 590–596.

Sellers, P.J., Berry, J.A., Collatz, G.J., Field, C.B. & Hall, F.G. (1992). Canopy reflectance, photosynthesis, and transpiration. III. A reanalysis using improved leaf models and a new canopy integration scheme. *Remote Sensing of Environment*, 42, 187–216.

Smith, H. (1986). The perception of light quality. In *Photomorphogenesis in Plants*. ed. R.E. Kendrick & G.H.M. Kronenberg, pp. 187–217. Dordrecht: Martinus Nijhoff.

Smith, H., Samson, G. & Fork, D.C. (1993). Photosynthetic acclimation to shade: probing the role of phytochromes using photomorphogenic mutants of tomato. *Plant, Cell and Environment*, 16, 929–937.

Suits, G.H. (1972). The calculation of the directional reflectance of a vegetative canopy. *Remote Sensing of Environment*, 2, 117–125.

Verhoef, W. (1984). Light scattering by leaf layers with application to canopy reflectance modelling: the SAIL model. *Remote Sensing of Environment*, 16, 125–141.

Wang, Y.-P. & Jarvis, P.G. (1990a). Description and validation of an array model – MAESTRO. *Agricultural and Forest Meteorology*, 51, 257–280.

Wang, Y.-P. & Jarvis, P.G. (1990b). Effect of incident beam and diffuse radiation on PAR absorption, photosynthesis and transpiration of Sitka spruce – A simulation study. *Silva Carelica*, 15, 167–180.

Wang, Y.-P., Jarvis, P.G. & Benson, M.L. (1990). Two-dimensional needle-area density distribution within the crowns of *Pins radiata*. *Forest Ecology and Management*, 32, 217–237.

J.P. GRIME, K. THOMPSON
and C.W. MACGILLIVRAY

Scaling from plant to community and from plant to regional flora

Introduction

At each scale, from local to global, there is a long-established botanical tradition of seeking causal explanations for vegetation pattern by reference to the characteristics of component species (MacLeod, 1894; Raunkiaer, 1934; Ramenskii, 1938; Holdridge, 1947; Grime, 1974; Noble & Slatyer, 1979; Box, 1981; Woodward, 1987). With the passage of time and with increasing concern about the impacts of land-use and climate change, there has been a shift in emphasis from phytogeographical correlation to predictive modelling. However, the recent emphasis upon computer projections of the future course of vegetation development has not brought about any fundamental change to the core research needs in this field. The dual challenge since the time of Raunkiaer has remained:

1. To identify individual plant traits or sets of traits that recur widely and provide consistent clues to the processes controlling vegetation structure and dynamics
2. To assemble data bases of species distributions and relevant traits matched to the scale of enquiry.

It is the second of these two that provides the main obstacle to progress. As Keddy (1992) and Hendry & Grime (1993) have pointed out, severe limitations arise from the fragmentary form of the species × trait matrices available to plant ecologists and plant geographers. Incomplete data bases not only restrict opportunities to extend models to unfamiliar vegetation but they also limit the initial review processes required to identify useful traits and to test their statistical reliability as predictors. Faced with these limitations, researchers have adopted various stratagems largely dictated by data availability and specific objectives. Information on plant morphology is available on a world scale as a by-product of taxonomy and, following the rationale of Raunkiaer (1934), this source continues to be used as a basis for predicting the determining effect of climate on vegetation, either by correlative methods (Holdridge, 1947; Box, 1981) or by mechanistic modelling of vegetation structure (Monteith, 1973; Woodward, 1987). In order for this approach to become more widely applicable to natural ecosystems, further development will be required to take account of soil factors that frequently override the influence of climate in controlling productivity and vegetation type. More serious still is a growing requirement, in all parts of the world, to deal with circumstances where 'natural' or anthropogenic disturbances usurp climate as the proximal control on vegetation composition. The accelerating rate of destruction of mature ecosystems and their replacement by early-successional and relatively labile plant communities presents a severe test to our predictive capabilities. We cannot ignore disturbed vegetation; it is expanding rapidly over the land surface and, as we shall suggest later, it is likely to include floristic elements that are highly responsive to climate changes.

This paper illustrates progress in a formal search (The Integrated Screening Programme or ISP), conducted in northern England with the objective of recognising plant attributes useful for the interpretation of plant community structure and vegetation dynamics. A protocol will be described in which predictions from the ISP are tested against independently derived information from field monitoring and experimentation. Although our main purpose is to bridge the gap from plant to community, at the end of the paper we explore the possibilities of extrapolation to regional floras.

A Research protocol

Figure 1 describes a set of procedures that appears to be gaining acceptance as an approach to understanding plant community responses to environmental change. An important step in devising the protocol has been the development in various parts of the world of standardised screening procedures in which comparatively large numbers of plants of contrasted ecology and drawn from regional floras have been grown under controlled conditions and their responses to individual factors measured. Examples of such experiments are available from Canada (Keddy, 1992), Australia (Jurado, Westoby & Nelson, 1991), Japan (Washitani & Masuda, 1990) and northern England (Grime & Hunt, 1975; Grime *et al.*, 1981; Hunt *et al.*, 1991; 1993a) and recently a manual of screening methods has been published (Hendry & Grime, 1993). The purpose of such screening is to allow objective review of variation in basic attributes of plant morphology, physiology and biochemistry in order to recognise recurring functional types. The perspective emerging from this screening

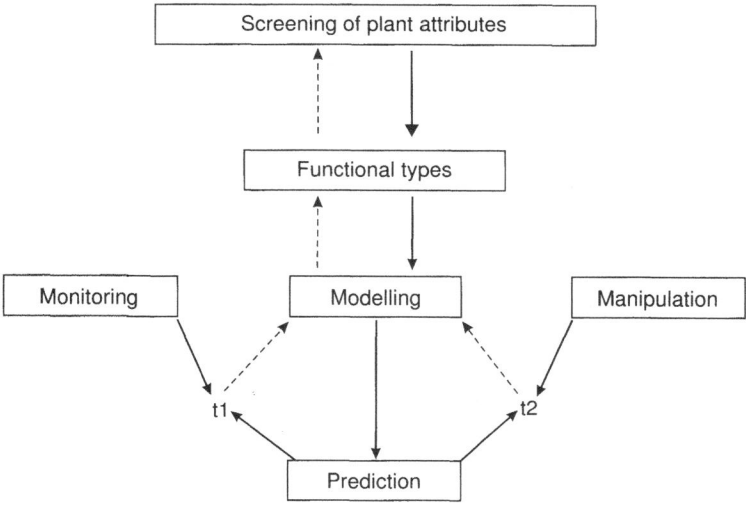

Fig. 1. Protocol for development and testing of predictions of vegetation response to environmental change. Discrepancies revealed at t1 and t2 initiate further modelling cycles, each of which may necessitate refinement of the functional types or even additional screening.

activity is that, despite the apparently infinite variety in plant design, evolutionary and ecological specialisation in plants throughout the world is tightly constrained with respect to certain key attributes such as seed size (Hodgson & Mackey, 1986; Jurado *et al.*, 1991), the apportionment of captured resources between growth, storage and reproduction (Grime, 1977), genome size (Grime & Mowforth, 1982), leaf longevity (Reich, Walters & Ellsworth, 1992) and root penetration (Reader *et al.*, 1992). As we shall see later in this paper, there is growing evidence that certain of the attributes used singly (Noble & Slatyer, 1979; van der Valk, 1981) or as sets (Grime, Hunt & Krzanowski, 1987a) can provide a basis for predicting plant responses to specific changes in climate, soils or management. Where sufficient data are available to classify all component species into functional types, it may be possible, as indicated in Fig. 1, to predict vegetation responses to specific scenarios of changed climate or land-use.

In the protocol of Fig. 1, it is suggested that there are two alternative methods by which hypotheses concerning community and ecosystem function may be tested and refined. The first, illustrated in the left-hand side of the figure, involves comparison of model predictions against data collected by monitoring. With respect to both land-use and climate change this probably represents the most realistic mechanism for recognising the imperfections of models in that, as indicated by the arrows in Fig. 1, discrepancies can stimulate not only changes in the model but also, where necessary, further data inputs from new screening procedures.

Unfortunately, monitoring studies on communities and ecosystems are an exceedingly scarce resource and, for many purposes, alternative mechanisms of hypothesis testing (illustrated on the right-hand side of Fig. 1) must be sought. These follow a logical pathway similar to that involving monitoring but rely upon manipulative experiments. Some of these experiments can be conducted by the synthesis of vegetation under controlled conditions (e.g. Grime *et al.*, 1987b; Lawton *et al.*, 1993). For many purposes, however, it is necessary to perform replicated manipulations of climate and land-use at the field scale, often necessitating an alliance between ecologists and engineers.

Examples

Example 1. Predicting plant community response to year-to-year variation in temperature

From extensive screening operations (e.g. Stebbings, 1956; Bennett & Smith, 1976; Levin & Funderburg, 1979) and various other investigations including the ISP (Hartsema, 1961; Bennett, 1971; 1976; Grime & Mowforth, 1982; Grime, 1983; Grime, Shacklock & Band, 1985), it has been established that there is more than a thousand-fold variation in nuclear DNA amount in vascular plants, and that differences in DNA amount in cool temperate regions such as the British Isles coincide with differences in the timing of shoot growth. A mechanism interpreting variation in DNA amount, cell size and the length of the cell cycle (the three attributes are inextricably linked) as a consequence of climatic selection has been proposed (Grime & Mowforth, 1982) but will not be reviewed here. In the context of the ISP and the search for functional types in the British flora, it is pertinent to focus upon one specific prediction concerning the potential impact of global warming. An essential precursor to this prediction is the hypothesis that the delayed phenology of many British plants is imposed by the greater sensitivity of small-celled, low DNA content species to the inhibitory effects of low temperature on cell division. If this hypothesis is correct, we may expect that a rise in temperature would confer an advantage on plants with low DNA content by differentially lengthening their growing seasons (Fig. 2). Conversely, plants with low DNA contents would be expected to be more strongly inhibited than those with high contents in years with an unusually cold spring or an unusually dry summer. This leads to the prediction that in response to year-to-year variation in climate, variance in contribution to the shoot biomass will be greatest in plant species of low DNA amount. Following the protocol of the left-hand side of Fig. 1, this hypothesis has been tested by examining fluctuations in the species composition of roadside vegetation at Bibury in Gloucestershire, which has been monitored continuously for more than 30 years (Hunt *et al.*, 1993b). Vegetation bulk was estimated non-destructively every year in the second half of July, by which time most species have achieved their peak seasonal biomass (Yemm & Willis, 1962). The results (Fig. 3) confirm that variance is highest in the biomass component made up of low DNA content plants.

Example 2. Predicting plant community responses to elevated CO_2

Recently a comprehensive screening of responses to elevated CO_2 has been conducted under non-limiting conditions of mineral nutrient supply and temperature, in natural summer daylight in the comparatively high-radiation environment of Littlehampton on the south coast of England. Growth of seedlings for six to eight weeks at each of four levels of CO_2 (ambient, 500, 650 and 800 μmol mol^{-1}) has allowed curve-fitting procedures to be used to characterise the response of each species. Experimental methods and results for 40 native C3 species are described by Hunt *et al.* (1991; 1993a). From these data, it is evident that, under the conditions of the screening experiment, there were large species-specific differences in response to elevated CO_2, varying from zero in some ephemerals and slow-growing perennials to very large increases in dry matter production in some species. Among the latter, potentially large, fast-growing, clonal perennials of productive habitats (e.g. *Chamerion angustifolium* and *Urtica dioica*) are conspicuous, prompting the hypothesis that responsiveness to elevated CO_2 may be dependent upon the existence of strong carbon sinks. Here it is interesting to note that fast-growing clonal herbs are prominent among the life forms identified as currently expanding in abundance in the British flora (Grime, Hodgson & Hunt, 1988). In previous attempts to explain this phenomenon (e.g. Hodgson, 1989), emphasis has been placed on changes in land-use and eutrophication by mineral fertilisers and atmospheric deposition of nitrogen. The data suggest that a stimulatory effect of elevating CO_2 could be contributing to the observed shift in functional types.

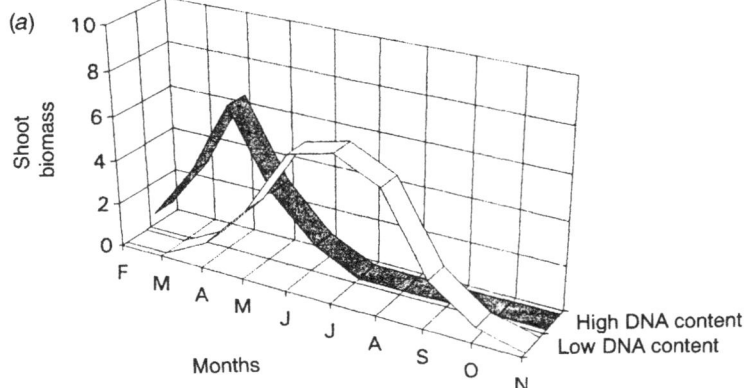

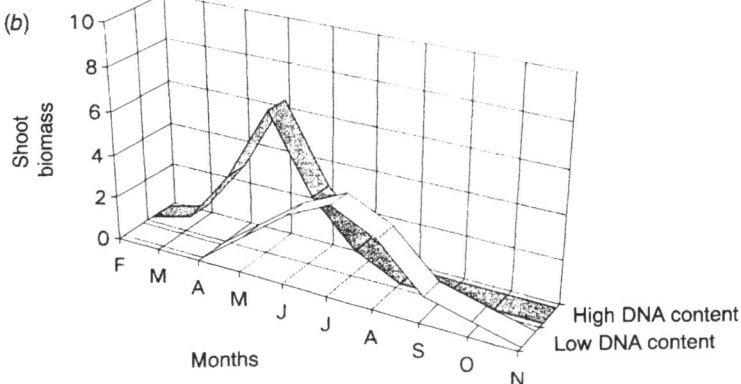

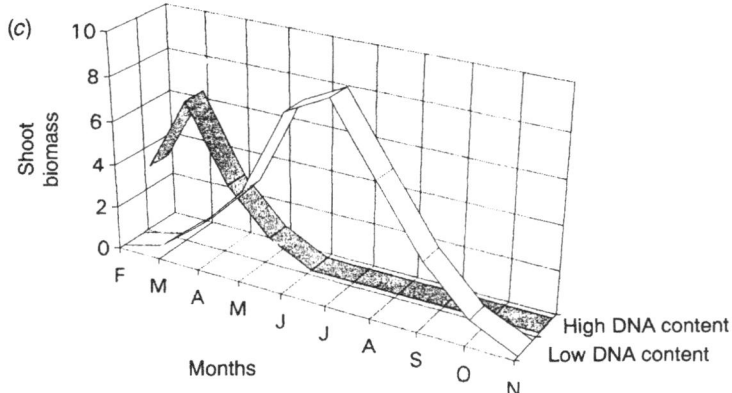

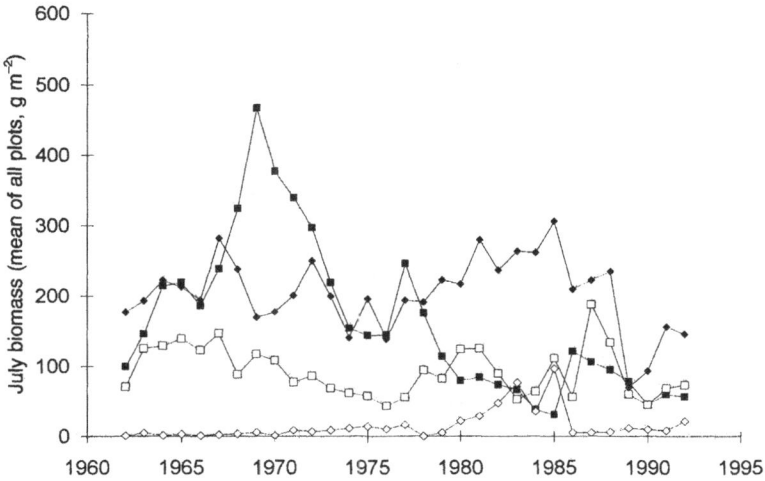

Fig. 3. Variation in shoot biomass in components of the road verge vegetation at Bibury, Gloucestershire, classified with respect to nuclear DNA amount. CV = coefficient of variation. Group 1, < 5 pg (■), CV = 0.67, are *Urtica dioica, Convolvulus arvensis*; group 2, 5.1–10.0 pg (□), CV = 0.38, common species *Agrostis stolonifera, Dactylis glomerata*; group 3, > 10 pg (◆), CV = 0.26, common species *Arrhenatherum elatius, Elytrigia repens*; group 4, unclassified, CV = 1.43.

Fig. 2. Diagram illustrating the predicted differential effect of winter and spring temperatures upon the shoot biomass of two plants of contrasted DNA content (*a*) 'normal' year; (*b*) year with cooler winter and spring; (*c*) year with warmer winter and spring. In the determinate phenology of the vernal geophyte (high DNA content), shoot expansion is retarded or advanced but is terminated by the limited supply of preformed tissue. The indeterminate summer green herb (low DNA content), although always relatively delayed by its dependence upon warm temperatures and current cell division, experiences a shortened or lengthened growing season and is predicted to show an immediate reduction or increase in shoot biomass. Species with low nuclear DNA content are also expected to be more immediately responsive to unusually dry summer conditions.

The screening is conducted on plants grown for a limited period as isolated individuals under non-limiting conditions. As many workers in this field have recognised (e.g. Korner, 1993), it is vital that any hypotheses arising from differential responses observed in such experiments are subjected to tests in more natural conditions. Following the protocol in the right-hand side of Fig. 1, hypotheses have been tested by examining the effect of elevated CO_2 on the structure of herbaceous communities. These were allowed to regenerate from natural seed banks on disturbed, fertile soils transferred to laboratory microcosms. The results of these manipulative experiments with natural communities (Díaz et al., 1993) are strongly at variance with the laboratory-derived predictions in that promotory effects of CO_2 were not confirmed in potentially responsive species. It is particularly interesting to note that under a doubling of CO_2 concentration the early successional, fast-growing and potentially large *Rumex obtusifolius*, although attaining a higher shoot biomass, did not increase its relative abundance in the community and showed marked symptoms of leaf stunting, reduced levels of foliar nitrogen and carbohydrate accumulation (Fig. 4). These patterns were associated with a stimulation of microbial sequestration of carbon and nitrogen in the soil. Whilst these results do not invalidate the patterns of response evident in the screening data, they are a salutary reminder of the complexities that distinguish the screening laboratory from a multispecies ecosystem. They suggest that sink strength is not the only trait accounting for CO_2 responsiveness. Other characteristics, such as the nutritional requirements of species and the nature of their interactions with the microbial flora, may be important in predicting responses to high CO_2 at the community level.

Example 3. Predicting plant community responses to extreme events

Anecdotal evidence indicates that events such as severe frosts, droughts and fire have ecological impacts that are large relative to their duration (Hopkins, 1978; Woodward, 1987). They are, therefore, likely to be important determinants of the speed and nature of the responses of plant species to climatic change (Walker, 1991). Westman (1978) identified two key aspects of response. *Resistance* is the ability of a species' biomass to resist displacement from control levels. *Resilience* is the speed of recovery back to control levels.

Leps, Osbornova-Kosinova & Rejmánek (1982) built on earlier speculation by Levitt (1978; 1980) and Grime (1974) in suggesting that particular variable traits of life history and resource allocation are of

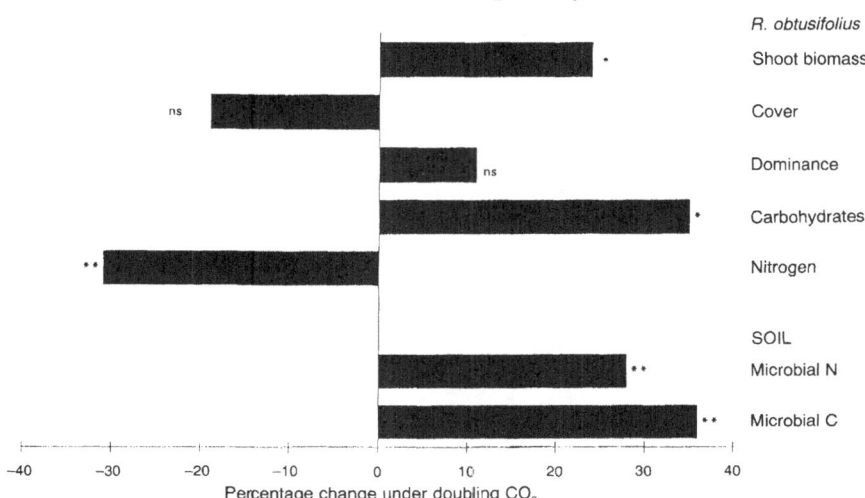

Fig. 4. Responses of *Rumex obtusifolius* and soil microflora grown in microcosms to a doubling of atmospheric CO_2 (700 µmol mol^{-1}) compared with controls at 350 µmol mol^{-1}. Vegetation was allowed to develop for 84 days by natural recruitment from seed banks in soils removed from a tall herb community in Derbyshire and placed in microcosms (six replicates per treatment) in cabinets without nutrient addition. Shoot biomass was measured as milligrams dry weight, cover as number of touches in a point-quadrat analysis, dominance as biomass of *R. obtusifolius*: total community biomass, carbohydrates (starch + glucose + sucrose) as milligrams per gram fresh weight and nitrogen as milligrams per gram dry weight of fully expanded young leaves, microbial C and N as milligrams per gram dry soil. ns, non-significant; *, $p < 0.05$; **, $p < 0.01$ (ANOVA).

universal value as predictors of resistance and resilience. They argue that traits which promote the tolerance of mineral nutrient stress (long-lived organs, low rates of nutrient turnover) lead to correlated resistance to other forms of damage. It was further predicted that the existence of within-plant tradeoffs leads to an association of these traits with low growth rates and, therefore, with low rates of recovery after damage (low resilience). These predictions were broadly supported by the field observations of Leps *et al.* (1982) on the impacts of a severe drought on two contrasting grasslands and by Lambers & Poorter's review (1992) of the ecological consequences of inter-specific variation in growth rate.

In a recent investigation (Fig. 5) closely adhering to the protocol

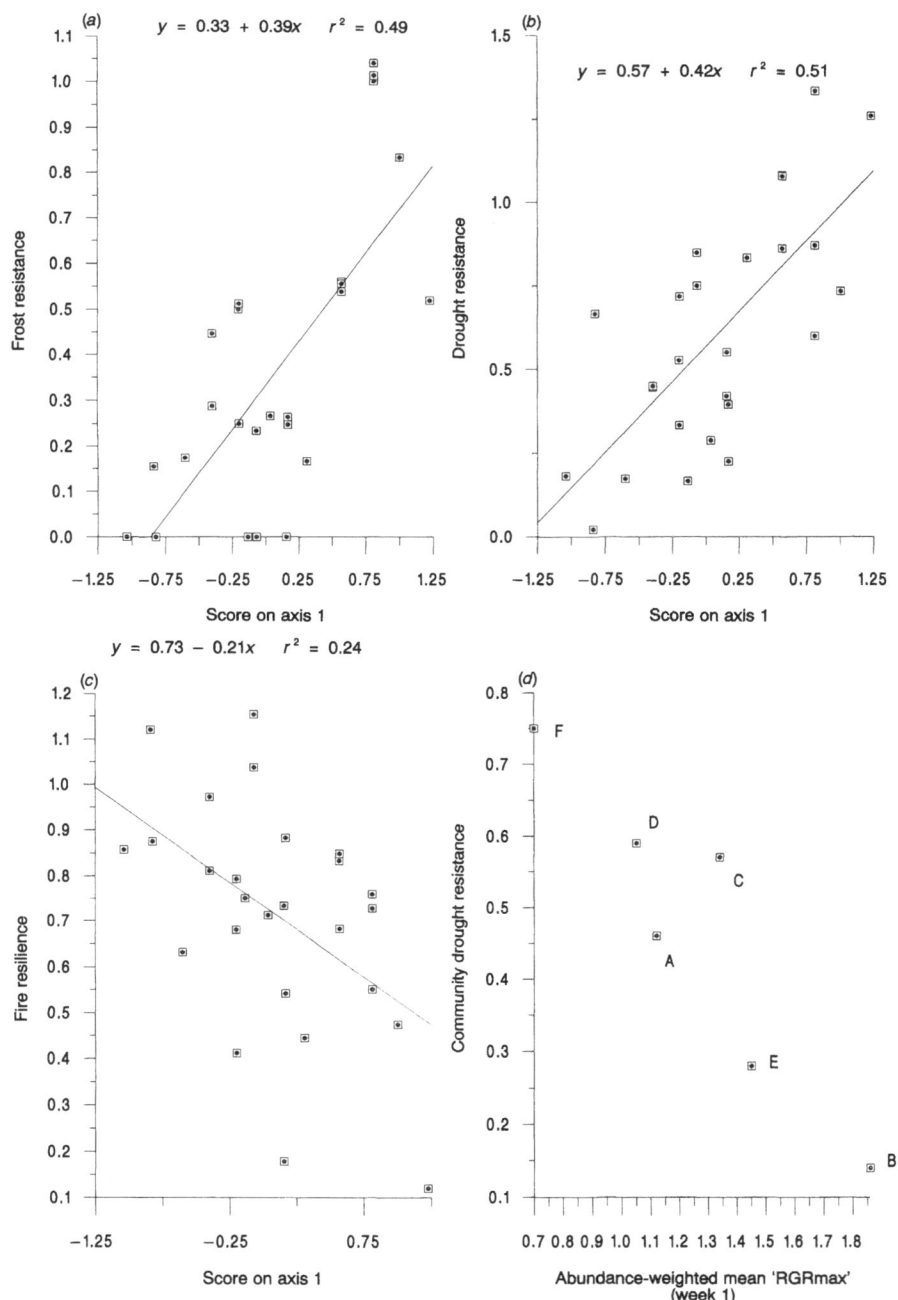

of Fig. 1, MacGillivray *et al.* (1995) have tested the ability of species traits related to nutrient stress tolerance to predict resistance and resilience of five types of herbaceous vegetation subjected to three types of extreme event (frost, drought and fire). The frost treatment was conducted by placing pieces of turf in the freezer compartment of a commercial food refrigeration vehicle on 11 and 12 June 1991. Mean minimum night temperature of leaf surfaces at 15 cm above the soil surface was reduced by the treatment to −6.5 °C (range: ± 0.5 °C, $n = 10$; control temperature = 5.5 °C); mean maximum daytime temperatures were 1.1 °C (frost) and 8.6 °C (control). Turves were removed from six plant communities (A–F) at Harpur Hill, near Buxton, Derbyshire, UK. Surface turf dimensions were either 50×50 cm (communities A–D) or 15×15 cm (communities E and F); turf depth was determined as the minimum necessary to avoid obvious damage to root systems. Five turves from each community were subjected to the frost treatment and five served as untreated controls. Abundance was estimated for each turf as the total number of contacts with 64 pins passed through the vegetation perpendicular to the soil surface in a regular 8×8 grid arrangement over the centre of each turf, leaving the perimeter unmonitored to avoid possible edge effects. Turves were assessed on 15–16

Fig. 5. Relationship between estimates of (*a*) frost resistance, (*b*) drought resistance and (*c*) fire resilience of above ground biomass and the scores of the species on the ISP-derived principal component analysis axis 1. This axis, fully described in MacGillivray *et al.* (1995), incorporates a wide range of ecological attributes but can very briefly be interpreted as one from fast-growing species typical of productive habitats (negative values) to slow-growing species tolerant of nutrient stress (positive values). (*d*) Relationship between drought resistance of whole communities and abundance-weighted mean maximum potential relative growth rate (RGR_{max} *sensu* Grime & Hunt, 1975) of component species. Resistance of individual speices was calculated as the ratio of the mean number of point-quadrat hits per turf immediately after the treatment in June 1991 to that in the control at the same time. Resilience was calculated as the same ratio one year after the fire treatment. Resistance of whole communities was calculated as the ratio of estimated mean above-ground biomass in treated turf to that in control turf. Biomass was estimated using previously established regression relationship between point-quadrat contacts and biomass (MacGillivray, 1993).
The methodology is described in the text.

June 1991. The slope (1.02) of a linear regression of the mean abundance of species in the untreated control turves on that recorded in further *in situ* control areas was not significantly different from 1 (*t* test: $p < 0.05$) indicating that turf removal *per se* did not affect species abundance. For descriptions of communities see MacGillivray *et al.* (1995).

The results provide strong support for the hypothesis that the syndrome of plant traits associated with nutrient stress tolerance is positively correlated with resistance to extreme events and has a negative relationship with resilience. In the case of drought resistance, the trend observed at the community level (Fig. 5*d*) is entirely consistent with the trend observed between species (Fig. 5*b*), indicating that community drought resistance is largely a function of the properties of the component species.

Example 4. Extrapolations to regional and national floras

As progress occurs in explaining and predicting plant community properties as a function of the attributes of component species, it seems inevitable that similar methods will be used to analyse vegetation processes operating on regional and national floras. An example of this approach is provided by the investigations of Hodgson (1986) in which various species traits were used to analyse the causes of commonness and rarity in a regional flora in northern England. Extending this approach, Thompson (1994) has recently compared the attributes of expanding and declining components of the vascular floras of several countries in Western Europe in an attempt to identify the dominant processes driving floristic change at this regional scale.

In Thompson's study, increasing and decreasing species were examined separately in each of six countries by discriminant analysis. The recently completed Botanical Society of the British Isles (BSBI) monitoring scheme (Rich & Woodruff, 1990) provided lists of species that have recently increased and decreased in each of the four separate countries of the UK (England, Wales, Scotland and Northern Ireland) and in the Republic of Ireland. The BSBI monitoring scheme was a 10×10 km square sample survey during 1987 and 1988 to assess the current status of the flora and, specifically, to compare it with data collected for the *Atlas of the British Flora* (Perring & Walters, 1962). The deficiencies of the monitoring scheme, particularly as a means of discovering changes since 1960, have been discussed by Rich & Woodruff (1992). The data analysed in Thompson (1994) derive from a revision (T.C.G. Rich, personal communication) that attempts to eliminate apparent changes caused by various forms of recording bias, and to list only those species showing genuine increases or declines since 1960. Additional lists of species that have increased or decreased in the Netherlands between 1940 and 1990 were provided by the Dutch Centraal Bureau Voor de Statistiek. Table 1 lists and explains the plant attributes used and Fig. 6 provides details of the system by which each species was classified according to CSR plant strategy theory (Grime, 1974).

CSR plant strategy theory (Grime, 1974) classifies plants into functional types according to their responses to gradients of productivity and disturbance. The three main functional types are *competitors*, *stress-tolerators* and *ruderals*. Competitors are robust perennials that thrive in conditions of high productivity and low disturbance. Stress tolerators are slow-growing species tolerant of nutrient stress and low productivity. Ruderals have a short life span and can survive by rapidly reproducing from seed on any fertile, open ground that may temporarily be produced by the action of disturbance. Ruderals, however, are sensitive to low productivity.

The results (Table 2) show that the discriminant function was equally effective in classifying increasing and decreasing species in each of the six countries, but the importance of the individual discriminating variables varied enormously. The ranking of the discriminating variables in Table 2 serves to highlight three features. First, the order of the variables is almost the same in the Netherlands and in England, and it is very different from the other four countries. S class, R class and canopy height or C class occupy the first three places in both countries. Second, regenerative variables (seed bank, wind dispersal, flowering period and seed weight) make little or no contribution to the discriminant function. Third, in the Netherlands and England, a single variable, S class, stands out as much more significant than all the others. In contrast, no single discriminating variable stands out in either Northern Ireland, the Republic of Ireland, Scotland or Wales, and no consistent pattern emerges from these four countries.

The results of the analysis for England confirm earlier work which suggested that fast-growing species (i.e. species of low S class) are at an advantage in the modern English landscape (Hodgson, 1986). In this respect, the Netherlands is clearly very similar to England. The other four countries show no clear pattern. Apparently in these less intensively disturbed countries, the causes of success and failure of individual species vary locally both within and between countries, and no single attribute or suite of attributes confers success. The difference

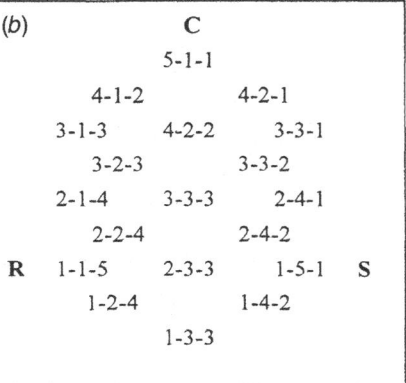

Fig. 6. The location of species in the stress/disturbance matrix can be quantified by dividing the matrix into 19 zones. Each of these corresponds to one of the recognised CSR strategy types (*a*), the most extreme ones being the pure competitors (C), stress-tolerators (S) and ruderals (R). These three are located at the corners of a triangular framework within which the intermediate strategic types are distributed according to a nodal network. The combination of attributes displayed by a species occupying any given position can be represented as a CSR co-ordinate (*b*). A value 5 represents an extreme position most like the pure C, S or R strategy, and a value 1 represents the opposite position, most unlike the pure strategy.

Table 1. *Autecological attributes used in the analysis of increasing and decreasing plant species*

Attribute	Range of values
Life history	1–9
C class	1–5
S class	1–5
R class	1–5
Canopy height	1–8
Lateral spread	1–5
Flowering period	number of months
Clonal growth	0 or 1
Seed bank	1–3
Wind dispersal	0 or 1
Dispersule weight	0–5

The data are essentially those in Grime *et al.* (1988), but some attributes differ. Life history is expressed on a scale from 1 (summer annual) to 9 (polycarpic perennial); higher numbers, therefore, represent increasing longevity. C, S and R class are components of CSR strategy (Fig. 6) and, therefore, represent, on a scale of 1 to 5, the extent to which the species is adapted to competition, stress or disturbance, respectively (Grime 1974; 1977). C class is, therefore, primarily, though by no means exclusively, a function of plant height (Gaudet & Keddy 1988). S class is largely an inverse function of maximum potential relative growth rate, while R class has much in common with life history. Canopy height ranges from 1 (< 100 m) to 8 (> 15 m), and lateral spread from 1 (annuals) to 5 (perennials > 1 m in diameter). Flowering period is simply the number of months the species is normally in flower. Dispersule weight varies from 0 (too small to be easily measured, e.g. orchid seeds, fern spores) to 5 (> 10 mg). The seed bank classes are 1 (seeds persist for < 1 year), 2 (1–4 year) and 3 (> 4 year). Wind dispersal and clonal growth are both either present (1) or absent (0). Wind dispersal is here defined as the presence of obvious morphological adaptations for wind dispersal (pappus or wing) and does not include species such as orchids or ferns, which are often assumed to be wind dispersed by virtue of their very tiny seeds or spores.

between the two groups of countries can be clearly seen in Fig. 7, which relates the mean S class of increasing and decreasing species in the six countries to their human population densities.

In the less populous countries, mean S class hardly differs, while in the two densely populated countries the S class of decreasing species is almost twice that of the increasing species. Human population density is a convenient metaphor for a wide range of human impacts on the

Table 2. *Results of a discriminant analysis of increasing and decreasing plant species in each of six European countries*[a]

	Netherlands	England	Northern Ireland	Republic of Ireland	Scotland	Wales
N	282	157	73	112	53	57
N included in analysis	178	105	43	70	32	27
S class	0.873	0.868	0.239	0.127	0.306	0.002
R class	−0.431	−0.293	0.181	−0.122	−0.080	−0.005
Canopy height	−0.399	−0.438	−0.252	0.094	0.012	0.221
C class	−0.389	−0.436	−0.495	−0.331	−0.132	−0.139
Life history	0.256	0.194	−0.335	−0.150	0.097	0.037
Seed bank	−0.172	−0.099	−0.055	−0.075	0.317	0.098
Wind dispersal	−0.085	−0.127	−0.441	−0.089	−0.303	−0.112
Clonal growth	0.080	0.010	−0.227	0.031	−0.155	−0.121
Dispersule weight	−0.078	−0.090	−0.038	−0.246	0.015	0.101
Flowering period	−0.025	-0.088	0.017	−0.007	0.032	0.442
Lateral spread	0.013	−0.001	−0.214	−0.199	−0.270	0.017
Percentage of species correctly classified	83.15	79.05	83.72	77.14	81.25	88.89

[a]Values shown are correlations between canonical discriminant functions and discriminating variables. N is total number of species and N in analysis is those species with no missing values for any of the discriminating variables. Percentage of species correctly classified as 'increasing' or 'decreasing' by the discriminant function is also shown. A positive correlation indicates a larger value of the variable in decreasing species, and *vice versa*. For details of analysis see text, and for a description of the discriminating variables see Table 1.

landscape, both urban and agricultural, but two widespread processes can be recognised. The first is the increasing restriction of slow-growing plants of infertile, relatively undisturbed habitats to fragmented islands of suitable habitat, many of them in nature reserves, surrounded by a sea of unsuitable landscape. Continued attrition of these fragments by road and house building, pollution and agriculture, combined with inevitable chance extinctions from small, isolated fragments, have

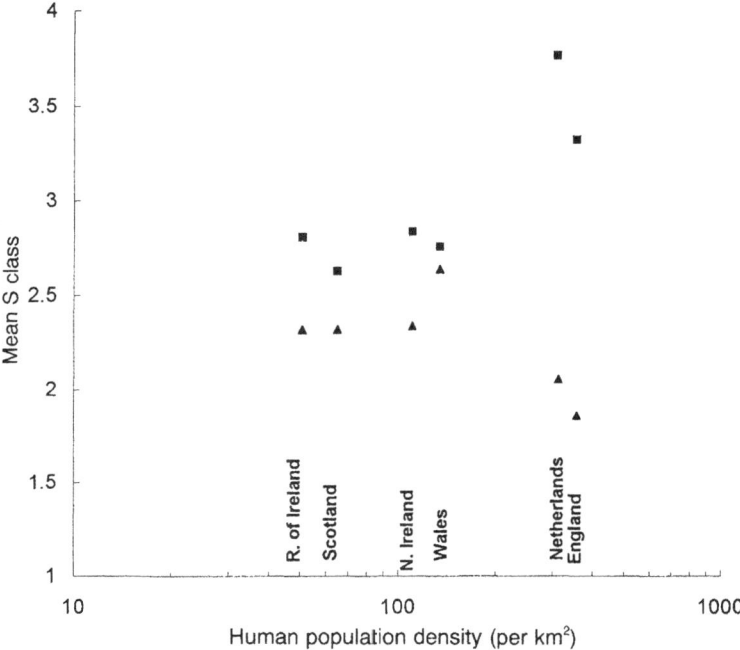

Fig. 7. Relationship between mean S class of increasing (▲) and decreasing (■) plant species and human population density in six European countries. The S class of these two groups were not significantly different in Scotland, Northern Ireland or Wales. The two groups are significantly different in R for Ireland ($p = 0.049$), England and the Netherlands (both $p < 0.001$). For description of CSR theory and interpretation of S class see the text and Table 1. Modified from Thompson (1994).

inescapably led to the contraction of slow-growing species. Second, the increasing abundance of disturbed fertile, artificial habitats, which support communities of fast-growing ruderal and competitive species, has led to the continuing expansion of these species. The result, graphically demonstrated in Fig. 7, is an increasing polarisation of the English and Dutch floras into an expanding, fast-growing component and a declining, slow-growing component characteristic of an older, seminatural landscape. In the less populous northern and western fringe of the British Isles this dichotomy can hardly be discerned.

A surprising feature of the results is the apparent lack of importance of regenerative attributes in characterising increasing and decreasing species in England and the Netherlands. If we had not excluded species with dust-like seeds or spores (nearly all orchids or ferns, and nearly all declining), wind dispersal might even have turned out to be a predictor of *decreasing* status. Perhaps the movement of propagules (both seeds and vegetative fragments) by construction, mining, quarrying, agriculture and the sheer volume of human traffic is now so pervasive that dispersal has ceased to be a barrier to the spread of species in densely populated landscapes. The example of a few wind-dispersed aliens that have achieved high rates of migration in the UK (e.g. *Senecio squalidus*, *Epilobium ciliatum*) should not obscure the fact that others have spread as quickly while producing little or no seed (e.g. *Veronica filiformis*, *Fallopia japonica*).

Conclusions

Several of the examples described in this review provide evidence of success in answering the challenges identified at the beginning of the paper. In particular a functional classification of plants based upon CSR strategy theory appears to provide strong clues to the processes controlling vegetation structure and dynamics, at least in so far as these reflect responses to changes in productivity and in the intensities of vegetation disturbance. Perhaps we should not be surprised that CSR theory has emerged as a good predictor of changed land use; deployment of life histories, growth rates and other traits related to resource capture and utilisation made this an inevitable consequence. Scaling-up to plant communities and regional floras by means of CSR theory has advantages from an operational perspective. A large data base of CSR strategies already exists, and at least for temperate species strategies can be readily allocated to new species using a minimum of easily measured variables (Grime *et al.*, 1988). At the level of the landscape and the regional flora, therefore, where ecology must deal with hundreds of species, CSR theory has no serious rival as a predictive tool. The *validity* of CSR theory may remain in doubt in some quarters, but its *utility* is beyond question.

Difficulties remain in the practical extension of CSR theory to floras outside western Europe, especially to rich floras of relatively unknown ecology. Here progress will come from the development of simple predictors of strategy that can be measured easily by non-specialists and do not depend on familiarity with the autecology of the species concerned.

The need for caution and for experimental tests in our attempts to scale-up from individual plants to larger units of vegetation are well exemplified by the evidence relating CSR theory to CO_2 responses, reviewed earlier in this paper. The association between responsiveness to elevated CO_2 and the strong carbon sinks present in fast-growing perennials of fertile soils remains a powerful basis for extrapolation. It is quite apparent, however, that sink strengths are subject to modifying factors in the context of the plant community and its attendant rhizosphere (Díaz *et al.*, 1993). We need to know much more about the potential of mycorrhizal fungi to influence the fate of the extra carbon fixed by vascular plants growing in enhanced concentrations of CO_2.

Care is also needed in defining the usefulness of CSR theory in relation to interpretation and prediction of vegetation responses to climate. This paper contains evidence of the capacity of the CSR classification to predict differences in the resistance and resilience of plant communities exposed to extreme events. This is explicable in terms of the strong involvement of disturbance and resource release in such events (Leps *et al.*, 1982). However, many less disruptive effects of climate are related only tangentially, or not at all, to the CSR axes of variation. Responses to more subtle shifts in climate, e.g. slight changes in seasonal patterns of temperature and rainfall, may be largely neutral in terms of CSR strategies and require other predictors, e.g. frost tolerance, tailored to specific circumstances.

In this paper, we have illustrated the potential of nuclear DNA amount as a simple and widely available predictor of sensitivity to year-to-year variation in temperature and rainfall. We forecast that eventually this variable, used carefully and in tandem with other selected plant attributes, will be widely used as a predictive tool in climate change research. In the ISP, plasticity in rooting depth has emerged as a useful predictor of drought response and evasion (Z. Liu, K. Thompson, R.E. Spencer & R.J. Reader, unpublished data), but data limitations seem destined to restrict the application of this attribute. In the longer term, anatomical measurements reflecting xeromorphy and water transport appear to offer the prospect of a practicable basis for prediction and scaling.

In order to predict the future course of vegetation development it is essential to take account of the processes of regeneration and dispersal. Earlier in this paper, we concluded that in heavily populated and exploited landscapes the fate of plant species appears to be determined by life history and resource dynamics rather than by specific attributes related to seed or spore dispersal. The possibility remains, however, that in less-intensively disturbed landscapes, propagule characteristics may reliably discriminate between rapid and slow dispersers. The convention that plants can be divided simply into those with adaptations for wind dispersal and those without (Wilson, Rice & Westoby, 1990) is clearly inadequate, and is tolerated only because a data base of real potential for wind dispersal is lacking for all but a tiny minority of species. Similarly, capacity for seed dispersal in time (i.e. seed persistence) is largely unknown for most species, although it is now possible to predict this attribute from simple measurements of dispersules (Thompson, Band & Hodgson, 1993).

Acknowledgements

This work was supported by the Natural Environment Research Council, partly through its TIGER (Terrestrial Initiative in Global Environmental Research) programme, award number GST/02/635. Analysis of the Bibury road verge data was supported by Nuclear Electric plc. Thanks also to Stuart Band, Sandra Díaz, Rod Hunt and Arthur Willis. Professor Bob Crawford and an anonymous referee made valuable comments on the manuscript.

References

Bennett, M.D. (1971). The duration of meiosis. *Proceedings of the Royal Society of London, Series B*, 178, 277–299.

Bennett, M.D. (1976). DNA amount, latitude and crop plant distribution. *Environmental and Experimental Botany*, 16, 93–108.

Bennett, M.D. & Smith, J.P. (1976). Nuclear DNA amounts in angiosperms. *Philosophical Transactions of the Royal Society of London, Series B*, 274, 227–274.

Box, E.O. (1981). *Macroclimate and plant forms: an introduction to predictive modelling in phytogeography*. The Hague: Junk.

Díaz, S., Grime, J.P., Harris, J. & McPherson, E. (1993). Effects of elevated atmospheric carbon dioxide on plant communities during the initial stages of secondary succession. *Nature*, 364, 616–617.

Gaudet, C.L. & Keddy, P.A. (1988). A comparative approach to predicting competitive ability from plant traits. *Nature*, 334, 242–243.

Grime, J.P. (1974). Vegetation classification by reference to strategies. *Nature*, 250, 26–31.

Grime, J.P. (1977). Evidence for the existence of three primary strategies in plants and its relevance to ecological and evolutionary theory. *American Naturalist*, 111, 1169–1194.

Grime, J.P. (1983). Prediction of weed and crop response to climate based upon measurements of nuclear DNA content. *Aspects of Applied Biology*, 4, 87–98.

Grime, J.P. & Hunt, R. (1975). Relative growth rate; its range and adaptive significance in a local flora. *Journal of Ecology*, 63, 393–422.

Grime, J.P. & Mowforth, M.A. (1982). Variation in genome size – an ecological interpretation. *Nature*, 299, 151–153.

Grime, J.P., Mason, G., Curtis, A.V. *et al.* (1981). A comparative study of germination characteristics in a local flora. *Journal of Ecology*, 69, 1017–1059.

Grime, J.P., Hunt, R. & Krzanowski, W.J. (1987a). Evolutionary physiological ecology of plants. In *Evolutionary Physiological Ecology*, ed. P. Calow, pp. 105–126. Cambridge: Cambridge University Press.

Grime, J.P., Mackey, J.M.L., Hillier, S.H. & Read, D.J. (1987b). Floristic diversity in a model system using experimental microcosms. *Nature*, 328, 420–422.

Grime, J.P., Hodgson, J.G. & Hunt, R. (1988). *Comparative Plant Ecology: A Functional Approach to common British Species.* London: Unwin Hyman.

Grime, J.P., Shacklock, J.M.L. & Band, S.R. (1985). Nuclear DNA contents, shoot phenology and species coexistence in a limestone grassland community. *New Phytologist*, 100, 435–444.

Hartsema, A.M. (1961). Influence of temperature on flower formation and flowering of bulbous and tuberous plants. In *Handbuch der Pflanzenphysiologie. 16. Ansenfaktoren in Wachstum und Entwicklung*, ed. W. Ruhland, pp. 123–167. Berlin: Springer.

Hendry, G.A.F. & Grime, J.P. (1993). *Comparative Plant Ecology – A Laboratory Manual.* London: Chapman & Hall.

Hodgson, J.G. (1986). Commonness and rarity in plants with special reference to the Sheffield flora. II. The relative importance of climate, soils and land use. *Biological Conservation*, 36, 253–274.

Hodgson, J.G. (1989). What is happening to the British flora? An investigation of commonness and rarity. *Plants Today*, 2, 26–32.

Hodgson, J.G. & Mackey, J.M.L. (1986). The ecological specialization of dicotyledenous families within a local flora: some factors constraining optimization of seed size and their possible evolutionary significance. *New Phytologist*, 104, 479–515.

Holdridge, L.R. (1947). Determination of world formations from simple climatic data. *Science*, 105, 367–368.

Hopkins, B. (1978). The effects of the 1976 drought on chalk grassland in Sussex, England. *Biological Conservation*, 14, 1–12.

Hunt, R., Hand, D.W., Hannah, M.A. & Neal, A.M. (1991). Response to CO_2 enrichment in 27 herbaceous species. *Functional Ecology*, 5, 410–421.

Hunt, R., Hand, D.W., Hannah, M.A. & Neal, A.M. (1993a). Further responses to CO_2 enrichment in British herbaceous species. *Functional Ecology*, 7, 661–668.

Hunt, R., Willis, A.J., Ward, L.K. *et al.* (1993b). *Vegetation and Climate: A Thirty-five Year Study in Road Verges at Bibury, Gloucestershire, and a Twenty-one Year Study in Chalk Grassland at Aston*

Rowant, Oxfordshire. The NERC Unit of Comparative Plant Ecology, Terrestrial Ecology Research on Global Warming: Phase 2. Contract report to Nuclear Electric plc.

Jurado, E., Westoby, M. & Nelson, D. (1991). Diaspore weight, dispersal, growth form and perenniality of central Australian plants. *Journal of Ecology*, 79, 811–828.

Keddy, P. (1992). A pragmatic approach to functional ecology. *Functional Ecology*, 6, 621–626.

Korner, C. (1993). CO_2 fertilization: the great uncertainty in future vegetation development. In *Vegetation Dynamics and Global Change*, ed. A.M. Solomon & H.H. Shugart, pp. 53–70. London: Chapman & Hall.

Lambers, H. & Poorter, H. (1992). Inherent variation in growth rate between higher plants: a search for physiological causes and ecological consequences. *Advances in Ecological Research*, 23, 187–261.

Lawton, J.H., Naeem, S., Woodfin, R.M. *et al.* (1993). The Ecotron: a controlled environmental facility for the investigation of population and ecosystem processes. *Philosophical Transactions of the Royal Society of London, Series B*, 341, 181–194.

Leps, J., Osbornova-Kosinova, J. & Rejmánek, M. (1982). Community stability, complexity and species life-history strategies. *Vegetatio*, 50, 53–63.

Levin, D.A. & Funderburg, S.W. (1979). Genome size in angiosperms: temperate versus tropical species. *American Naturalist*, 114, 784–795.

Levitt, J. (1978). An overview of freezing injury and its interrelationships to other stresses. In *Plant Cold Hardiness and Freezing Stress*, ed. P.H. Li & A. Sakai. London: Academic Press.

Levitt, J. (1980). *Responses of Plants to Environmental Stresses*. Vol. II. *Water, Radiation, Salt and Other Stresses*. New York: Academic Press.

MacGillivray, C.W. (1993). Extreme climatic events and plant community structure. PhD thesis, University of Sheffield.

MacGillivray, C.W., Grime, J.P. & the ISP Team (1995). Testing predictions of the resistance and resilience of vegetation subjected to extreme events. *Functional Ecology*, 9, 640–649.

MacLeod, J. (1894). Over de bevruchting der bloemen in het Kempisch gedeelte van Vlaanderen. *Deel II. Botanisch Jaarboek Dodonaea*, 6, 119–511.

Monteith, J.L. (1973). *Principles of Environmental Physics*. Edward Arnold: London.

Noble, I.R. & Slatyer, R.O. (1979). The use of vital attributes to predict successional changes in plant communities subject to recurrent disturbances. *Vegetatio*, 43, 5–21.

Perring, F.H. & Walters, S.M. (1962). *Atlas of the British Flora.* London: Nelson.

Ramenskii, L.G. (1938). *Introduction to the Geobotanical Study of Complex Vegetations.* Moscow: Selkzgiz.

Raunkiaer, C. (1934). *The Life Forms of Plants and Statistical Plant Geography: being the Collected Papers of C. Raunkiaer, Translated into English by H.G. Carter, A.G. Tansley and Miss Fansboll.* Oxford: Clarendon Press.

Reader, R.J., Jalili, A., Grime, J.P., Spencer, R.E. & Matthews, N. (1992). A comparative study of plasticity in seeding rooting depth in drying soil. *Journal of Ecology,* 81, 543–550.

Reich, P.B., Walters, M.B. & Ellsworth, D.S. (1992). Leaf life-span in relation to leaf, plant and stand characteristics among diverse ecosystems. *Ecological Monographs,* 62, 365–392.

Rich, T.C.G. & Woodruff, E.R. (1990). *The BSBI monitoring scheme 1987–1988. Report to the Nature Conservancy Council, CSD report 1265.* Peterborough: Nature Conservancy Council.

Rich, T.C.G. & Woodruff, E.R. (1992). Recording bias in botanical surveys. *Watsonia,* 19, 73–95.

Stebbins, G.L. (1956). Cytogenetics and evolution of the grass family. *American Journal of Botany,* 43, 890–905.

Thompson, K. (1994). Predicting the fate of temperate species in response to human disturbance and global change. In *Biodiversity, Temperate Ecosystems and Global Change,* ed. T.J.B. Boyle & C.E.B. Boyle, pp. 61–76. Berlin: Springer Verlag.

Thompson, K., Band, S.R. & Hodgson, J.G. (1993). Seed size and shape predict persistence in soil. *Functional Ecology,* 7, 236–241.

van der Valk, A.G. (1981). Succession in wetlands: a Gleasonian approach. *Ecology,* 62, 688–696.

Walker, B.H. (1991). Ecological consequences of atmospheric and climate change. *Climatic Change,* 18, 301–316.

Washitani, I. & Masuda, M. (1990). A comparative study of the germination characteristics of seeds from a moist tall grassland community. *Functional Ecology,* 4, 543–558.

Westman, W.E. (1978). Measuring the inertia and resilience of ecosystems. *Bioscience,* 28, 705–710.

Willson, M.F., Rice, B.L. & Westoby, M. (1990). Seed dispersal spectra: a comparison of temperate plant communities. *Journal of Vegetation Science,* 1, 547–562.

Woodward, F.I. (1987). *Climate and Plant Distribution.* Cambridge: Cambridge University Press.

Yemm, E.W. & Willis, A.J. (1962). The effects of maleic hydrazide and 2,4-dichlorophenoxyacetic acid on roadside vegetation. *Weed Research,* 2, 24–40.

J.D.B. WEYERS, T. LAWSON and Z.Y. PENG

Variation in stomatal characteristics at the whole-leaf level

Introduction

A full understanding of the role of stomata in higher plant physiology and ecology will not be forthcoming without due consideration of models and scaling processes. However, modelling and attempts to scale-up must pay due regard to the intricacy of real life, while the values used in them should be unbiased and have quantified precision. This chapter deals with the real-life complexity relevant to models describing water loss from stomata at the whole-leaf level and below (Table 1). Stomatal dimensions, stomatal frequency and stomatal or leaf water vapour conductance (g_s and g_1, respectively) are the characters chosen for analysis as they are variables used in several frequently cited models. Stomatal dimensions can be used to predict the conductance of individual pores (Equation 2, Table 1), and when combined with stomatal frequency, they can be used to estimate g_s for patches of leaf (Equation 5, Table 1), which can also be measured by porometry or infrared gas analysis. Stomatal or leaf conductances are also important variables in more complex models such as those of Jarvis & McNaughton (1986) and McNaughton & Jarvis (1991), which deal explicitly with sensitivity of transpiration to stomatal control at scales from the whole leaf and above (Equations 7–9, Table 1). Variability in stomatal frequency is also of interest in the context of acclimation to changes in CO_2 partial pressure (Woodward, 1987; Körner, 1988).

Methods of studying variability in stomatal characteristics

Measurements of stomatal aperture and conductance

Observations of individual pores

Accurate measurements of individual pores are required to study pore characteristics and inter-pore variability. The greatest degree of

Table 1. *Simple models used for estimating water vapour fluxes at different scales*[a]

Scale and equation	Meaning of symbols (units)
Single pore	
1. $E_p = g_p D_1 / P_a$	E_p = pore water vapour flux (mol s^{-1})
	g_p = pore conductance (mol s^{-1} pore^{-1})
where	D_1 = water vapour saturation deficit at leaf surface (Pa)
	P_a = atmospheric pressure (Pa = J m^{-3})
	D_w = water diffusivity in air (m^2 s^{-1})
2. $g_p = \dfrac{P_a D_w A_s}{RT(d+2c)}$	A_s = mean pore area (m^2)
	R = gas constant (J mol^{-1} K^{-1})
	T = mean of leaf and air temperature (K)
	d = pore depth (m)
	c = 'end correction' (m)

Patch (one surface, assuming g_b is constant)	
3. $E_1 = D_a g_t / P_a$	E_1 = leaf water vapour flux density (mol m^{-2} s^{-1})
	D_a = water vapour saturation deficit outside boundary layer (Pa)
where	g_x = conductance of pathway component x (mol m^{-2} s^{-1})
	where subscripts indicate:
	t = total (leaf + boundary layer)
4. $g_t = \dfrac{g_l g_b}{g_l + g_b}$	l = leaf
	b = boundary layer
	i = intercellular air spaces
	s = stomata
5. $g_1 \dfrac{g_i g_s}{g_i + g_s} + g_c$	c = cuticle
	v = stomatal frequency (m^{-2})
and	
6. $g_s = \dfrac{P_a D_w A_s v}{RT(d+2c)}$	

Patch, leaf and above (two equal surfaces, where g_b is influenced by g_s)	
7. $E_1 = E_{eq}(\Omega) + E_{imp}(1-\Omega)$	E_{eq} = 'equilibrium' transpiration rate (mol m^{-2} s^{-1})
	Ω = decoupling coefficient (dimensionless, $0 < \Omega < 1$)
where	E_{imp} = 'imposed' transpiration rate (mol m^{-2} s^{-1})
	c_p = molar heat capacity of air (J mol^{-1} K^{-1})
8. $E_{imp} = c_p 2 g_1 D_a / (\lambda \gamma)$	λ = molar latent heat of evaporation of water (J mol^{-1})
	γ = psychometric constant (Pa K^{-1})
and	ε = s/γ (dimensionless), where
	s = slope of curve relating saturation vapour pressure to temperature (Pa K^{-1})
9. $E_{eq} = \varepsilon \Phi / (\lambda(\varepsilon + 1))$	x = net radiation flux density (J m^{-2} s^{-1})

[a] For history of the development of models, assumptions made and further explanation, see Jarvis & McNaughton, (1986); Nobel, (1991); Jones, (1992).

resolution is possible with scanning electron microscopy (SEM); for example, van Gardingen, Jeffree & Grace (1989) used low-temperature SEM to measure dimensions of *Avena fatua* stomata, but this method is probably impractical for whole-leaf sampling. While light microscopy may be more convenient, its limit of resolution will affect the accuracy of measurements, particularly when pore widths are narrow. A further problem with *in vivo* measurement with light microscopy is that leaves may be exposed to high light intensities on the microscope stage. It may, therefore, be preferable to work from photomicrographs or leaf replicas. The use of silicone rubber impressions (Fig. 1; Smith, Weyers & Berry, 1989) has allowed sampling of whole leaves using light microscopy, but the technique can be subject to artefact, and care must be taken in determining the mean aperture of a given sample (Weyers & Meidner, 1990).

Studies at the whole-leaf level can utilise measurements of individual pore dimensions, but the number of observations required is large: even

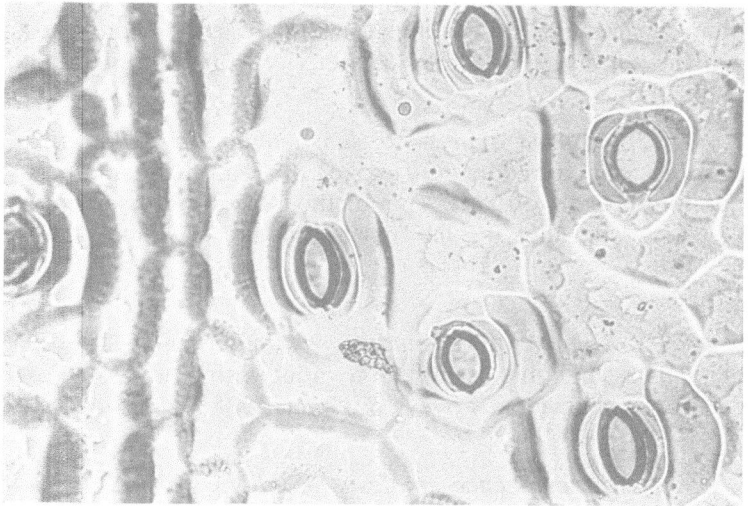

Fig. 1. Silicone rubber impression of an area of a *Commelina communis* leaf. The method was described in Weyers & Meidner (1990). Note: (i) the stomata are not randomly distributed – the distance between pores is relatively constant and there are no pores over the vein, the position of which is indicated by the elongate epidermal cells; (ii) the stomata are not of uniform aperture and appear narrower nearer the vein; (iii) each pore is more-or-less, but not exactly, symmetrical.

sampling 5% of the stomata on a leaf of area 2000 mm^2 and stomatal frequency 50 mm^{-2} involves looking at and/or measuring some 5000 stomata. If the whole population of stomata cannot be measured, then due regard must be given to sampling strategy. In either light or electron microscopy, image analysis can be used to reduce the tedium of making large numbers of measurements, as described by Omasa, Hashimoto & Aiga (1983) and van Gardingen *et al.* (1989).

Indirect methods

Certain indirect methods are capable of resolving variation in g_s (or variables correlated with g_s) over whole leaves or substantial areas of leaf and are generally more convenient than microscopy, although there is inevitably a loss of information about the state of individual pores.

1. *Gas exchange measurements.* Diffusion porometers and infrared gas analysers provide information about the movement of water vapour out of leaves (see Weyers & Meidner, 1990; Hall *et al.*, 1993). Many instruments utilise cuvettes that are small enough to be used to generate an image of whole-leaf activity from multiple readings.

2. *Infrared analysis of leaf temperature differences.* Infrared imaging can provide an image of leaf temperature differences (Hashimoto *et al.*, 1934). This method can yield a series of images of the same leaf, allowing the dynamic nature of stomatal movements to be studied.

3. *Uptake of ^{14}C and starch iodine tests.* These destructive methods assume that assimilation varies with g_s under the experimental conditions (Downton, Loveys & Grant, 1988; Terashima *et al.*, 1988).

4. *Fluorescence analysis of photosynthetic activity.* In this technique, light flashes of < 1 s duration are used to excite leaf areas. Fluorescence images are obtained and converted to non-photochemical quench coefficients, which are used to derive the rate of photosynthesis and gas exchange (Daley *et al.*, 1989). This method also allows stomatal movements to be followed indirectly (Cardon, Mott & Berry, 1994).

5. *Vacuum infiltration.* This rapid, simple method involves exposing detached leaves to varying pressures under water

(Beyschlag & Pfanz, 1990). The resulting infiltration of water (Fig. 2) is related linearly to whole-leaf conductance and presumably to that of patches.

Measurements of stomatal frequency, index and size

These characteristics may be observed microscopically on intact leaves or epidermal strips, but it is generally more convenient to make a replica of the leaf surface (Fig. 1). Stomatal frequency can be estimated from counts of pores in a defined area, with due regard for proper sampling procedure (see Kubínová 1994). Here, as with stomatal dimensions, some degree of automation in recording data is desirable. In most if not all species, there are no stomata in the epidermis covering veins (Fig. 1); hence, truly random or stratified random samples result in skewed frequency distributions leading to difficulty with statistical analysis. For some investigations, such as those concentrating on cell development, it may be beneficial to consider inter-veinal stomatal frequencies instead. However, stomatal frequency may not be the best parameter for determining treatment effects on stomatal development; for this, the stomatal index (stomata as a percentage of stomata plus epidermal plus subsidiary cells) or total number of stomata per leaf may be more appropriate. In view of the pronounced variability in stomatal frequency over the leaf surface (see below), an unbiased sampling method is required to derive total stomatal numbers from stomatal frequency and leaf area measurements.

Analysis and presentation of data

Noise, patches and trends

Stomatal characteristics are not constant over a leaf. They can exhibit three types of variation: noise, patches and trends. Noise is the scatter about a local mean value defined with reference to the assumed 'macro' pattern of variance. Patches and trends are alternative forms of macro-variance at the areola to whole-leaf scale. A general definition of patchiness is 'non-random behaviour' and for stomatal aperture it is commonly understood to apply to areas (usually areolae or groups of areolae) having a mean pore width or conductance that is different from those of other neighbouring areas. A trend in stomatal aperture can be defined as a continuous and smooth transition in pore width or conductance over an area of leaf. In theory, patchy behaviour could also exhibit trends in the mean patch values.

A standard technique for quantifying the degree of non-uniform or

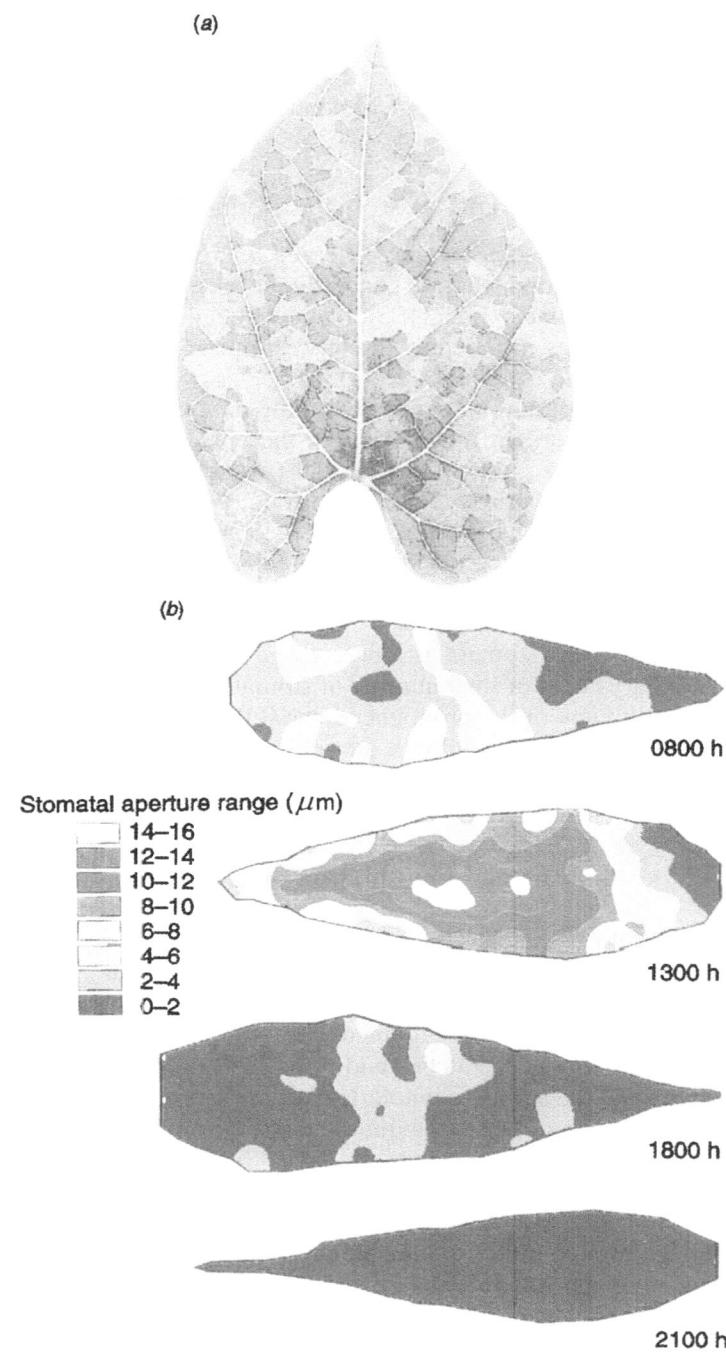

(a)

(b)

Stomatal aperture range (μm)

14–16
12–14
10–12
8–10
6–8
4–6
2–4
0–2

0800 h

1300 h

1800 h

2100 h

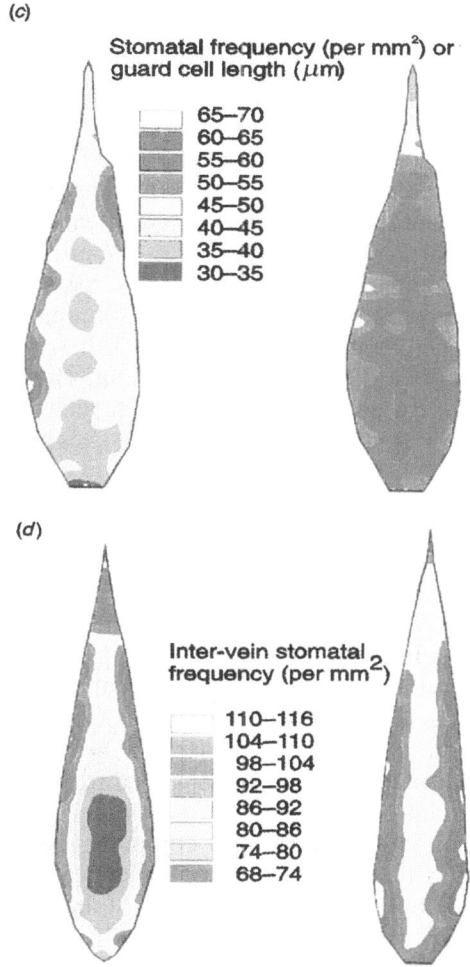

Fig. 2. Colour representations of variation in stomatal characteristics. (*a*) Patchy stomatal opening in a *Phaseolus vulgaris* leaf following excision and wilting, as visualised by vacuum infiltration (leaf length 79 mm). (*b*) Maps of mean stomatal aperture over the surface of *Commelina communis* leaves in a glasshouse at different times of day (new presentation of data of Smith *et al.*, 1989). The leaves were, in order, 90, 87, 96 and 115 mm long. (*c*) Maps of mean stomatal frequency (left) and guard cell length (right) over the surface of a *C. communis* leaf (new presentation of data of Smith *et al.*, 1989). (*d*) Maps of inter-vein stomatal frequency in fully expanded leaves from *C. communis* plants provided with continuous watering (left) or no water for 24 days (right).

A colour version of this figure is available for download from www.cambridge.org/9780521471091

patchy stomatal closure would be desirable (van Kraalingen, 1990). Clearly, the existence of patches or trends on the leaf surface needs to be considered in relation to noise (equal to within patch) variability. The mean patch aperture value is expected to be relatively constant over the patch area, and comparisons between areas should take into account within-patch variation via appropriate statistical procedures. In transects across a leaf, patches and trends will differ in plots of the variance of the sample against distance (H.G. Jones, personal communication). There should be pronounced peaks in the running standard deviation of patchy data, whereas in the case of trends, the profile of running standard deviation should be relatively constant. However, this difference might not be obvious if the difference in mean patch aperture were small or if the aperture variance within patches were high. An alternative would be to change the area of averaging and plot variance against sample area; peaks in variance would then represent scales of patchiness.

The same principles ought to apply to other stomatal characteristics, although in general, these are treated as exhibiting trends.

Presentation of data

Samples of individual pore characteristics can be presented as frequency distributions (Laisk, Oja & Kull, 1980; van Gardingen *et al.*, 1989; Fig. 3). Trends in stomatal characteristics can be visualised using suitable iso-trait contour maps or three-dimensional representations where x and y represent spatial co-ordinates and z is the character of interest (Figs. 2 and 4). An important element of computer mapping programs is the ability to overlay the map with the co-ordinates of the leaf outline, such that data interpolation stops with a 'fault' at the leaf margin. The design of the mapping program may constrain the sampling pattern if it requires a regular data array, whereas a stratified sampling pattern may be more efficient (T. Lawson & J.D.B. Weyers, unpublished data).

Results from certain techniques, like leaf infiltration, are best recorded by photography (Fig. 2), although treatments might be compared gravimetrically (Beyschlag & Pfanz, 1990) or via image analysis (Daley *et al.*, 1989). Where a method produces data that are digitised, as in indirect temperature or fluorescence measurements, the use of grey levels or false-colour for data presentation is a possibility. The relationships between selection of false-colour and mapping resolution and the interpretation of images are complex. Too many divisions of the range produce an overly complex image that is difficult to interpret. In this case, genuine trends could appear as patches; if there were too few divisions, a genuinely patchy distribution might appear as trends.

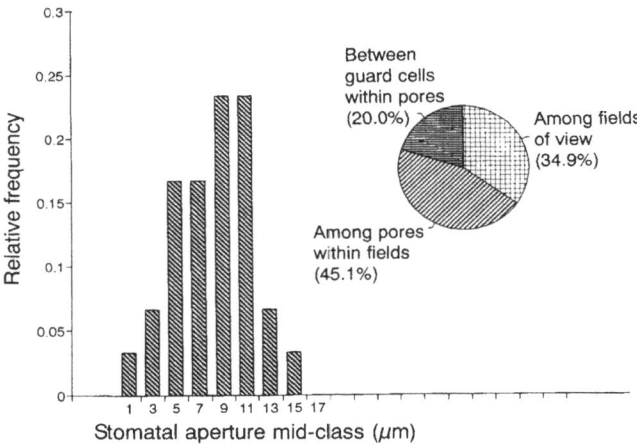

Fig. 3. Statistical analysis of a sample of 30 *Commelina communis* stomata. Measurements were taken from photographs of five fields of view within a 3×3 mm area of a silicone rubber impression, as shown in Fig. 1. Main histogram: frequency distribution of mean pore widths in the area. Inset: pie chart showing partitioning of variance in guard cell half-apertures. Analysis of variance showed (i) that there were significant differences among stomata in the different fields of view ($p < 0.01$), indicating that, even in this small area, there were trends or patches evident; and (ii) that the variation in guard cells between pores in the same field of view is significant ($p < 0.01$), compared with the difference between the half-apertures of a pair, indicating a higher degree of symmetry than would be expected from the local variability in pore widths. Data analysis kindly performed by Dr R. A'Brook.

When choice of colours for the contour bands is also considered, it seems that a subjective element to data presentation is inevitable.

Observations of spatial heterogeneity in stomatal characters

Stomatal aperture/conductance

Individual pores and small groups of stomata
A noticeable aspect of the individual stomatal pore is its apparent symmetry when open (Fig. 1). As shown in Fig. 3, careful measurement reveals that the half-widths of a pair of guard cells are significantly less

(a)

Leaf conductance
(mmol m⁻² s⁻²)

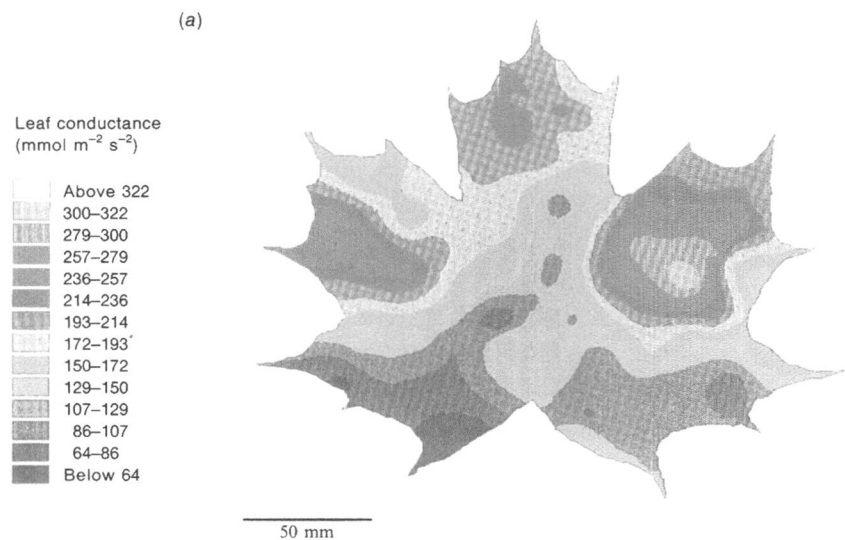

	Above 322
	300–322
	279–300
	257–279
	236–257
	214–236
	193–214
	172–193ʹ
	150–172
	129–150
	107–129
	86–107
	64–86
	Below 64

50 mm

Fig. 4. Colour representations of variation in leaf conductance. (*a*) Map of leaf water vapour conductance over an *Acer pseudoplatanus* leaf in the field, based on 58 readings over about 0.8 h. Data obtained by J.D.B. Weyers and A. Yool. (*b*) Map of leaf water vapour conductance in a variegated *Ficus elastica* leaf in a glasshouse (left), based on 61 readings over about 0.3 h, and a photograph of the leaf itself (right). Data obtained by J.D.B. Weyers and A. Hunter. All data were obtained using a Mk 3 automatic porometer, Delta-T devices, Cambridge.

A colour version of this figure is available for download from
www.cambridge.org/9780521471091

variable than would be expected on the basis of local variation among pores. Another striking feature of any small sample of stomata is the difference in aperture among adjacent pores (Figs. 1 and 3). For *Commelina communis*, the relative variation of stomatal apertures among adjacent pores is greater than that of other stomatal traits such as size and distance apart: Weyers & Meidner (1990) reported the respective coefficients of variation for these characteristics to be 23.0%, 7.2% and 15.4%. Spence (1987) concluded that the dispersion of stomatal aperture data (in epidermal strips) depends both on plant species and pretreatments. Laisk *et al.* (1980) noted that the frequency distributions of stomatal apertures in *Vicia faba* and *Hordeum vulgare* were symmetrical and bell-shaped, except when the mean aperture approached the extremes of zero, when the distribution was skewed right, and close to

(b)

Leaf conductance
(mmol m^{-2} s^{-2})

Above 237
224–237
212–224
200–212
189–200
175–189
163–175
151–163
139–151
126–139
114–126
102–114
90–102
78–90
65–78
53–65
41–53
29–41
Below 29

50 mm

the maximum possible aperture, when it was skewed left. Between these limits, they suggested that differences in apertures among individual stomata might be conserved as the mean value altered and the frequency distribution was translated along the aperture axis.

Patchiness and trends

Francis Darwin was among the first to realise that variation in stomatal activity occurred at the whole-leaf level. He developed a horn hygroscope porometer to measure rates of transpiration and noted that 'in some leaves, particularly those of monocotyledons, the hygroscope gives very different readings in different parts of the leaf' (Darwin, 1898). However, Knight (1916), using the more sophisticated technique of viscous-flow porometry, concluded that differences in conductivity between different areas of a *Ficus elastica* leaf were small. Knight's report possibly influenced subsequent research, which did not focus on the phenomenon again until the 1960s when a series of reports described spatial variation at the whole-leaf level in measurements of stomatal conductance, leaf temperature and stomatal frequency (Slavik, 1963; Cook, Dixon & Leopold, 1964; Raschke, 1965). None of these studies, however, involved a large number of samples, and none was analysed for spatial pattern. Therefore, in a major review published in 1983, Solárová and Pospíšilová concluded that 'little attention had been paid to heterogeneity of the leaf blade with respect to gas exchange'.

Fresh impetus was provided by Downton *et al.* (1988) and Terashima *et al.* (1988), who showed that the assumption of uniform aperture had given rise to the erroneous conclusion that abscisic acid affected photosynthesis directly rather than acting solely via stomatal closure. They concluded that 'patchy' g_s was the cause of these errors, as had been predicted by Laisk (1983). Their observations were confirmed using a variety of methods and treatments (see, for example, Daley *et al.*, 1989; Sharkey & Seemann, 1989; Beyschlag & Pfanz, 1990; Mott, Cardon & Berry, 1993; Cardon *et al.*, 1994). The general conclusion can be made that traumatic treatments lead to patchy stomatal movements in species with heterobaric leaves (i.e. those where inter-vein areas are effectively gas-tight compartments sealed by the bundle sheath extensions to the veins, as opposed to homobaric leaves where there is relatively free diffusive exchange of gases between areolae).

In the homobaric leaves of *C. communis*, Smith *et al.* (1989) found little evidence for patches in the unstressed state, but they did find that stomatal aperture varied over the leaf surface in a complex non-uniform manner. They found that stomata were generally more open in the centre and closed nearer the margins (Fig. 2b). The results of Hashimoto *et al.* (1984), obtained via thermal imaging of the heterobaric leaves of *Helianthus annuus*, also indicate trends rather than patches in both control and water-stressed leaves, although this could be an artefact of the false-colour scales which were used.

Few reports to date deal with leaves under natural conditions, although Beyschlag & Pfanz (1990) showed that patchy movements can occur naturally before and after mid-day stomatal closure in a glasshouse. Figure 4a shows the trends in g_1 over an *Acer pseudoplatanus* leaf in the field. The source data were obtained with a diffusion porometer and so do not allow patchy behaviour to be identified; nevertheless, the map indicates that there were up to five-fold differences in g_1 over the leaf surface. A shadefleck was present in the lower left part of this leaf, accounting for the low conductances in this region. Under controlled conditions, patterns of conductance can be manipulated by creating simulated sunflecks or shadeflecks and by altering wind speed (results not shown). They are strongly influenced by photosynthesis in the underlying mesophyll, as can be seen from Fig. 4b; in this set of observations on a variegated *F. elastica* leaf, analysis of variance revealed a significant difference ($p < 0.05$) in conductance between yellow areas and dark or pale green areas that were matched for distance away from the leaf margin.

It is well established that stomata on upper and lower leaf surfaces exhibit different stomatal frequencies, dimensions and conductance (see Weyers & Meidner, 1990). Mott *et al.* (1993) have further shown that the pattern of patchy fluorescence in *Xanthium strumarium* is different on either leaf surface. These observations add further complexity when modelling transpiration (see Jarvis & McNaughton, 1986).

Stomatal frequency, index and size

Around each guard cell pair there is a zone where other stomata appear *less* frequently than would be expected by chance; this has been interpreted as showing that stomatal initials are capable of influencing the development of neighbouring cells (see Sachs, 1991). Hence, at the scale of individual areolae, the stomata are relatively evenly spaced (Fig. 1). However, it should be noted that the lack of stomata over vascular tissue imposes a form of patchiness in their distribution.

Tichá (1982) reviewed many studies of changes in stomatal frequency and size during leaf ontogeny and noted in passing that frequency on any given leaf blade is 'extremely variable'. Most of the studies quoted do not provide enough data to allow a two- or three-dimensional representation. However, Smith *et al.* (1989) examined trends in frequency and size over mature *C. communis* leaves and presented their data in this way (Fig. 2*c*). For these maps, the variables exhibit a significant negative correlation ($r = -0.434$, $n = 210$, $p < 0.01$), suggesting that the effects on stomatal conductance of variation in these two factors may partially cancel out (Equation 6, Table 1).

In *C. communis*, maps of stomatal frequency (Fig. 2*c*,*d*) indicate that the stomata are most dense at the margins and tips of leaves. Stomatal frequency is greatly affected by treatments that influence leaf area, such as water deficit stress. The maps shown in Fig. 2*d* represent leaves from control and stress treatments where the weighted mean inter-vein frequency increased significantly by 16.5% ($p < 0.01$) following withdrawal of water for 24 days. However, this was not the only effect of the treatment, since the stomatal index decreased from 10.26 to 9.60 ($p < 0.05$). Such effects on stomatal development could possibly affect the potential value of g_s when stresses are relieved.

As discussed by Körner (1988), the extent of variability in stomatal frequency and index needs to be considered in studies of how they may have varied in response to changing partial pressures of CO_2 (Woodward, 1987). M.D. Davidson, J.D.B. Weyers & J.A. Raven (unpublished) studied stomatal frequency in *Ulmus glabra* leaves and found that variability among sites within leaves (coefficient of variation = 12.8%) and among leaves (coefficient of variation = 15.3%) was so high that great care would be required in choosing both the number and position of samples for a study of changes in frequency through time. As a very minimum, it seems essential that the location of samples for frequency estimates should be stated in such studies.

Observations of temporal heterogeneity in stomatal conductance

Space precludes a detailed discussion of this topic, but it must be borne in mind both when taking samples and when scaling-up. The fact that g_s varies through the day has been known for many years. Similarly, short-term cycling in g_s with periods of 4–120 min are well documented and frequently follow 'shock' treatments (Barrs, 1971). The observations of Beyschlag & Pfanz (1990) illustrate the complexity of spatial heterogeneity and its interaction with time of day: not only was a daily cycle in leaf conductance evident, but also a change in the *pattern* of conductance. Patchiness appeared most evident when the stomata were opening or closing; otherwise the pattern of infiltration was constant or showed trends. In their study, Smith *et al.* (1989) also found that the pattern as well as the absolute values of stomatal apertures varied over a 13 hour period (Fig. 2*b*). The report of Cardon *et al.* (1994) characterising the dynamics of patchy stomatal movements is extremely valuable because it shows clearly that the behaviour of individual patches is dynamic and sometimes independent. Two contrasting sets of results are presented: lowered humidity or light either elicited patches that oscillated in concert, such that the overall gas exchange for the area oscillated in phase, or it caused independent patchy behaviour that cancelled out to give a relatively constant rate of gas exchange. However, the leaf temperature patterns in control leaves observed by Hashimoto *et al.* (1984) remained very similar over a 2.5 hour period under constant, defined conditions.

Physiological and anatomical bases for heterogeneity

Stomatal aperture and conductance

Variability among stomata
We can conclude from the data presented in Fig. 3 that the main 'unit of variability' is the stomatal pore rather than the individual guard cells. The reason for this is unclear, but the greater-than-expected degree of pore symmetry could reveal some mechanism that tends to equilibrate the individual guard cells in a pair. The reason for inter-pore variability also remains obscure. Spence (1987) identified variability in guard cell size as a possible factor leading to aperture variance in certain species (*V. faba, Datura stramonium*). In *C. communis*, this is unlikely to be the whole story, because the coefficient of variation of stomatal aperture is at least three-fold greater than that of guard cell length. Laisk *et al.* (1980) proposed that variability in aperture reflected differences in guard cell solute potentials (as observed in sigmoidal plasmolysis curves); they accounted for changes in the shape of the aperture–frequency distribution by assuming guard cells below a certain turgor threshold would all have zero aperture, while those close to the upper extreme would encounter wall stiffening. Guard cell solute levels alter as part of their physiological responses to external and internal stimuli; their turgors are also affected by the vapour pressure deficit and leaf water potential, ψ_w. If the stimuli and physical conditions alter spatially or temporally, then so will pore widths, especially if each stomate responds independently. The observed noise may reflect out-of-phase 'hunting' towards some control value set by a feedback system within each guard cell.

Patchiness and trends

At the 'macro' level, patterns of conductance are likely to be related to changes in external and internal conditions over the leaf surface. Examples of important factors include windspeed, through its effect on the boundary layer (van Gardingen & Grace, 1991), and sunflecks, through their direct effect on guard cells and indirect effect on photosynthesis and the substomatal CO_2 concentration, C_i (Pearcy, 1990). As Fig. 4b shows, stomatal activity is strongly correlated with underlying variation in the capacity for photosynthesis. Guard cell turgor also depends on the local ψ_w, which is known to vary over the leaf surface (Slavik, 1963). The proximity and position of a particular stoma in relation to the vascular system should be important for this reason; the vascular system may also influence behaviour through physiologically active solutes delivered via the transpiration stream, such as abscisic acid, K^+ and Ca^{2+} (Mansfield, Hetherington & Atkinson, 1990). It should be noted that these influences cannot be predicted simply on the basis of the cell water relations equation ($\psi_w = \psi_p + \psi_s$, where ψ_p is the pressure potential and ψ_s the solute potential), because of the optimum water potential for stomatal opening (see Weyers & Meidner, 1990). There is also the possibility of spatial and temporal differences in sensitivity to stimuli and of synergistic or antagonistic interactions among stimuli when influencing responses. Furthermore, the cells of a leaf differ in maturity, especially in monocots (Miranda, Baker & Long, 1981).

There can be little doubt that patchiness in g_s exists as a phenomenon as it has been shown in numerous species and with several different methods. However, control leaves from many reported experiments show steady conductance or smooth transitions between areas on the leaf. Patchiness is apparently a response to a sudden change in conditions, where the stomata in a particular area respond in concert but essentially independently from neighbouring areas. A possible reason for this lies in the fact that the patches themselves are generally bounded by veins of medium size. Terashima *et al.* (1988) suggested that patchiness might be a consequence of a heterobaric leaf anatomy, where bundle sheath extensions provide zones of leaf whose inter-cellular air space is isolated from other zones. Bundle sheath extensions are restricted to leaves of certain species but occur in monocots and dicots; in a given heterobaric leaf, they are not universal, being normally associated with larger veins; the proportion of the total vein length with extensions differs among species (see Mauseth, 1988). The role of these extensions appears to be to provide structural support and to supply water to the epidermis, so areolae may also be hydraulically isolated from each other and possibly sensitive to cavitation in feeder xylem elements.

Observations of patchy stomatal or assimilatory behaviour are generally in accord with the above, regions of high or low 'conductance' or 'assimilation' being bounded by slightly larger veins and including subareas bounded by smaller veins (see Fig. 2a). Moreover, the isolation of areolae in terms of inter-cellular gas phase can explain why anatomical patches become physiological patches when environmental conditions and stimuli are expected to vary smoothly over the leaf surface. If stomata respond to a *secondary* stimulus transmitted rapidly via the gas phase (that is, C_i), it follows that responses of stomata within patches will be similar but responses among patches may be dissimilar and asynchronous. Many of the data so far obtained represent snapshots of behaviour, but patches could represent groups of stomata acting in unison to 'hunt' some optimum stomatal response. The elegant work of Cardon *et al.* (1994) provides experimental support for this notion. Patches themselves might be out of phase, but, assuming that the new environmental condition is even, they will eventually all reach some optimum state.

Stomatal dimensions and frequency

Little is known about the reasons why stomatal dimensions and frequencies vary over leaf surfaces, although it is probably related to the same factors that control leaf expansion. Hence, although the total number of stomata per leaf varies ontogenetically and stomatal index may alter following certain treatments (Tichá, 1982; Woodward, 1987), stomatal frequency is generally at its highest where leaves may not have fully expanded (leaf base, margins, tip). Moreover, treatments affecting leaf expansion, such as water deficit stress, affect frequency greatly (Fig. 2d). If leaf expansion is achieved via cell expansion, then this may also explain why there is a negative correlation between stomatal frequency and stomatal size.

Conclusions: relevance to scaling-up

Stomatal function is vitally important to plants because it affects the key processes of photosynthesis and transpiration. Agronomists, ecologists and physiologists are interested in these processes at several scales and in the relationship between activity at one scale and another. Several reviewers have concluded that heterogeneity of stomatal behaviour is important from the perspective of methodology and physiology (Spence, 1987; Weyers & Meidner, 1990; Mansfield *et al.*, 1990) and many have pointed out its importance for scaling-up and scaling-down (Laisk, 1983; Jarvis & McNaughton, 1986; Downton *et al.*, 1988; Terashima *et al.*, 1988). This chapter highlights the problem of stomatal heterogeneity in relation to predicting transpiration rates and concentrates on three aspects that can influence the rate of transpiration: stomatal dimensions, stomatal frequency and stomatal conductance. The main conclusions in relation to extrapolating these estimates up to larger areas are as follows:

1. There is a high degree of spatial variability in stomatal dimensions, stomatal frequency and stomatal conductance over leaf surfaces. Among-patch variance is greatly increased following traumatic treatments. Hence, the use of subleaf sampling will probably give biased results unless a weighted mean value is used in models. This requires an adequate sampling protocol.

2. The g_s value for a given leaf patch varies in space *and in time*. The high degree of temporal variation in g_s means that steady-state models may be unrealistic.

3. It is desirable to quantify the effects of sampling errors when scaling-up. Difference in sampling position could give rise to scaled-up values differing by several-fold. A representative sampling scheme should be used both to quantify and to limit these errors. It should, however, be noted that the effect of such errors will diminish as larger and larger scales are considered, because of the reduction of coupling of g_s to transpiration rate (Jarvis & McNaughton, 1986).

4. Further observations and measurements are required to establish fully the nature of stomatal heterogeneity and the reasons for it. A statistico–physiological approach to the following fascinating questions is required. How is pore symmetry achieved? Why is there so much inter-pore variability? Is patchy stomatal behaviour confined to heterobaric species? How prevalent is patchy behaviour in the field? What conditions influence the spatial and temporal pattern of stomatal conductance in leaves? What effects do within- and among-patch variance have on overall gas exchange?

Finally, it must be emphasised that, depending on the level of scale (Table 1), other information could be relevant for determining transpiration rates and may, indeed, exhibit as much if not more variability as the factors considered here. Examples include boundary layer conductance (van Gardingen & Grace, 1991), cuticular conductance and mesophyll anatomy (Miranda *et al.*, 1981) and water relations (Slavik, 1963). For some potentially important factors, such as the depth of the stomatal pore, virtually nothing is known about within-leaf variability. This illustrates one of the benefits of modelling (Jones, 1992), in that it serves to drive research, in this case highlighting which factors *could* be important for accurate predicting.

Acknowledgements

We thank the UK BBSRC for supporting T.L. via a quota studentship. The following have directly assisted our research for this paper: Stephen Barr, Richard A'Brook, Martin Davidson, Karen Findlay, Audra Hunter, Yvonne Lindsay, Alison Roberts, Susan Smith, Andrew Yool and Iain Tennant. Our colleagues Bill Berry, John Hillman, Roy Oliver, Richard Parsons, Neil Paterson, John Raven and Janet Sprent are thanked for material support and advice.

References

Barrs, H.D. (1971). Cyclic variations in stomatal aperture, transpiration, and leaf water potential under constant environmental conditions. *Annual Review of Plant Physiology*, 22, 223–236.

Beyschlag, W. & Pfanz, H. (1990). A fast method to detect the occurrence of nonhomogenous distribution of stomatal aperture in heterobaric plant leaves. Experiments with *Arbutus unedo* during the diurnal course. *Oecologia*, 82, 52–55.

Cardon, Z.G., Mott, K.A. & Berry, J.A. (1994). Dynamics of patchy stomatal movements, and their contribution to steady-state and oscillating stomatal conductance calculated with gas-exchange techniques. *Plant, Cell and Environment*, 17, 995–1008.

Cook, G.D., Dixon, J.R. & Leopold, A.C. (1964). Transpiration: its effects on leaf temperature. *Science*, 144, 546–547.

Daley, P.F., Raschke, K., Ball, J.T. & Berry, J.A. (1989). Topography of photosynthetic activity of leaves obtained from video images of chlorophyll fluorescence. *Plant Physiology*, 90, 1233–1238.

Darwin, F. (1898). Observations on stomata. *Philosophical Transactions of the Royal Society of London*, 190, 531–621.

Downton, W.J.S., Loveys, B.R. & Grant, W.J.R. (1988). Stomatal closure fully accounts for the inhibition of photosynthesis by abscisic acid. *New Phytologist*, 108, 263–266.

Hall, D.O., Scurlock, J.M.O., Bolhàr-Nordenkampf, H.R., Leegood, R.C. & Long, S.P. (1993). *Photosynthesis and Production in a Changing Environment*. London: Chapman & Hall.

Hashimoto, Y., Ino, T., Kramer, P., Naylor, A.W. & Strain, B.R.

(1984). Dynamic analysis of water stress of sunflower leaves by means of a thermal imaging system. *Plant Physiology*, 76, 266–269.

Jarvis, P.G. & McNaughton, K.G. (1986). Stomatal control of transpiration: scaling up from leaf to region. *Advances in Ecological Research*, 15, 1–49.

Jones, H.G. (1992). *Plants and Microclimate*. Cambridge: Cambridge University Press.

Knight, R.C. (1916). On the use of the porometer in stomatal investigation. *Annals of Botany*, 30, 57–76.

Körner, C. (1988). Does global increase of CO_2 alter stomatal density? *Flora*, 181, 253–257.

Kubínová, L. (1994). Recent stereological methods for measuring leaf anatomical characteristics: estimation of the number and sizes of stomata and mesophyll cells. *Journal of Experimental Botany*, 45, 119–127.

Laisk, A. (1983). Calculation of leaf photosynthetic parameters considering the statistical distribution of stomatal apertures. *Journal of Experimental Botany*, 34, 1627–1635.

Laisk, A., Oja, V. & Kull, K. (1980). Statistical distribution of stomatal apertures of *Vicia faba* and *Hordeum vulgare* and the *Spannungsphase* of stomatal opening. *Journal of Experimental Botany*, 31, 49–58.

Mansfield, T.A., Hetherington, A.M. & Atkinson, C.J. (1990). Some current aspects of stomatal physiology. *Annual Review of Plant Physiology and Plant Molecular Biology*, 41, 55–75.

Mauseth, J.D. (1988). *Plant Anatomy*. Menlo Park: Benjamin/Cummings.

McNaughton, K.G. & Jarvis, P.G. (1991). Effects of spatial scale on stomatal control of transpiration. *Agricultural and Forest Meteorology*, 54, 279–301.

Miranda, V., Baker, N.R. & Long, S.P. (1981). Anatomical variation along the length of the *Zea mays* leaf in relation to photosynthesis. *New Phytologist*, 88, 595–605.

Mott, K.A., Cardon, Z.G. & Berry, J.A. (1993). Asymmetric patchy stomatal closure for the two surfaces of *Xanthium strumarium* L. leaves at low humidity. *Plant, Cell and Environment*, 16, 25–34.

Nobel, P.S. (1991). *Physicochemical and Environmental Plant Physiology*. San Diego: Academic Press.

Omasa, K., Hashimoto, Y. & Aiga, I. (1983). Observation of stomatal movements of intact plants using an image instrumentation system with a light microscope. *Plant and Cell Physiology*, 24, 281–288.

Pearcy, R.W. (1990). Sunflecks and photosynthesis in plant canopies. *Annual Review of Plant Physiology and Plant Molecular Biology*, 41, 421–453.

Raschke, K. (1965). *Das Siefenblasenporometer (zur messung der stomaweite an amphistomatischen blättern). Planta,* 66, 113–120.

Sachs, T. (1991). *Pattern Formation in Plant Tissues.* Cambridge: Cambridge University Press.

Sharkey, T.D. & Seemann, J.R. (1989). Mild water stress effects on carbon-reduction-cycle intermediates, ribulose bisphosphate carboxylase activity, and spatial homogeneity of photosynthesis in intact leaves. *Plant Physiology,* 89, 1060–1065.

Slavik, B. (1963). The distribution pattern of transpiration rate, water saturation deficit, stomata number and size, photosynthetic and respiration rate in the area of the tobacco leaf blade. *Biologia Plantarum,* 5, 143–153.

Smith, S., Weyers, J.D.B. & Berry, W.G. (1989). Variation in stomatal characteristics over the lower surface of *Commelina communis* leaves. *Plant, Cell and Environment,* 12, 653–659.

Solárová, J. & Pospíšilová, J. (1983). Photosynthetic characteristics during ontogenesis of leaves. 8. Stomatal diffusive conductance and stomata reactivity. *Photosynthetica,* 17, 101–151.

Spence, R.D. (1987). The problem of aperture variability in stomatal responses, particularly aperture variance, to environmental and experimental conditions. *New Phytology,* 107, 303–315.

Terashima, I., Wong, S.-C., Osmond, C.B. & Farquhar, G.D. (1988). Characterisation of non-uniform photosynthesis induced by abscisic acid in leaves having different mesophyll anatomies. *Plant and Cell Physiology,* 29, 385–394.

Tichá, I. (1982). Photosynthetic characteristics during ontogenesis of leaves. 7. Stomata density and sizes. *Photosynthetica,* 16, 375–471.

van Gardingen, P.R. & Grace, J. (1991). Plants and wind. *Advances in Botanical Research,* 18, 189–253.

van Gardingen, P.R., Jeffree, C.E. & Grace, J. (1989). Variation in stomatal aperture in leaves of *Avena fatua* L. observed by low-temperature scanning electron microscopy. *Plant Cell and Environment,* 12, 887–888.

van Kraalingen, D.W.G. (1990). Implications of non-uniform stomatal closure on gas exchange calculations. *Plant Cell and Environment,* 13, 1001–1004.

Weyers, J.D.B. & Meidner, H. (1990). *Methods in Stomatal Research.* Harlow, UK: Longman.

Woodward, F.I. (1987). Stomatal numbers are sensitive to increases in CO_2 from pre-industrial levels. *Nature,* 327, 617–618.

D. ATKINSON and R. FOGEL

Roots: measurement, function and dry-matter budgets

Introduction

Scaling-up, the conversion of data collected at one spatial or temporal scale to a value at a larger scale, is normally carried out for a specific purpose. The purpose will influence the confidence limits that are acceptable for the scaled value(s). This is the case for scaled-up measurements of plant roots and plant root systems, which are usually required either as part of a carbon budget or to aid the understanding of an agronomic process, a physiological effect or an ecological consequence. Scaling the results of small-scale processes that occur in soils to a field scale is arguably more difficult than carrying out similar operations for the above soil-surface environment. The soil is physically and chemically more complex than the above-ground atmosphere and the range of processes that influence biological activity is more diverse (O'Toole & Bland, 1987). To cover all the issues involved in scaling soil processes is not possible in this chapter. The combination of rate measurements made in the laboratory with state measurements made in the field, the indirect estimation of root mass and the use of field data to construct a simple ecosystem root dry-matter budget have been selected as examples. These are used to identify the current limits to our ability to scale-up measurements of the mass and activity of plant roots and their associated arbuscular mycorrhizal fungi (AMF) from a laboratory to a field scale.

Roots and the carbon budget

The need to quantify root biomass and turnover has become more important as a result of current interest in global climate change and in the role of carbon-containing compounds in influencing observed effects. Cannell, Dewar & Thornley (1992), summarising data from a range of sources, suggested that the terrestrial biosphere contains 550 Gt carbon and soils contain around 1500 Gt. Soil, therefore, represents a major component of the global carbon budget. Root systems, as part of the biosphere embedded within the soil, clearly represent a key interface and are arguably one of the largest means by which carbon moves from the atmosphere to the soil and to other components of the soil biota (Fogel & Hunt, 1979). Data to validate this point are rare, at least in part because of the difficulty of scaling-up measurements of plant root systems to produce data that can be related to above-ground measurements.

Estimates of the amount of photosynthate passing to the root system are variable. A woodland in Northern Michigan, USA dominated by mature *Populus* had an annual production budget of 16.5 Mg ha^{-1} of which leaf production accounted for 2.9 Mg ha^{-1} and small roots for 5.1 Mg ha^{-1} (R. Fogel, unpublished data). Cannell *et al.* (1992) estimated litter production from the leaves of Sitka spruce, eucalyptus and beech as 2.4, 1.8 and 1.6 Mg ha^{-1} per year, and from fine roots at 2.8, 1.8 and 2.2 Mg ha^{-1} per year, respectively. For a perennial fruit crop, Buwalda & Lenz (1992) estimated that 2% of annual photosynthesis was used for root growth, 3 to 10% to replace losses owing to root turnover, 5 to 15% was passed to other organisms and 15 to 25% was used for maintenance, most of which represented respiration. This gave a range of 25–52% of total photosynthate moving below ground. From the data given by these authors, it was unclear whether their estimate of 'total' photosynthesis included above-ground respiration. 'Whether questions concern carbon sequestration in vegetation and ecosystems or process level responses to a complex suite of environmental variables, root systems must be a significant component of the analysis' (Norby, 1994). Roots have been suggested as the source of the carbon that is often found to be 'missing' from the global carbon cycle (Dyson, 1992). Clearly, the construction of a carbon budget must include accurate estimates of the amounts of carbon contained within and moving to roots and their associated micro-organisms. A review, by Rogers, Runian & Krupa (1994), of the effects of elevated CO_2, a key component in global-change scenarios, indicated that increasing atmospheric CO_2 levels usually increase root biomass and so had the potential to increase rhizodeposition. In addition, several other authors (for example, Pregitzer, 1993; Tinker, 1993) have indicated the importance of CO_2 levels for root growth and dynamics. A detailed understanding of the global carbon cycle will, therefore, require improved understanding of carbon flows into and within soil. An understanding of the amounts and activities of roots will be key to this. Similar information is required to aid the design of more sustainable land-use systems. Such systems will make more effective use of non-renewable resources by enhancing the functioning of the crop root system and improving our management of soil micro-organisms and the chemical transformations for which they are responsible, such as the mineralisation of organic sources of nitrogen and phosphorus.

Linking field and laboratory measurements

The growth and functioning of the root system have been the subject of many studies. Most studies of processes, for example the uptake of a specific mineral nutrient, have necessarily been carried out at a scale involving the use of small seedlings or excised roots under controlled conditions. To scale these measurements up to the level of either a whole field situation or an ecosystem, a number of issues must be accommodated. These include variation in the rate of the process in different parts of the plant, variation owing to size and age, and the changes in activity that may result from the more variable natural environment.

Studies of the growth or activity of roots can be carried out under a range of conditions. Some features of studies carried out in solution culture, microcosms (a laboratory system allowing many of the features of real environments in a controlled situation), or at a field scale are identified in Table 1.

The simplicity of the environment created in solution culture makes it an ideal means of understanding the effects of a single factor, for example the effects of solution temperature upon the uptake of an individual ion by a root of a given and limited age. The reasons for using this type of experimental system for basic-process studies, however, represent the major difficulties in attempts to use information of this type at a field scale. Microcosms and field experiments represent increasing degrees of reality but also an increasing level of complexity and lack of precision. The difficulties of scale resolve into categories.

1. The appropriate multipliers: the factors to be applied to convert state, for example the amount of root, or rate, for example the phosphorus inflow rate, variable from one scale unit to another
2. The variability associated with any biological measurement
3. The modifiers that need to be applied to the multiplier so as to encompass reality. The most difficult areas involved in the scaling of root processes seem likely to relate to the potential of a root system for plastic responses, the impact of other organisms and the variation which is normal for large areas of land.

Table 1. *The characteristics of experimental systems used to assess the functioning of plant roots*

Attribute	Scale of activity		
	Solution culture	Microcosm	Field plots
Homogeneity	Spatially homogeneous, at least initially	Spatially semihomogeneous	Very heterogeneous
Climate	Usually controlled and uniform	Controlled and documented	Very variable both in cyclic and in unpredictable ways
Resources	Theoretically unlimited other than when part of experimental design	Less controlled than in solution especially where soil-based media is used, but outwith treatments relatively unlimited	Potentially limited by interactions of physical and mechanical factor with biological and chemical activity (unless the subject of large inputs)
Potential for influencing performance through a plastic response (e.g. stress avoidance)	Limited	Some potential to change performance by exploiting a more varied physical and chemical environment	Major potential for plants to optimise growth to physical environment and to gain additional resources by modifying growth
Other organisms	Microflora and fauna usually absent	Limited microflora but usually no fauna	A wide range of microflora and fauna present and responsible for major modifications to growth

Elements of these problems are illustrated in the following case.

The extent to which the whole or only part of the root system functions in the absorption of water and mineral nutrients has been a subject of debate for a number of years. Evidence has recently been reviewed by Kozinka (1991) for water and by Kolek & Holobrada (1991) for mineral nutrients. The difficulty in assessing the ability of a large root system to absorb has meant that water uptake is usually assessed using small plants, or a small segment of a large intact root system using a potometer (Clarkson & Sanderson, 1971; Sanderson, 1983) or excised roots (Steudle, Oren & Schulze, 1987). Recent developments in the heat pulse method (Higgs, 1994), to-date mainly used for measurements on stems, may in the future allow direct measurements of flow to be made in roots. Currently, precise measurements depend upon a variant of the potometer technique. Despite advances in isotope techniques, this remains essentially true also for nutrients such as phosphorus. The above studies showed that the rate of water uptake per unit length is highest close to the root tip, but despite this tips and laterals are responsible for only 25% of total water uptake. Brouwer (1953) showed that the rate of uptake by older roots varies with the plants total rate of water use. Atkinson & Wilson (1979) used a solution culture system, similar to that devised by Russell & Sanderson (1967), to assess the ability of either white or woody *Prunus* roots to absorb ^{32}P. They obtained relative rates of 5.4 ± 1.3 and 4.4 ± 0.2 pmol phosphorus $cm^{-2} s^{-1}$, respectively, expressed on the basis of total root area. Under the controlled conditions of a solution culture experiment, the measurement of the uptake of isotopes, such as ^{32}P, will be relatively precise. Other parts of the procedure contain more errors. The estimation of root surface area, which is needed to compare similar lengths of roots of very different diameters, is much less dependable. Surface area is normally calculated on the basis of a root being a smooth cylinder, which clearly it is not. This will be even less true for AMF-infected roots. Root activity is usually modelled on the basis either of the whole root system functioning in uptake (Atkinson, 1983) or of only the root tips being active (Graham, Clarkson & Sanderson, 1974). Attempts to estimate total uptake from measurements made in solution culture tend to underestimate the significance of environmental variation in soil. Cooper (1973) documented the magnitude of the variation that occurs in soil temperature with depth. A consequence of this variation is that the younger zones of the root, those estimated as being most active in studies using excised roots, usually occur in the deeper and, therefore, colder soil. This will modify both relative and absolute activity. Soil water potential also varies within the soil profile and will modify activity (Atkinson, 1983). The combination of these, and other environmental factors, suggests that the scaling of root functions will require careful specification of the inherent assumptions (Kozinka, 1991). Phosphate inflow data has, however, been combined with measurements of the amount of white and woody root on tree root systems of the same species to provide estimates of potential activity. Measurement of the root system of a field-grown tree, with many metres of both white and woody root, is more complex than the estimation of the surface area of a root system confined within a rigid potometer. Errors in the estimation arise from the separation of the root system from soil, the division of the root system into white and woody components and from the estimation of the length of substantial field samples on the basis of sub-samples. A range of inflow rates for whole root systems have been obtained (Atkinson, 1986) with average values per unit root length of between 1.3 and 2.8 pmol cm^{-1} s^{-1}. Comparison of these values with phosphate diffusion characteristics in soil (Nye & Tinker, 1977) suggest that this potential rate of inflow could not be sustained for long in the soil situation. Potential values assessed in soil are likely to be much lower (Bhat, 1983). This relatively simple case study illustrates some of the difficulties in scaling-up. The difficulty of combining data obtained at different scales, for example nutrient inflow and root length, and with different patterns of standard errors is an area now of considerable interest to biological statisticians.

Many of the scaling problems related to measurements of the root system and associated microorganisms relate to technical difficulties. They are, however, compounded by the ability of the root system to adapt its growth strategy (Jones & Jones, 1989). In the example given above, the trees studied had the possibility of obtaining their total phosphorus needs through a range of different rates of inflow distributed over the whole root system. The trees also had the ability to grow roots in new soil areas and to take advantage of the activity of mycorrhizal fungi and the more extensive volume of soil which they exploit.

The remainder of this paper concentrates on how scaling is influenced by biological variability. The effects of level of resource and homogeneity, which have been more extensively studied, see Linder & Flower-Ellis (1992) for an example, are not treated here.

Indirect estimates of the amount of root in soil

Measuring the amount of root presents difficulties that relate principally to removing roots from soil, visualising roots within soil, spatial variation within the soil volume and the rate of temporal variation (Hooker *et al.*, 1995). An interesting comment on this is that in a recent volume on allometric relationships in plants (Niklas, 1994) roots are wholly excluded. In many studies, attempts have been made to estimate the amount of root present in a particular soil type or in an ecosystem by using generalised allometric relationships. This mainly involves the linking of an easily measured above-ground parameter, such as tree girth, with root mass. This practice has been discussed by Dewar & Cannell (1992), King (1993) and Vogt *et al.* (1995). Where only an imprecise measure of the standing crop is needed, this approach will produce an estimate, although one with wide confidence limits. The limitations to this approach need to be examined. Vogt *et al.* (1995) commented 'The ecotone of an individual plant must be examined separately for the above-ground and below-ground since they may not respond in a similar manner. Above-ground parts of plants do not necessarily reflect how the below-ground parts are responding'. The relationship between root and shoot mass may vary as a result of a range of factors, of which age, nutrient supply and soil water potential have been most extensively studied (see Davidson (1969) and Cooper (1973) for examples).

A number of other factors are often ignored. Some of these seem likely to become of increasing importance as more assessments are made of natural ecosystems. Inter-specific variation and its interaction with major environmental factors seem likely to be especially important. Most laboratory studies are based upon a single genetic source of the species. In nature, vegetation is more genetically diverse. Lavender, Atkinson & Mackie-Dawson (1993) grew a range of 30 genotypes of *Betula pendula* under standard conditions. The genotypes were selected as having relatively similar shoot growth. Under the conditions of the experiment, 90% of the clones fell within a band that was ± 15% of the mean for shoot growth. Only 60% of clones fell within a similar band for mean total root production. If variation was calculated as the difference between the highest and lowest average values within the range of clones and expressed as a percentage of the overall mean value for that characteristic, root mass (75%) was more variable than root length (64%) but less variable than specific root length per unit mass (SRL) (85%). Root : shoot ratio (R:S) (54%) was less variable than the above parameters but had a mean of 0.37 and a range of 0.28 to 0.48. Birch plants of relatively similar above-ground appearance on average partitioned one-third of the amount of above-ground resource into the root system but could partition as little as a quarter or as much as a half of relative resources. This is a source of variation that the use of simple allometric relationships will ignore. A selection of four of the clones used by Lavender *et al.* (1993) were used to assess the affect of nitrogen supply on R:S (Lavender, 1992). Variation in nitrogen supply, of an order of magnitude, resulted in the plants exposed to high levels of nitrogen having R:S values in the range 0.27 to 0.51, while those exposed to low levels had ratios of 0.47 to 0.85. This type of effect has been found in a number of other studies (for example, Davidson, 1969; Ericsson, 1995). However, in the plants exposed to high levels of nitrogen, this range of R:S values was associated with plants where shoot weights varied from 3.5 to 6.7 g dry weight (Fig. 1). In plants with a low nitrogen supply, a much higher range of R:S values was associated with variation in shoot weight of only 0.7 to 1.2 g; that is, partitioning to the root system varied relatively little across a substantial range of shoot weight values for plants grown with a high supply of nitrogen while, with a low nitrogen supply, plants with a particular shoot weight could be associated with very different partitioning to the root system.

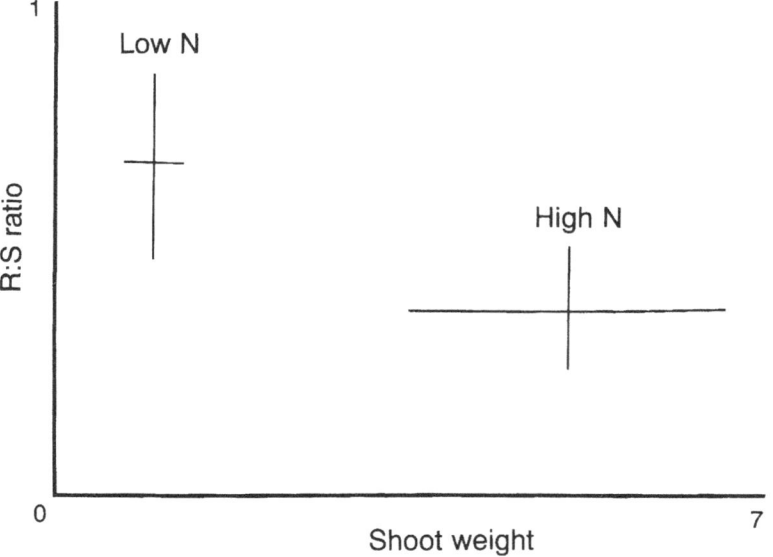

Fig. 1. The relationship between shoot growth and the R:S ratio of seedlings of *Betula pendula* plants grown under varying nitrogen conditions. Vertical lines indicate the range of R:S ratio and horizontal lines indicate the range of shoot weight (g). (Redrawn using data from Lavender, 1992.)

Estimating the mass of below-ground material from an above-ground measurement would not have been possible in the low nitrogen situation. A similar result was obtained for variation in the level of phosphorus supply. The R:S values for plants of *Lolium perenne* given a low nitrogen supply under a range of soil moisture conditions indicated a relatively similar picture (Davidson, 1969). Plants grown with low nitrogen supply had an R:S value of around 2, whereas those with high levels of nitrogen achieved a ratio of 0.4 to 0.8 and over a wider range of shoot growth. With the exception of plants that were acutely deficient in nitrogen, *Trifolium repens* showed a constant R:S value of 0.2 to 0.4. Under conditions of limited nutrient supply, the estimation of a root mass by the use of an allometric relationship seems likely to give only a crude estimate of the root system. Rather than the use of a standard value for R:S, the approach will gain if this value is modified in relation to factors such as soil nitrogen status. Estimated R:S values should also allow for variation in root mass within a season, and for the variation being out of synchrony in root and shoot.

Effects of microorganisms

In natural ecosystems and many other non-sterile soil-based systems, most roots are infected with mycorrhizal fungi, usually AMF. Infection with AMF influences the functioning of the root system (Berta, Fusconi & Trotta, 1993) but also modifies its form and development (Hooker, Munro & Atkinson, 1992). The effects of *Glomus* on root system development has been summarised by Hooker *et al.* (1992) and Atkinson (1992). In these studies, infection with AMF had little effect on the growth rate of primary roots but increased the length growth rate (mm per day) of most other root orders. The branching interval was reduced, giving a root system that was more highly branched and with more root orders. As a consequence, a much higher proportion of the AMF root system was made up of high-order laterals (Table 2). As lateral roots have been reported to have different patterns of activity (Kozinka, 1991) and have different patterns of longevity (Atkinson, 1992; Hooker *et al.*, 1995), then the infection of the root system with AMF will influence the functioning of the total root system. This will occur irrespective of the contribution of AMF to nutrient supply. Hooker *et al.* (1995) found that where plants of *Populus* were infected with AMF only 16% of colonised roots survived for longer than 49 days. For comparable uninfected roots, 49% survived over 49 days.

In addition to the effects of AMF on root functioning, different fungal species vary in their effects on the same plant species. Hooker *et al.*

Table 2. The effect of Glomus E3 on the development of the root system of Populus sp.

Root order	Length of individual roots (mm)			No. lateral roots per unit root length			Percentage of the root system made up by different orders		
	Control	AMF	LSD ($p < 0.05$)	Control	AMF	LSD ($p < 0.05$)	Control	AMF	LSD ($p < 0.05$)
2°	33	66	16	356	646	178	59	21	15
3°	2.6	5.7	1.5	18	129	70	38	58	14
4°	2.7	2.9	1.4	–	21	30	3	18	9
5°	–	4.9	11.3	–	–	–	0	2	2

After Hooker et al. (1992).

(1992) found secondary roots of *Populus* infected with *Glomus* E3 averaged 66 mm in length compared with roots infected with *Scutellispora calospora*, which averaged 47 mm. Infection with these two fungi resulted in the production of 129 and 39 lateral roots per unit length of tertiary root. In scaling any root properties that are likely to be influenced by AMF, it is necessary to be aware of the importance of both the probability of infection and the fungal species responsible.

Estimation of the root components of a carbon budget for a woodland system

Information presented in previous sections highlights the difficulties likely to be associated with attempts to produce estimates of root mass or activity within the variable environment of the field situation.

The University of Michigan Soil Biotron (Fogel & Lussenhop, 1991) enables measurements to be made of a range of root and soil parameters in an area of mixed deciduous woodland in northern Michigan. The observation panels of the rhizotron allow the estimation of the length of different classes of root in the range of species on the Biotron site. It allows the rate and timing of new root production and decomposition to be estimated through the use of techniques such as cohort analysis (Hendrick & Pregitzer, 1992a). Each of the 34 panels has the observational surface of five of the mini-rhizotron tubes most commonly in use (Mackie-Dawson & Atkinson, 1991; Hendrick & Pregitzer, 1992b). In addition, the use of cameras with higher magnification than is currently possible with mini-rhizotrons allows the quantification of organisms such as soil fungi. The use of these measurements in a root system carbon-flow budget requires the scaling-up of measurements from small plot to ecosystem. This project raises many of the more general issues associated with the need to scale detailed soil parameters to the ecosystem level. The use of Biotron data in this way represents a useful model system (Atkinson & Fogel, 1992).

Many of the issues related to the scaling of root measurements relate to the difficulties of visualising roots in, or extracting roots from, a material with the density and complexity of soil. Simply estimating the amount of root, even in a simple ecosystem, with an acceptable degree of accuracy is complex. In addition to technical difficulties, the spatial variability of roots in soil mean that most root estimates (Mackie-Dawson & Atkinson, 1991) are associated with high coefficients of variation. Coefficients of variation for estimates of root mass, estimated by washing roots from soil core samples, are typically greater than 20% and can be as high as several hundred per cent, even for well-replicated studies (Atkinson & Mackie-Dawson, 1991). Similar coefficients of variation of root length, estimated using mini-rhizotron methods, have been identified as around 50% (Hendrick et al., 1992b) and over 100% (Upchurch, 1987). In addition, even within a single ecosystem, there will be spatial variation in the vegetation. Where species composition has a significant influence on the amount or timing of root activity, which is the case for this ecosystem (Atkinson & Fogel, 1992), then factors relating to species composition, as well as the above sources of variation, must be allowed for in deriving an acceptable mean with suitable confidence limits. Interactions of spatial and temporal variation further complicate the problem. Root budgets commonly ignore micro-organisms such as AMF, whose close association with roots would make them difficult to physically separate. Recent studies, for example Fogel (1991), suggest, in addition, that mycorrhizas are a significant component. There are, however, major technical difficulties in estimating the amount of AMF hyphae in soil (Sylvia, 1991).

A useful starting point in defining the need for data for a root budget is to ask why we cannot use a well-established estimate based on core sampling. Subject to the qualifications given above and control of the efficiency of extraction, core sampling can give useful values for the standing crop. In the recent past, measurements made using soil coring methods have been regarded as being absolute. Few published studies, however, include estimates of amounts being lost from cores during the extraction of roots. Where studies use 2 mm diameter mesh sieves and fine roots occur at diameters of around 60 μm, then losses seem likely to occur. Estimates of root production and carbon flow are more difficult to obtain using core sampling. Rhizotron and mini-rhizotron measurements allow the estimation of turnover, which is not possible using coring. The ability to re-observe the same area of soil for an extended period allows the use of cohort analysis methods. The equation detailing primary production can be expressed as $NPP = B_{t+1} - B_t + L$ where NPP is the net primary production, B_t is the standing crop at time t and L represents losses. The use of this approach has changed our concept of the importance of roots in ecosystem processes (Fogel, 1991). The size of the dynamic root component, 40 to 73% of net primary production (Fogel, 1985), has emphasised the need to fully estimate the root length. The turnover of fungal hyphae in soil can be as rapid. D. Atkinson, R. Fogel & O. Pauline (unpublished data) using a cohort analysis approach found that less than 50% of fungal hyphae in soil survived for more than seven days. Estimates of the quantity of mycorrhizal fungi hyphae in soil are rare, but re-calculation of values listed by Sylvia (1991) gave standing crop estimates of around 60 kg ha^{-1} for the surface 100 mm of soil. The key components of a root carbon budget, because of the size of state and rate variables, seem to relate to the amounts of new (white), suberised but living (brown), woody and dead root and the amounts of associated soil microbes (Table 3). The methods of estimating these in a soil biotron and the main assumptions in scaling estimates to ecosystem level are detailed in Table 3. On the Biotron site, parameters were estimated non-destructively. The intensive destructive sampling on a long-term site raises its own difficulties. An observation method gives the ability to quantify labile components and allows the sequential measurements needed for cohort analysis (Mackie-Dawson & Atkinson, 1991). Many of the problems associated with the scaling of quantitative estimates from the Biotron are similar to those involved in scaling mini-rhizotron estimates (Upchurch, 1987; Hendrick & Pregitzer, 1992a,b). Many published studies of mini-rhizotron and rhizotron studies in reviewing limitations have tended to concentrate upon potential effects of observation surfaces on the amount of root adjacent to that surface (for example, Harper *et al.*, 1991). The most severe problems, however, tend to relate to the ability to generalise from the data obtained. Upchurch (1987) identified coefficients of variation of >100% as being common for mini-rhizotron data. He considered the variance to be an inherent property of the soil–plant system. In addition, Glen, Brown & Tekeda (1987) found that variance was correlated with the mean for a range of crops, which violates the key assumptions that underpin the analysis of variance.

Many of the problems encountered in attempts to analyse root data are common to the analysis of other soil data for which a series of geostatistical methods, such as kriging, have been developed (Voltz & Webster, 1990; Goovaerts & Journel, 1995). These methods take advantage of spatial correlations between observations within and across a data set. The lack of communication between soil microbiologists and geostatisticians seems to have inhibited the uptake of methods, such as kriging, into biological studies. This must represent a necessary collaboration for the future. Hendrick & Pregitzer (1992b) concluded that 'There was no right way to analyse mini-rhizotron data'. They suggested the use of nested or pooled data as a means of overcoming high levels of spatial variation. The use of cohort analysis, i.e. following the fate (survival) of a population of roots produced at a given time has proved to be the key to obtaining estimates of both production and death and hence estimates of the standing crops of white and dead roots (Hendrick & Pregitzer, 1992a; Hooker *et al.*, 1995).

A key difficulty in obtaining estimates of root (or AMF) remains the conversion of two-dimensional data to three dimensions, that is the

Table 3. *The components of a soil carbon budget that can be estimated using rhizotron methods of estimation, and problems related to scaling to an ecosystem level*

Component of soil budget	Method of estimation in the Biotron	Scaling-related difficulties
New root production (white root growth)	Direct measurement on observation windows to give lengths; visualisation using video recording of windows Cohort analysis to indicate longevity and thus net production per unit time	1. Estimation of length per unit volume from area measurements 2. Calculation of root mass from root length 3. Measurement of root production for each individual species, i.e. identifying the roots of individual species in a mixture of individuals
Senescence to brown (suberised) root	As for new root production	1. As for new root production 2. Estimation of tissue losses on suberisation 3. Distinguishing from dead root
Dead root 'production'	As for new root production	As for brown root
Woody root production	Length as for new root Volume increase by diameter change using a sample	1. As for new root 2. Variation associated with the relative scarcity of this root type 3. Time period needed for observations
Mass of fungal hyphae	*In situ* visualisation and measurement via video recording Cohort analysis to estimate survival, turnover, etc.	1. Problems of visualisation leading to coarse initial estimates 2. Estimation of length per unit volume 3. Estimation of mass from length 4. Resolution of AMF and other soil fungi 5. Spatial variation
Quality of soil fauna	Visualisation using video recording	1. Identification of species present 2. Spatial and temporal variation

definition of the effective depth of field. Upchurch (1987) identified three main approaches that could be used to convert mini-rhizotron or rhizotron measurements to root length density (RLD). These were (i) an empirical relationship between rhizotron root length and that obtained by another method; (ii) the association of a particular soil volume with a rhizotron measurement; or (iii) the use of theoretical conversion factors. Upchurch (1987) indicated that (i) had been shown to give variable results and that 'recalibration of the technique with each experiment and perhaps with time within an experiment will be required.' Approaches (ii) and (iii) involve a number of practical problems in their operational use. The power of the length–volume factor means that it will have a major effect upon the absolute size of the estimate produced. An assumption of a 1 mm depth of field will give double the estimate of production obtained using a 2 mm depth of field. The real depth of field will differ for every root, so any estimate must be a weighted average.

An estimate of the total standing crop of the main root types for all species has been produced for the mixed deciduous woodland in northern Michigan, using the Biotron data. This is shown in Table 4. The estimates given here are in line with those derived from other studies of the area (Hendrick & Pregitzer, 1992a). The values are also in line with a number of other published measurements. Dewar & Cannell (1992) reviewed data for fine root density, which ranged from 1.4 to 3.6 Mg carbon per ha. Assuming an average carbon content of 50% for the roots from the Michigan woodland and fine root as being the sum of white plus brown root, this gave a value of 3.2 Mg carbon per ha for this study. This falls within the above range.

In assessing the errors likely to be associated with the above estimates, it may be helpful to look at the factors used in a sample calculation. The average root length visible on a window is around

Table 4. *Estimates derived from measurements made in the Biotron of the mean standing crop of the roots of all species in a mixed woodland ecosystem*

Component	Estimate (Mg ha^{-1})	Estimated variance (SE)
White root	1.2	0.28
Brown root	5.3	1.48
Dead root	2.2	0.46
Woody root	13.8	2.89

1.5 mm cm^{-2}. Scaling this to a volume basis (mm cm^{-3}) involves a factor of 2×10^{-1} (Table 5). The conversion of a 1 cm depth of soil to a typical profile of 100 cm requires multiplication by 10^{-2}. Conversion from a soil surface area of 1 cm^2 to 1 ha involves a factor of 10^8. Conversion of millimetre length to gram weight uses a factor of 3×10^{-4}. As a result, a mean window length of 1.5 mm cm^{-2} represents a root standing crop of approximately 3.0 Mg ha^{-1}. The variation (standard error) associated with this will be around 0.84 Mg. As many of the data would appear not to be normally distributed, the standard error must be regarded as only a guide.

The estimates given in Table 4 involve the use of a number of simplifying assumptions. These are listed in Table 5, together with the principal areas of uncertainty. Effects related to area-to-volume conversion, species effects and sampling variation remain the major problems. Estimates produced in this way have the potential to be made more dynamic than those from allometric relationships.

Conclusions

Norby (1994) suggested that analysis of the significance of a root response to atmospheric CO_2 concentrations must consider (i) roots as a platform for nutrient acquisition and as a mediator of the whole-plant response; (ii) roots as a component of whole-plant carbon storage; and (iii) the implications of root turnover for soil organic matter. While the question of the effect of CO_2 on below-ground response is topical and important, the ability to estimate the magnitude of the root system, its directly associated soil flora and their activities is important also in relation to the ability of crops to access nutrients and water and to the maintenance of soil resources. This chapter has attempted to examine issues associated with scaling of indirect estimations of root activity, the application of laboratory data to the field scale and the use of detailed field data to produce a root biomass budget for temperate woodland ecosystem.

Studies that attempt to model the amount of carbon within a perennial root system often derive the root components of their models from above-ground productivity or even from a single variable, such as trunk diameter. The information discussed here suggests that this approach will be less effective for a natural ecosystem, with its attendant biodiversity, than for more genetically homogenous vegetation types such as a crop. Even in such simple systems, attempts to estimate root functioning commonly involve combining functional information on properties that can only be estimated in the laboratory with root-quantity data

Table 5. *Assumptions made in the estimation of the components of a root system budget, using a rhizotron method*

Component	Assumptions	Areas of uncertainty
White root mass	1. Root density at the window is similar to that at a similar position not visible near the window 2. All roots within 0.6 mm of the glass surface are visible 3. Estimated values of length per unit volume can be converted to mass using average values of SLR (mg g^{-1}) 4. The windows measured are representative of the ecosystem as a whole	1. Variation in the periodicity of root production between different species 2. The effects of using average SRL values for populations of roots that vary in diameter by an order of magnitude 3. The 'real' average depth of field 4. The representativeness of the window recorded 5. The effect of the presence of the window on the values being recorded 6. The criteria to define 'white' root for many species 7. Uniformity of criteria and accuracy used by recorders
Brown root mass	1. As for white root 2. Brown root can be distinguished on the basis of appearance from dead and woody root	1. Variation in the rate of suberisation for the roots of different species 2. Points 2, 3, 4, 5 and 7 for white root
Dead root mass	1. As for white root 2. Dead root can be estimated from white root and a measured survival factor 3. Dead root disappears from the system within a season	1. Variation between species related to the timing of suberisation, distribution with depth and decomposition 2. Points 2, 3, 4, 5 and 7 identified for white root
Woody root mass	1. All roots within 4 mm of the window surface can be visualised 2. The mean SRL for this ecosystem was 35 cm g^{-1} 3. Points 1, 4 identified for white root	1. Points 2, 3, 4 and 5 identified for white root 2. The extent to which any window sample can represent an acceptable sample of a spatially uncommon root type

SRL, specific root length

obtained in the field. The use of laboratory data obtained under relatively uniform conditions to estimate field-level activity involves the
implicit assumption that the more variable conditions found in the field
will not significantly influence the basic response. Interaction of factors
such as soil temperature seem likely to affect processes such as root-
system absorption of water and nutrients. Although complex laboratory
systems that provide varied environments and allow changes in strategy
are being used (Grime, Crick & Rincon, 1986), most laboratory experiments prevent the plant from responding by a change in its behaviour
or allocation strategy, that is by exhibiting a plastic response. Under
field conditions, these may become the dominant component, for
example, the allocation of additional resource to permit accessing a
more distant nutrient supply. These effects are further complicated by
interactions with organisms such as AMF. Mycorrhizas, in addition to
representing an additional sink for assimilate, modify the form, development and internal allocation of assimilates within the root system, and
different AMF species have different effects. These biological and
microbial factors illustrate some of the factors that are commonly significant in the more complex field situation. The model developed by
Dewar & Cannell (1992) assumes that fine root biomass turns over once
during a season. Assuming that turnover occurs twice in a season
doubled soil carbon storage (50% of the forest carbon budget) and
increased total storage by 19%. Despite the size of the soil carbon pool,
this component is the least well documented. This suggests that scaling
to a field level needs more detailed information on root activity and the
factors influencing it.

Scaling estimates of root production, made at a small-plot scale, to
an ecosystem scale appears potentially to be simpler than that of scaling
root functional attributes. The problems of scaling estimates from the
Soil Biotron installed in deciduous woodland to an ecosystem scale
have been discussed as a means of illustrating the assumptions involved
in this type of exercise. The estimation of real sample volume, the
identification of species of root being measured and the most appropriate sampling strategy continue to influence our ability to scale root
measurements derived from observational methods. Molecular biology,
and especially fingerprinting techniques, are being suggested as the
means of solving many of the identification problems of soil biology.
In combination with destructive methods, such methods should eliminate some of our current taxonomic problems but would have little
impact upon key issues relating to sample volume or sampling strategy.
At the present time, the problems involved in defining the dimension
of the soil volume being sampled, the spatial relationship of such

samples to the rest of the soil volume and the biological identity of the structures being measured remain the key challenges to our ability to scale measurements of roots from laboratory studies to the ecosystem.

References

Atkinson, D. (1983). The growth, activity and distribution of the fruit tree root system. *Plant Soil*, 71, 23–35.

Atkinson, D. (1986). The nutrient requirements of fruit trees: some current considerations. *Advances in Plant Nutrition*, 2, 93–128.

Atkinson, D. (1992). Tree root development: the role of models in understanding the consequences of arbuscular endomycorrhizal infection. *Agronomie*, 12, 817–820.

Atkinson, D. & Fogel, R. (1992). The use of a Soil Biotron to quantify the flow of carbon to plant root systems in forest soils. In *Root Ecology and its Practical Application*, ed. L. Kutschera, E. Hubl, E. Lichtenegger, H. Persson & M. Sobotik, pp. 731–734. Klagenfurt, Austria: Verein fur Wuizelforschung.

Atkinson, D. & Mackie-Dawson, L.A. (1991). Root growth: methods of measurement. In *Soil Analysis: Physical Methods*, ed. K.A. Smith & C.E. Mullins, pp. 447–509. New York: Dekker.

Atkinson, D. & Wilson, S.A. (1979). The root soil interface and its significance for fruit tree roots of different ages. In *The Soil Root Interface*, ed. J.L. Harley & R.S. Russell, pp. 259–271.

Berta, G., Fusconi, A. & Trotta, A. (1993). VA mycorrhizal infection and the morphology and function of root systems. *Environmental Experimental Botany*, 33, 159–173.

Bhat, K.K.S. (1983). Nutrient inflows into apple roots. *Plant and Soil*, 71, 371–380.

Brouwer, R. (1953). Water absorption by the roots of *Vicia faba* at various transpirational strengths. *Proceedings Kon Netherlands Academy Wet Series C*, 56, 106–115.

Buwalda, J.G. & Lenz, F. (1992). The carbon costs of root systems of perennial fruit crops. In *Root Ecology and its Practical Applications*, ed. L. Kutschera, E. Hubl, E. Lichtenegger, H. Persson & M. Sobotik, pp. 285–289. Klagenfurt, Austria: Verein fur Wuizelforschung.

Cannell, M.G.R., Dewar, R.C. & Thornley, J.H.M. (1992). Carbon flux and storage in European forests. In *Responses of Forest Ecosystem to Environmental Changes*, ed. A. Teller, P. Mathy & J.N.R. Jeffers, pp. 256–271. London: Elsevier.

Clarkson, D.T. & Sanderson, J. (1971). Relationship between the anatomy of cereal roots and the absorption of nutrients and water. *Annual Report of ARC Letcombe Laboratory for 1970*, 16–25.

Cooper, A.J. (1973). *Root Temperature and Plant Growth*. Commonwealth Agricultural Bureau.

Davidson, R.L. (1969). Effects of soil nutrients and moisture on root/ shoot ratios in *Lolium perenne* L. and *Trifolium repens* L. *Annals Botany*, 33, 571–577.

Dewar, R.C. & Cannell, M.G.R. (1992). Carbon sequestration in the trees, products and soils of forest plantations: an analysis using UK examples. *Tree Physiology*, 11, 49–71.

Dyson, F. (1992). *From Eros to Gaia*. New York: Panthean Books.

Ericsson, T. (1995). Growth and shoot: root ratio of seedlings in relation to nurtrient availability. *Plant Soil*, 168, 205–214.

Fogel, R. (1985). Roots as primary producers in below-ground ecosystems. In *Ecological Interactions in Soil*, ed. A.H. Fitter, D. Atkinson, D.J. Read & M. Usher, pp. 23–36. Oxford: Blackwell.

Fogel, R. (1991). Root system demography and production in forest ecosystems. In *Plant Root Growth*, ed. D. Atkinson, pp. 89–101. Oxford: Blackwell.

Fogel, R. & Hunt, G. (1979). Fungal and arboreal biomass in a Western Oregon Douglas-fir ecosystem: distribution patterns and turnover. *Canadian Journal of Forest Research*, 9, 245–256.

Fogel, R. & Lussenhop, J. (1991). The University of Michigan Soil Biotron. In *Plant Root Growth*, ed. D. Atkinson, pp. 61–68. Oxford: Blackwell.

Glen, D.M., Brown, M.W. & Takeda, F. (1987). Statistical analysis of root count data from minirhizotrons. In *Minirhizotrons Observation Tubes: Methods and Application for Measuring Rhizosphere Dynamics*, pp. 81–88. Madison, WI: American Society of Agrenomy.

Goovaerts, P. & Journel, A.G. (1995). Integrating soil map information in modelling the spatial variation of continuous soil properties. *European Journal of Soil Science*, 46, 397–414.

Graham, J.P., Clarkson, D.T. & Sanderson, J. (1974). Water uptake by the roots of marrow and barley plants. *Annual Report of ARC Letcombe Laboratory for 1973*, 9–12.

Grime, J.P., Crick, J.C. & Rincon, J.E. (1986). The ecological significance of plasticity. In *Plasticity in Plants*, ed. D.H. Jennings & A.J. Trewavas, pp. 5–29. London: Society for Experimental Biology.

Harper, J.L., Jones, M. & Sackville-Hamilton, N.R. (1991). The evaluation of roots and the problems of analysis of their behaviour. In *Plant Root Growth: An Ecological Prospective*, ed. D. Atkinson, pp. 3–24. Oxford: Blackwell.

Hendrick, R.L. & Pregitzer, K.S. (1992a). The demography of fine roots in a Northern hardwood forest. *Ecology*, 73, 1094–1104.

Henrick, R.L. & Pregitzer, K.S. (1992b). Spatial variation in tree root distribution and growth associated with mini rhizotrons. *Plant Soil*, 143, 283–288.

Higgs, K.A. (1994). Water stress and water use in broad-leaved seedlings. Evaluation of sap flow gauges in water related research. In

Efficiency of Water Use in Crop Systems, Aspects of Applied Biology, Vol. 38, pp. 153–164. Warwick, UK: Association of Applied Biologists.

Hooker, J.E., Black, K.E., Perry, R.L. & Atkinson, D. (1995). Arbuscular mycorrhizal fungi induced alteration to root longevity of poplar. *Plant Soil*, 172, 327–329.

Hooker, J.E., Munro, M. & Atkinson, D. (1992). Vesicular–arbuscular mycorrhizal fungi induced alteration in poplar root system morphology. *Plant Soil*, 145, 207–214.

Jones, H.G. & Jones, M.B. (1989). Introduction: some terminology and common mechanisms. In *Plants Under Stress*, ed. H.G. Jones & T.J. Flowers, pp. 1–10. Cambridge: Cambridge University Press.

Kolek, J. & Holobrada, M. (1991). Ion uptake and transport. In *Physiology of the Plant Root System*, ed. J. Kolek & V. Kozinka, pp. 204–284. Dordrecht: Kluwer.

Kozinka, V. (1991). Uptake and transport of water. In *Physiology of the Plant Root System*, ed. J. Kolek & V. Kozinka, pp. 129–202. Dordrecht: Kluwer.

King, D.A. (1993). A model analysis of the influence of root and foliage allocation of forest production and competition between trees. *Tree Physiology*, 12, 119–135.

Lavender, E.A. (1992). Genotype variation in the root system of *Betula pendula* Roth. PhD Thesis, University of Aberdeen.

Lavender, E.A., Atkinson, D. & Mackie-Dawson, L.A. (1993). Variation in root development in genotypes of *Betula pendula. Aspects of Applied Biology*, 34, 183–192.

Linder, S. & Flower-Ellis, J. (1992). Environmental and physiological constraints to forest yield. In *Resource of Forest Ecosystems to Environmental Change*, ed. A. Teller, P. Mathy & J.N.R. Jeffers, pp. 149–164. London: Elsevier.

Mackie-Dawson, L.A. & Atkinson, D. (1991). Methodology for the study of roots in field experiments and the interpretation of results. In *Plant Root Growth*, ed. D. Atkinson, pp. 25–47. Oxford: Blackwell.

Niklas, K.J. (1994). *Plant Allometry*. Chicago: University of Chicago Press.

Norby, R.J. (1994). Issues and perspectives for investigating root responses to elevated atmospheric carbon dioxide. *Plant Soil*, 172, 327–329.

Nye, P.H. & Tinker, P.B. (1977). *Solute Movement in the Soil–Root System*. London: Blackwell.

O'Toole, J.C. & Bland, N.L. (1987). Genotypic variation in crop plant root systems. *Advances in Agronomy*, 41, 91–145.

Pregitzer, K.S. (1993). Impact of climate change on soil processes and soil biological activity. In *Global Climate Change*, ed. D.

Atkinson, pp. 71–82. Farnham, UK: British Crop Protection Council.

Rogers, H.H., Runion, G.B. & Krupa, S.V. (1994). Plant responses to atmospheric CO_2 enrichment with emphasis on roots and the rhizosphere. *Environmental Pollution*, 38, 155–189.

Russell, R.S. & Sanderson, J. (1967). Nutrient uptake by different parts of the intact roots of plants. *Journal of Experimental Botany*, 18, 491–508.

Sanderson, J. (1983). Water uptake by different regions of the barley root. Pathways of radial flow in relation to development of the endodermis. *Journal of Experimental Botany*, 34, 240–253.

Steudle, E., Oren, R. & Schulze, E.D. (1987). Water transport in maize roots: measurement of hydraulic conductivity, solute permeability and of reflection coefficients in excised roots using the root pressure probe. *Plant Physiology*, 84, 1220–1232.

Sylvia, D.M. (1991). Quantification of external hyphae of vesicular arbuscular mycorrhizal fungi. *Methods in Microbiology*, 24, 53–66.

Tinker, P.B. (1993). Climatic change and its implications. In *Global Climate Change*, ed. D. Atkinson, pp. 3–12. Farnham, UK: British Crop Protection Council.

Upchurch, D.R. (1987). Conversion of mini-rhizotron–root interactions to root length density. In *Mini-rhizotron Observation Tubes: Methods and Application for Measuring Rhizosphere Dynamics. ASA, Publication No. 50*, pp. 51–65. Madison, WI: American Society of Agronomy.

Vogt, K.A., Vogt, D., Asbjornsen, H. & Dahlgren, R. (1995). Roots, nutrients and their relationship to spatial patterns. *Plant Soil*, 168, 113–123.

Voltz, M. & Webster, R. (1990). A comparison of kriging cubic splenes and classification for predicting soil properties from sample information. *Journal of Soil Science*, 41, 473–490.

M.J. BARNSLEY, S.L. BARR and T. TSANG

Scaling and generalisation in land cover mapping from satellite sensors

Introduction

This chapter is divided into three sections. The first section briefly examines the need for information on land cover and the limitations of existing map-based data sets. It also reviews the role of satellite remote sensing in providing improved information on land cover and land cover change at regional and global scales. The second section considers some of the scaling issues involved in the production of land cover maps from coarse spatial resolution satellite sensor images. Particular attention is given to the impact of sensor spatial resolution on both the number and the nature of the categories that can be identified consistently. The final section investigates various techniques that can be used to generalise raster-format land cover maps to smaller cartographic scales. These techniques have two functions. First, they provide a means of validating land cover maps derived from satellite sensor images acquired at a spatial resolution of 1 km or coarser. Second, they can be used to transform these data to still smaller cartographic scales (larger area), appropriate to meso-scale and macro-scale environmental models.

Land cover and the role of satellite remote sensing

The need for regional and global land cover maps

Information on the spatial pattern and temporal dynamics of land cover is critical to many studies of climatological, hydrological and ecological processes operating at the Earth surface (Townshend, 1992a). In this context, land cover is important both as a key parameter in its own right (for example, as an indicator of human activity and longer-term climate change) and as a surrogate for other variables of interest (such as albedo, leaf area index (LAI) and surface roughness) (Tucker & Sellers, 1986; Dorman & Sellers, 1989; Townshend, 1992b). This information is required at spatial scales ranging from the local to the global, and at temporal frequencies ranging from days to years.

Several large-area (i.e. regional or global) land cover data sets have been produced to meet these needs (Townshend *et al.*, 1991). Many of these have been developed specifically for use with global climate models: examples include Matthews (1982), Olsen, Watts & Allison (1983) and Wilson & Henderson-Sellers (1985). However, most of these data sets have a number of important limitations, notably:

1. They are of variable or unknown accuracy, having been compiled from disparate sets of (often dated) paper maps
2. They are essentially static – that is, they provide little or no information on changes in land cover over time
3. They tend to be highly generalised, both spatially and categorically
4. The classes that they depict often relate to climax vegetation communities, rather than to actual land cover or functional land cover types
5. They are inconsistent: studies by Townshend *et al.* (1991) and DeFries & Townshend (1994a) indicate that these and other global data sets vary not only in terms of the proportion of the land surface occupied by each broad cover type but also with respect to the total land area.

The role of satellite remote sensing

Satellite remote sensing represents an alternative source of land cover information. At regional and global scales, it is probably the only viable means by which this information can be obtained in an accurate, timely and consistent manner. Satellite remote sensing also offers the important advantage that changes in land cover can be monitored over time. Thus, it may be possible to study both the response of land cover to various dynamic environmental and climatological processes, and any feedback mechanisms that might exist (Cess, 1978; Dickinson & Hanson, 1984).

The satellite sensor employed in most previous studies has been the relatively coarse spatial resolution AVHRR (Advanced Very High Resolution Radiometer) instrument mounted on board the NOAA (National Oceanic and Atmospheric Administration) series of satellites (Townshend, 1994). Although this sensor was designed primarily for meteorological purposes, it has been widely adopted for studies of the land surface. The advantages that NOAA–AVHRR offers in this respect include:

1. The provision of global coverage at daily intervals
2. The low cost of its data compared to those acquired by fine spatial resolution Earth-resources satellite sensor systems, such as those carried on board Landsat and SPOT (Systeme Probatoire pour l'Observation de la Terre)
3. The smaller quantities of data that it generates over a given study area (Malingreau & Belward, 1992; Townshend, 1992b).

NOAA–AVHRR can provide data at three nominal spatial resolutions, namely: High Resolution Picture Transmission (HRPT)/Local Area Coverage (LAC) data at 1.1 km, Global Area Coverage (GAC) data at 4 km and Global Vegetation Index (GVI) data at 15–20 km (Townshend & Justice, 1988; 1990; Belward, 1992). The GAC and GVI sets are produced by undersampling HRPT/LAC data both spatially and temporally (Belward & Lambin, 1990; Malingreau & Belward, 1992). Most global-scale studies of the land surface have tended to make use of these last two formats, since complete global coverage of HRPT/ LAC data is currently unavailable (Townshend *et al.*, 1991). Similarly, most global-scale studies have been based on a simple transformation of data acquired at red and near-infrared wavelengths (near-infrared − red/near-infrared + red), known as the normalised difference vegetation index (NDVI). The NDVI is sensitive to photosynthetically active vegetation within the sensor's Instantaneous Field-Of-View (IFOV) (Curran, 1983; Sellers, 1985; Tucker & Sellers, 1986; Goward, Tucker & Dye, 1987). It can, therefore, be used to provide an indication of vegetation type, amount (biomass) and primary production (Justice *et al.*, 1985; Townshend, Goff & Tucker, 1985). Moreover, since it combines information from two spectral wavebands, it further reduces the volume of data that has to be handled.

Several studies have used NDVI data derived from NOAA–AVHRR to monitor land cover at regional and global scales, notably Justice *et al.* (1985), Tucker, Townshend & Goff (1985), Townshend, Justice & Kalb (1987), Lloyd (1990), Justice, Townshend & Kalb (1991), Loveland *et al.* (1991) and DeFries & Townshend (1994b). In the main, these studies have made use of multitemporal data sets to distinguish land cover types or ecological communities on the basis of differences in their phenological cycles, as indicated by changes in their NDVI values over time. For example, Lloyd (1990) used a binary decision-tree approach to produce a phytophenological classification of land cover, based on observations of the date of onset, the duration and the intensity of photosynthetic activity.

As a result of these studies, together with the recent increased awareness of global environmental issues, several important developments are currently taking place. First, under the auspices of the International Geosphere–Biosphere Programme (IGBP), a global 1 km land cover data set is being produced from NOAA–AVHRR HRPT/LAC data (Townshend, 1992a; Townshend *et al.*, 1994). Second, a number of new 'medium' (or 'moderate') spatial resolution sensors are planned for launch before the end of the 1990s. These include the AATSR (Advanced Along-Track Scanning Radiometer), VEGETATION, MODIS (Moderate Resolution Imaging Spectrometer), MISR (Multi-angle Imaging Spectroradiometer) and MERIS (Medium Resolution Imaging Spectrometer) instruments (Barnsley *et al.*, 1994). Each will acquire data at or around a spatial resolution of 1 km, and each will produce data that can be used to generate regional and global land cover maps. As an example of this, scientists on the MODIS Land Team aim to produce global data sets based on a combined structural and phenological classification of land cover (Running *et al.*, 1994).

Scaling issues and the role of generalisation

Although satellite remote sensing is already being used to produce regional and global land cover data sets, several issues deserve further attention: specifically, what classification schemes should be adopted to map land cover at grid scales of 1 km or coarser; how can land cover maps produced at these spatial scales be validated; and how should the land cover data derived from medium spatial resolution satellite sensors be generalised to the still smaller cartographic scales required by, for example, global climate models? Each of these issues will be examined briefly in the remainder of this chapter.

Scaling issues in the production of multiclass land cover maps

The effect of sensor spatial resolution on the information content of remotely sensed images is well documented (Townshend, 1981; Forshaw *et al.*, 1983; Irons *et al.*, 1985; Woodcock & Strahler, 1987; Justice *et al.*, 1989). As the spatial resolution of the sensor becomes coarser, the likelihood that several different land cover types will fall within the sensor's IFOV increases. This produces what are commonly known as 'mixed pixels' (Irons *et al.*, 1985). The spectral response associated with a mixed pixel is a complex function of the relative proportions, spatial distribution and spectral response of the individual land cover types within the IFOV, and of the sensor's Modulation Transfer Function (MTF) (Townshend, 1981; Woodcock & Strahler, 1987).

As the spatial resolution of the sensor becomes coarser still, the number of mixed pixels in the image and the degree of spatial mixing within them increase. This has a number of important implications. First, there will be some loss of spatial information, as a result of the spatial averaging that takes place within the pixel. In particular, the ability to distinguish accurately the boundaries between discrete parcels of land cover will clearly diminish (Townshend & Justice, 1988; 1990). Second, there will be some loss of spectral separability between the land cover classes (defined and identified at some finer spatial scale), as a result of the composite spectral response associated with the mixed pixels. This will have an impact on the accuracy of any land cover map produced using conventional per-pixel, multispectral classification algorithms, although it may be offset, up to a point, by a concomitant reduction in within-class variance resulting from the spatial averaging effect within the larger IFOVs (Markham & Townshend, 1981; Woodcock & Strahler, 1987). The combined effect of these two processes is increasing uncertainty about the number, nature and the relative proportions of different land cover types present within the scene (Moody & Woodcock, 1994). The severity of this effect will vary between land cover types, depending on their typical parcel size and spatial distribution (fragmentation) within the scene (Moody & Woodcock, 1994).

Figure 1 gives an indication of the impact of the mixed-pixel effect at different sensor spatial resolutions for an area of intensive agriculture in south-east England, centred on Ashford in Kent. The diagram shows both the frequency and the cumulative frequency distributions of land-parcel sizes in this area. These data have been derived from a land cover map produced using a 1024×1024 pixel subscene extracted from a SPOT HRV (Systeme Probatoire pour l'Observation de la Terre/Haut Resolution Visible) multispectral image. It should be noted that approximately 99% of the land parcels in this area are smaller than the nominal spatial resolution of NOAA–AVHRR HRPT/LAC data (1.1 km $\times$ 1.1 km; 121 ha) and so are likely to form mixed pixels in such data. Similarly, approximately 85% of the land parcels are smaller than the finest spatial resolution data that will be produced by the National Aeronautic and Space Administration's (NASA) MODIS instrument (250 m; 6.25 ha). While these figures relate to a scene with

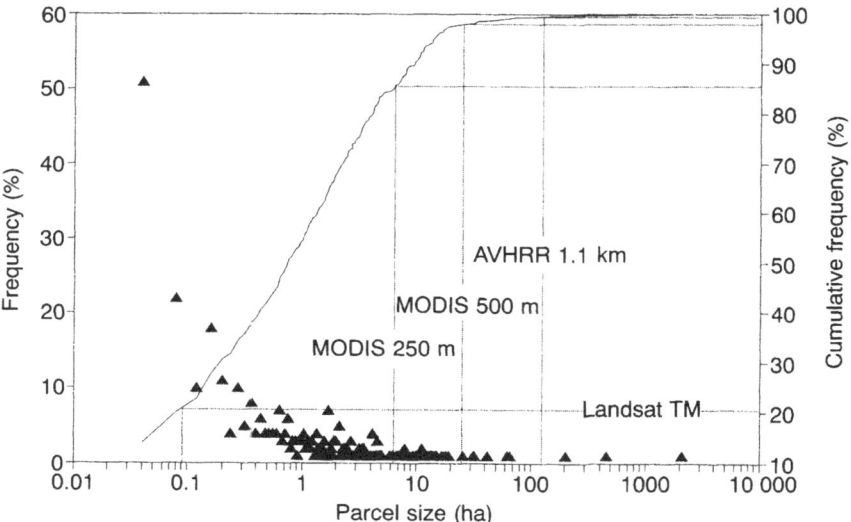

Fig. 1. Frequency (▲) and cumulative frequency (—) distribution of land parcel sizes for an area of intensive agriculture in south-east England, centred on Ashford, Kent. The data have been derived from a supervised multispectral classification of a 1024×1024 pixel sub-scene extracted from a SPOT HRV multispectral image.

a relatively complex spatial structure, they nevertheless provide an indication of the general magnitude of the mixed-pixel effect.

Given the problems presented by mixed pixels in coarse spatial resolution image data, the analyst has two options. The first is to select an alternative approach to image segmentation that may be more appropriate to coarse spatial resolution data (cf. per-pixel multispectral classification; Woodcock & Strahler, 1987). The second is to re-define the classification scheme – in terms of the number and the nature of the classes – so that it is more closely matched to the scale of the data.

Alternative approaches to image segmentation

Where there is a need to identify land cover classes whose typical parcel sizes are generally much smaller than the nominal spatial resolution of the sensor, one approach is to use techniques that estimate the proportion of each pixel covered by each class. Two such techniques are currently receiving attention in the literature, namely *mixture modelling* and *subpixel calibration* (Iverson, Cook & Graham, 1989; Cross *et al.*,

1991; Quarmby, 1992; Quarmby *et al.*, 1992; Ripple, 1994; Zhu & Evans, 1994).

Mixture modelling is an attractive option in that it is physically based. The signal associated with each pixel is considered to be some mixture of the spectral response of each of its constituent land cover types (Smith, Johnson & Adams, 1985); although multiple scattering between scene elements is usually ignored (Cross *et al.*, 1991). Each pixel in the output image is assigned several labels, one for each class, together with a corresponding set of numerical values indicating the proportions of that pixel assigned to each class (Quarmby *et al.*, 1992). Application of mixture modelling usually requires the identification of suitable 'end-members' ('pure' pixels of each class). This can be problematic, especially in very coarse spatial resolution images, with many studies relying on somewhat subjective, empirical methods, such as principal components analysis (Smith *et al.*, 1985; Cross *et al.*, 1991). Moreover, solution of the set of simultaneous equations used to estimate the class mixture in each pixel requires $N + 1$ logical channels of data (whether spectral or temporal); where N is the number of classes to be identified. Depending on the number of classes, this can be very demanding in terms of both data volumes and computational requirements.

An alternative to mixture modelling is subpixel calibration (Iverson *et al.*, 1989; Ripple, 1994; Zhu & Evans, 1994). This involves the use of least-squares regression to establish a relationship between the environmental variable of interest (usually the abundance/percentage cover of a given land cover type) and the detected reflectance in one or more spectral wavebands. The abundance values are determined separately, often through analysis of a finer spatial resolution data set. For example, Ripple (1994) identified closed-canopy conifer forests in Landsat MSS images (resampled to 50 m) using unsupervised multi-spectral classification. These data were subsequently used to estimate the percentage conifer cover within co-registered 1 km AVHRR pixels. A step-wise multiple regression procedure was then used to relate these values to various combinations of AVHRR wavebands. These relationships were employed to estimate percentage conifer cover throughout the AVHRR image.

There are, however, several problems with subpixel calibration techniques. First, they require additional image data; more specifically, they require *both* fine and coarse spatial resolution data sets, albeit only for certain sample sites, to develop an invertible model. Second, it is generally not possible to define a single, global relationship between the variable of interest and detected reflectance, especially over very large study areas, because of variations in factors such as edaphic and climatological conditions, topographic slope and aspect, and solar illumination and sensor view angles. Instead, the data must be stratified, with separate relationships determined for each subregion. This clearly increases the degree of intervention required by the analyst. Finally, even with stratification, most studies report relatively weak correlations between the environmental parameters and detected reflectance, with r^2 values often less than 0.5 (Iverson et al., 1989; Ripple, 1994; Zhu & Evans, 1994).

Hierarchical nested classification schemes and domains of scale

The second option open to the analyst is to define a new classification scheme that may be more appropriate to coarse spatial resolution data; that is, to re-define both the number and the nature of the candidate classes. The new classes may be related to those identified at finer spatial scales in either a simple (dominant class) or an abstract (barley → cereal crop → arable farmland) fashion. They may, therefore, form part of a nested set (or hierarchy) of classes (Strahler, Woodcock & Smith, 1986; Woodcock & Strahler, 1987). For example, the subject of the classification may change from individual species, through vegetation communities and ecosystems, to biomes.

One problem with hierarchical classification schemes is that they tend to promote the notion of an immutable relationship between cover type definitions at each level. More importantly, there are no objective guidelines to indicate the spatial scales at which classification should move from one level in the hierarchy to another. The question that this poses is whether a separate classification scheme should be produced for each new spatial resolution or whether there are 'domains of scale' (Wiens, 1989), separated by relatively sharp transitions in the spatial and spectral properties of the scene, within which the dominant environmental/climatological processes are constant and for which a single classification scheme will suffice.

It has been suggested that spatial variance increases as a transition is approached between two domains of scale in a hierarchical system (O'Neill et al., 1986; Wiens, 1989). Evidence for domains of scale might, therefore, be obtained by examining the spatial and spectral properties of a scene in remotely sensed images acquired over a wide range of different sensor spatial resolutions. In the absence of a true multiple spatial resolution data set, this can be achieved by degrading a fine spatial resolution image to simulate data acquired at successively coarser spatial scales. The spatial and spectral properties of the data can then be measured at each spatial resolution (scale) (Carpenter & Chaney, 1983; Carrere, 1990).

Numerous techniques have been used to quantify the spatial and spectral properties of digital remotely sensed images and other, related, raster data sets. These include the analysis of mean local image variance (Woodcock & Strahler, 1987; Belward, 1992), block variance (Greig-Smith, 1979; Weiler & Stow, 1991), scale variance (Moellering & Tobler, 1972; Townshend *et al.*, 1987; Townshend & Justice, 1988; 1990), fractal geometry (DeCola, 1989; Lam, 1990; LaGro, 1991; Milne, 1991; Lam & Quattrochi, 1992), semivariograms (Curran, 1988; Woodcock, Strahler & Jupp, 1988a,b; Bian & Walsh, 1993), Fourier transforms (Townshend & Justice, 1990; Weiler & Stow, 1991), spectral density functions (Renshaw & Ford, 1984; Weiler & Stow, 1991), trend surfaces (Davies, 1986; Turner *et al.*, 1991) and landscape indices (Murphy, 1985; Robinove, 1986; Turner, 1989; Turner, Dale & Gardner, 1989a; Turner *et al.*, 1989b). Most of these techniques operate on raw image values – commonly referred to, somewhat tautologously, as digital numbers (DN) – rather than class labels. As a result, somewhat different results may be obtained for the same scene in images acquired in different spectral wavebands.

It is beyond the scope of this chapter to provide a full discussion and evaluation of each of these techniques. Instead, attention will be focused on just one – scale variance analysis – while noting that the trends observed using this and other techniques are often very similar (Townshend & Justice, 1988; 1990).

In essence, scale variance analysis measures the percentage contribution of digital number variation at different spatial scales to the (total) variance of the image as a whole (Moellering & Tobler, 1972; Townshend & Justice, 1988; 1990). Figure 2 presents the results of scale variance analysis applied to a 2048 × 2048 pixel subscene extracted from a Landsat TM image centred on Ashford, Kent in south-east England. The analysis has been performed on four spectral wavebands. Although each waveband shows a slightly different response, all suggest that there are two distinct peaks in spatial variance for this scene: the first occurs at approximately 240 m; the second occurs between 4 km and 8 km. (It should be noted that there may be further peaks in spatial variance at both finer and coarser spatial scales, though these cannot be determined from analysis of the present data set.) While the former can readily be accounted for by variation in reflectance between land parcels (predominantly agricultural fields), the latter is more difficult to explain in terms of significant scene elements. Nevertheless, the important observation is that over this range of spatial scales, there appear to be

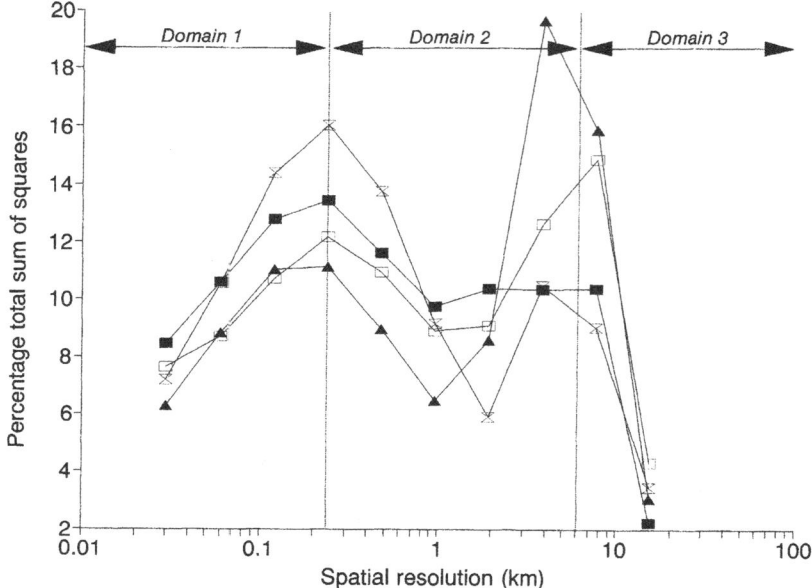

Fig. 2. Scale variance analysis applied to a 2048 × 2048 pixel sub-scene extracted from a Landsat TM image centred on Ashford, Kent in south-east England. The data indicate two peaks in spatial variance for this scene over the observed range of spatial scales. This, in turn, suggests that there may be three discrete 'domains of scale': namely, < 240 m, 240 m to 4 km, and > 4 km (see text for details). The four TM wavebands analysed were band 2 (green) (□); band 3 (red) (■); band 4 (near-infrared) (▲); and band 5 (shortwave infrared) (⊠).

three distinct 'domains' of spatial variation in the scene (< 240 m, 240 m to 4 km and > 4 km); although these results are likely to be both scene specific (Townshend *et al.*, 1987; Townshend & Justice, 1988; 1990) and time dependent (Belward, 1992; Malingreau & Belward, 1992).

The results of scale variance analysis do not by themselves provide incontrovertible evidence of the precise boundaries between domains of scale with respect to land cover, since they are based on untransformed digital numbers. Nevertheless, they do at least suggest that such domains exist. Even if specific domains of scale can be identified, however, there remain the more intractable problems of designing an appropriate classification scheme for each domain and how the validity/applicability of such schemes degrades towards the (spatial resolution) limits of the domain. Certainly, it is not at all clear how simple measures of spatial variance can be used to determine the number and the nature of land cover types that can be reliably identified in remotely sensed images acquired at a given spatial resolution. This remains an area for future research.

Generalisation of multiclass land cover maps

The second and third issues identified earlier in this chapter were the need to find some way of validating large-area, coarse grid-scale land cover maps and the requirement to convert these data to the still smaller cartographic scales appropriate to meso-scale and macro-scale climatological/environmental models. Each can be addressed, in part at least, through the use of categorical generalisation, which is the subject of this section. Since the literature on spatial data generalisation is particularly rich in the field of digital mapping, much more so than in remote sensing, extensive use will be made of the concepts and the terminology that have been developed therein (Buttenfield & McMaster, 1991).

Elements of spatial data generalisation

Generalisation is the process by which data are transformed systematically from larger (finer spatial resolution) to smaller (coarser spatial resolution) cartographic scales. In this context, the purpose of generalisation is to produce land cover maps that are appropriate to the investigation of environmental and climatological processes operating at a particular spatial scale. The aim is to reduce the volume of data to be handled while, as far as is possible, preserving its information content (for example, spatial and attribute accuracy).

Although generalisation can be performed at a number of different stages during data processing, the discussion here focuses on postclassification generalisation, sometimes known as *categorical generalisation* (McMaster, 1991). It is possible to identify a set of further stages and processes within categorical generalisation. These include *association* and *aggregation* (Nyerges, 1991). In the context of raster land cover maps, the process of aggregation involves the combination of several land cover classes identified at one spatial resolution into a set of new categories at some smaller cartographic scale (coarser spatial resolution). The aggregation process may be controlled by spatial and/ or aspatial criteria. The manner in which individual classes are aggregated is conditioned by the rules of association, that is, the member-of-set relationships between land cover classes/parcels.

Application of generalisation to land cover mapping

Generalisation has two main uses in the context of land cover mapping. The first is in the validation of land cover maps produced from coarse spatial resolution images. This might be achieved by generalising land cover maps produced from fine spatial resolution images, since these are comparatively easy to validate, to smaller cartographic scales. An estimate of the classification accuracy for the coarse spatial resolution data set can be obtained by cross-tabulating the class labels in that image against those for corresponding pixels in the generalised data set. By the same token, it should also be possible to generalise land cover data derived directly from coarse spatial resolution sensors to still smaller cartographic scales that are appropriate to, for example, global climate models. The key to both of these applications is to define appropriate mechanisms with which to generalise the data.

There are several ways in which classes identified at one spatial scale might be aggregated together to produce a new data set at some smaller cartographic scale. The optimum aggregation scheme will undoubtedly depend on the nature of the data that are to be generalised and their intended use. For instance, it may be appropriate to adopt a simple aggregation scheme involving a one-to-one mapping function, where each pixel in the coarse spatial resolution data set is assigned a single label from the set of classes, $C\{c_1, c_2, c_3, \ldots, c_n\}$, identified in the fine spatial resolution data. A simple majority filter would produce such a mapping function. The alternative is a many-to-one mapping function. This may be further subdivided. For instance, a composite category, $C_1c_2c_3$, might be derived from three categories c_1, c_2 and c_3, identified in the fine spatial resolution data. This might be the case where, for example, regions are classified on the basis of the dominant and subordinate vegetation species; thus, $C_1c_2c_3$ indicates that class C_1 is dominant, while classes c_2 and c_3 are subordinate within that region. Another many-to-one mapping function is where the initial categories map in an abstract fashion to an entirely new class at the coarser spatial resolution. For example, the classes *wheat* and *barley* might map to *arable farmland*.

These aggregation strategies can be implemented in the form of either kernel-based or region-based generalisation procedures. The former offer the advantage of computational simplicity, while the latter are much more flexible in terms of the nature of spatial postprocessing that can be performed. The remainder of this section will provide a brief review of a number of the techniques available under each of these headings, discussing their relative merits and disadvantages in the context of generalising large-area land cover maps.

Kernel-based categorical generalisation

The simplest form of categorical generalisation makes use of kernel-based procedures; that is, where a simple moving-window filter is passed across the land cover data set and, through the process of *convolution*, the input pixel values (class labels) are converted into a set of new class labels.

Perhaps the most widely used kernel-based generalisation technique is the simple majority filter (Gurney, 1981). This replaces the class label assigned to each pixel in the image with that of the most frequently occurring class in the region immediately surrounding it (Fig. 3*a*). In essence, the majority filter smooths the original data by relabelling isolated pixels of any given class. By the same token, however, this technique can result in the disproportionate loss of certain land cover classes. This is particularly true for classes that are highly fragmented throughout the scene, even where they constitute a significant percentage of the total area. As a result of this, important information on landscape heterogeneity may be lost using this procedure.

Although not originally intended for use in this context, a number of kernel-based techniques have been developed that can be used to perform many-to-one generalisation of raster-format land cover maps. These include the procedures developed by Wharton (1982), Gong & Howarth (1992) and Barnsley & Barr (1996). These techniques examine the frequency and/or the spatial arrangement of different class labels in the kernel and use this information to assign a new class label to each pixel in the image (Fig. 3*b,c*). With each of these techniques, generalisation can be performed either by comparing the extracted information against predefined training classes or through unsupervised classification of the observed spatial patterns (Wharton, 1982; Whitehouse, 1990; Gong & Howarth, 1992; Barnsley & Barr, 1996).

A further, related set of kernel-based techniques has been developed in the field of landscape ecology (Legendre & Fortin, 1989). These have been designed to characterise the spatial patterns of land cover in a landscape and to relate these to the dominant ecological processes operating in that area (Meentemeyer & Box, 1987; Turner *et al.*, 1989a,b; Wiens, 1989; Turner, 1990). The landscape ecology indices include relative richness, diversity, dominance, fragmentation and contagion (Table 1) (Murphy, 1985; Robinove, 1986; O'Neill *et al.*, 1988;

(a)

1	1	2	2	4
5	1	1	7	7
3	4	1	7	7
6	3	4	1	7
6	3	3	3	1

⟹ 1

(b)

1	1	2	2	4
5	1	1	7	7
3	4	1	7	7
6	3	4	1	7
6	3	3	3	1

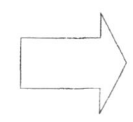

⟹

c	f
1	7
2	2
3	4
4	3
5	5
6	2

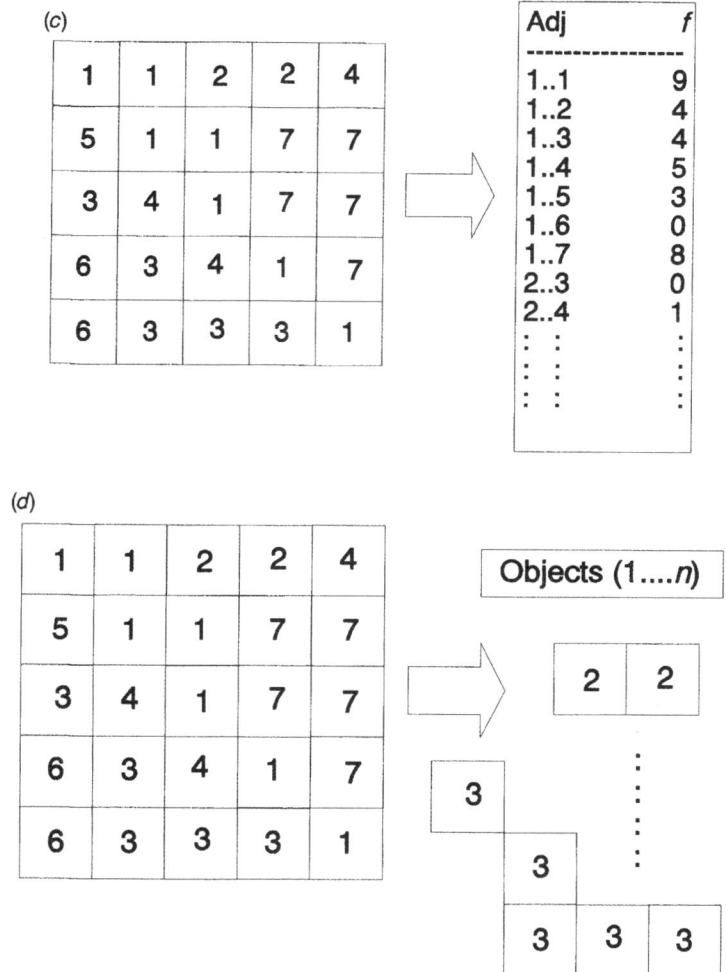

Fig. 3. Diagrammatic representation of various aggregation schemes. (*a*) Simple one-to-one (majority class) mapping function; (*b*) many-to-one mapping function (frequency based) (after Wharton, 1982); (*c*) many-to-one mapping function (using class frequency and spatial arrangement) (after Barnsley *et al.*, 1993); (*d*) region-based scheme (after Barr & Barnsley, 1995).

Table 1. *Formulae for five commonly used landscape indices (relative richness, diversity, dominance, fragmentation, and contagion; see Robinove 1986, Turner 1989 and Turner et al. 1989a, b). The spatial reclassification kernel (SPARK) is a related measure that examines both the frequency and spatial arrangement of class labels (Barnsley et al., 1993; Barnsley & Barr, 1996; see also Fig. 3c)*

Measure	Symbol	Formula
Relative richness	R	$R = 100 * (n / n_{\max})$
Diversity	H	$H = -\sum_{i=1}^{n} P_i \ln (P_i)$
Dominance	D_o	$D_o = H_{\max} - H$
Fragmentation	F	$F = (n-1)/(p-1)$
Contagion	C	$C = 2n \log n + \sum^{n} q(ij) \log q(ij)$
SPARK	Δ_k	$\Delta k = 1 - \sqrt{\dfrac{1}{2N^2} \sum_{i=1}^{n} \sum_{j=i}^{n} (A_{ij} - T_{kij})^2}$

n = number of classes in the kernel; $n_{\max}$ = number of classes in the image; i = class i; P_i = the number of pixels of class i in the kernel divided by the total number of elements in the kernel; $H_{\max}$ = the maximum diversity, when classes occur in equal proportions (i.e. $\ln(n)$); p = number of elements in the kernel; $q(i,j)$ = the probability of class i being adjacent to class j; ($q(i,j) \neq q(j,i)$); A_{ij} = element i,j of the adjacency-event matrix for the current kernel; T_{kij} = element i,j of the template matrix for composite category k.

Turner & Ruscher, 1988; Turner, 1989; Carrere, 1990; Turner *et al.*, 1991; Gustafson & Parker, 1992). Each provides a single output value that indicates the degree of spatial heterogeneity in land cover in a region or in the image as a whole. Although these indices are helpful in understanding scene structure, their application to image data generalisation requires some development. It might be possible to use them to indicate relatively uniform regions or, alternatively, spatial complexes of land cover types that are characteristic of certain ecosystems or biomes. In any case, work in this field is continuing with the aim of developing techniques that can be used to translate information across a range of scales (O'Neill *et al.*, 1988; Turner *et al.*, 1989a,b). One interesting aspect of this work is the use of neutral scene models to determine the influence of a given environmental process. These operate by generating the expected spatial pattern of land cover in the absence of the process, and then testing the goodness-of-fit between the expected and observed spatial patterns (Gardner *et al.*, 1987; Gustafson & Parker, 1992).

Whichever algorithm is adopted, kernel-based categorical generalisation has a number of limitations. First, the spatial mixing of land cover is examined within a somewhat arbitrary rectangular window the optimum dimensions of which are likely to vary throughout the image according to the size and shape of the local land cover parcels. Second, spatial relationships are examined at the level of individual pixels, as opposed to 'real world' entities, such as agricultural fields and vegetation communities. As a result, it is difficult to analyse more sophisticated morphological properties and spatial relationships, such as shape, containment, distance and direction. These may be useful in characterising the complex spatial assemblages of land cover that are appropriate to maps of large areas. Finally, in certain circumstances, it may be important to preserve the boundaries between key ecological/environmental units while removing or reducing spatial heterogeneity within them. In general, kernel-based techniques are poorly suited to this type of operation, because of the inherent spatial averaging that takes place within the kernel.

Region-based generalisation procedures

The alternative to kernel-based categorical generalisation is to examine the morphological properties and spatial arrangement of discrete regions within the land cover image (Fig. 3*d*; Mark, 1991). In the context of this study, a region is considered to be a section of the image exhibiting uniform land cover type, that is, a continuous block of pixels with the same class label or numeric identifier (Nichol, 1990; Barr, 1992; Barr & Barnsley, 1993). These regions are assumed to represent meaningful 'objects' within the corresponding scene, such as agricultural fields, urban areas, forests and water bodies.

The advantage of adopting a region-based approach to categorical generalisation is that a far richer set of spatial relationships can be exploited, including adjacency, containment, distance and direction (Freeman, 1975; Frank & Mark, 1991; Barr & Barnsley, 1995). Furthermore, information on the morphological properties of each region can also be derived, such as its area, perimeter, eccentricity, compactness and fractal dimension (Gonzalez & Wintz, 1987; Boyle & Thomas, 1988; Schalkoff, 1989; Mehldau & Schowengerdt, 1990; Sonka, Hlavac & Boyle, 1993). The key, however, is to utilise this information in deciding which regions to merge, how, and which class labels to assign to the merged regions (Barr & Barnsley, 1993; 1995; Goffredo, Wilkinson & Fisher, 1993).

A flexible, region-based categorical generalisation scheme requires the following components.

1. Some means of recognising and encoding discrete regions within the land cover image
2. A set of algorithms to derive the spatial relationships and morphological properties of interest
3. Some means of encoding and storing this information
4. A set of algorithms that allow specific categorical generalisation tasks to be performed, based on an analysis of the encoded spatial and morphological information
5. A set of techniques to structure and sequence the execution of these algorithms (Barr & Barnsley, 1995).

Several studies have begun to consider the development of region-based generalisation schemes. These studies can be divided into (i) those in which the original set of land cover classes remain unaltered after the regions have been generalised – that is, those that perform some form of one-to-one mapping or cartographic simplification, examples of which can be found in Nichol (1990) and Goffredo *et al.* (1993)) – and (ii) those in which new, composite classes or themes are derived (Gurney, 1981; Moller-Jenson, 1990). Most of these studies make use of a *single* spatial relationship or morphological property as the basis for generalisation. However, much greater flexibility and more general applicability can be obtained where *multiple* spatial relationships and *multiple* morphological properties can be used simultaneously to guide the generalisation process (Barr & Barnsley, 1995). An approach such as this has been adopted by Mehldau & Schowengerdt (1990) and by Barr & Barnsley (Barnsley *et al.*, 1993; Bar & Barnsley, 1995). This permits the construction of complex 'clauses' that describe the composition of the new, composite classes in terms of the expected spatial patterns and morphological characteristics of regions at the original spatial scale.

Region-based categorical generalisation schemes are still at a relatively early stage of development. Certainly, there is some way to go before they can be used operationally to transform land cover data from one spatial scale to another. The main requirement, however, is to define suitable aggregation schemes and rules of association that can be used to guide the generalisation process.

Summary and conclusion

Timely, consistent information on land cover has considerable value to a range of environmental and climatological models, both in its own right and as a surrogate for other variables of interest. The only viable means by which such data can be acquired over very large areas is via satellite remote sensing. At regional and global scales, 'medium' spatial resolution instruments such as the AVHRR and ATSR (Along-Track Scanning Radiometer) satellite sensors are appropriate for reasons of data costs, the frequency of global coverage and data volumes. These devices will be augmented in the near future by a range of new medium spatial resolution satellite sensors, including AATSR, VEGETATION, MODIS, MISR and MERIS. These should increase the frequency and the quality of global coverage.

There are, however, several important issues relating to the production of large-area land cover maps from satellite sensor images that require further attention. More specifically, greater attention needs to be given to:

1. The most appropriate classification schemes (in terms of the number and nature of classes) for mapping land cover at grid scales of 1 km or coarser
2. The mechanisms by which land cover maps produced at these spatial scales can be validated
3. The techniques that can be used to generalise land cover data derived from medium-scale spatial resolution satellite sensors to the still smaller cartographic scales required by, for example, global climate models.

This chapter has begun to explore some of these issues. In particular, it was noted that there is some evidence for domains of scale within which a single classification scheme may suffice. Evidence for the boundaries between these domains of scale can be obtained by analysing the spatial and spectral properties of a scene in a multiple spatial resolution data set. There is, however, still no objective method to determine the number and nature of the land cover categories that can be consistently identified in any given domain of scale.

It was also noted that there is considerable potential for the use of categorical generalisation techniques both to validate land cover maps derived from coarse spatial resolution image data and to transform these maps to still smaller cartographic scales. In this context, although kernel-based generalisation procedures offer advantages in terms of computational simplicity, region-based techniques provide greater flexibility, wider applicability and are better at preserving the boundaries between key ecological or environmental units.

Acknowledgements

The authors would like to acknowledge helpful discussions with Jan-Peter Muller (UCL), Philip Lewis (UCL) and Alan Strahler (Boston University). Thanks are also due to Peter Blamire for assistance with the data processing. M. Barnsley and S. Barr were supported in this work through the Natural Environment Research Council's TIGER (Terrestrial Initiative in Global Environmental Research) programme (GST/02/628 and T91/86-3.3/1d). T. Tsang was supported by a NERC postgraduate studentship (GT4/93/27/L).

References

Barnsley, M.J. & Barr, S.L. (1996). Inferring urban land use from satellite images using kernel-based spatial reclassification. *Photogrammetric Engineering and Remote Sensing*, 62, 949–958.

Barnsley, M.J., Barr, S.L., Hamid, A., Muller, J.-P., Sadler, G.J. & Shepherd, J.W. (1993). Spatial analytical tools to monitor the urban environment. In *Geographical Information Handling – Research and Applications*, ed. P.M. Mather, pp. 147–184. Chichester: Wiley.

Barnsley, M.J., Strahler, A.H., Morris, K.P. & Muller, J.-P. (1994). Sampling the Bidirectional Reflectance Distribution Function (BRDF): 1. Evaluations of current and future satellite sensors. *Remote Sensing Reviews*, 8, 271–311.

Barr, S.L. (1992). Object-based re-classification of high resolution digital imagery for urban land-use monitoring. In *Proceedings of XXIX Conference of the International Society for Photogrammetry and Remote Sensing, International Archives of Photogrammetry and Remote Sensing: Commission 7*, Washington DC, 1–14 August 1992, pp. 969–976. Washington, DC: International Society for Photogrammetry and Remote Sensing.

Barr, S.L. & Barnsley, M.J. (1993). Object-based spatial analytical tools for urban land-use monitoring in a raster processing environment. In *Fourth European GIS Conference (EGIS'93)*, pp. 810–822. Utrecht: EGIS Foundation.

Barr, S.L. & Barnsley, M.J. (1995). A spatial modelling system to process, analyse and interpret multi-class thematic maps derived from satellite sensor images. In *Innovations in GIS 2*, ed. P. Fisher, pp. 53–65. London: Taylor and Francis.

Belward, A.S. (1992). Spatial attributes of AVHRR imagery for environmental monitoring. *International Journal of Remote Sensing*, 13, 193–208.

Belward, A.S. & Lambin, E. (1990). Limitations to the identification of spatial structures from AVHRR data. *International Journal of Remote Sensing*, 11, 921–927.

Bian, L. & Walsh, S.J. (19930. Scale dependencies of vegetation and topography in a mountainous environment of Montana. *Professional Geographer*, 45, 1–11.

Boyle, R.D. & Thomas, R.C. (1988). *Computer Vision: A First Course*. Oxford: Blackwell Scientific.

Buttenfield, B.P. & McMaster, R.B. (eds.) (1991). *Map Generalization*. Harlow: Longman.

Carpenter, S.R. & Chaney, J.E. (1983). Scale of spatial pattern: four methods compared. *Vegetatio*, 53, 153–160.

Carrere, V. (1990). Development of multiple source data processing for structural analysis at regional scales. *Photogrammetric Engineering and Remote Sensing*, 56, 587–595.

Cess, R.D. (1978). Biosphere-albedo feedback and climate modeling. *Journal of Atmospheric Science*, 35, 1765–1768.

Cross, A.M., Settle, J.J., Drake, N.A. & Paivinen, R.T.M. (1991). Subpixel measurement of tropical forest cover using AVHRR data. *International Journal of Remote Sensing*, 12, 1119–1129.

Curran, P.J. (1983). Multispectral remote sensing for the estimation of green leaf area. *Philosophical Transactions of the Royal Society*, 309, 257.

Curran, P.J. (1988). The semi-variogram in remote sensing: an introduction. *Remote Sensing of Environment*, 24, 493–507.

Davis, J.C. (1986). *Statistics and Data Analysis in Geology*. New York: Wiley.

DeCola, L. (1989). Fractal analysis of a Landsat scene. *Photogrammetric Engineering and Remote Sensing*, 55, 601–610.

DeFries, R.S. & Townshend, J.R.G. (1994a). Global land cover: comparison of ground-based data sets to classifications of AVHRR data. In *Environmental Remote Sensing from Global to Regional Scales*, ed. G.M. Foody & P.J. Curran, pp. 84–110. Chichester: Wiley.

DeFries, R.S. & Townshend, J.R.G. (1994b). NDVI-derived land cover classifications at a global scale. *International Journal of Remote Sensing*, 15, 3567–3586.

Dickinson, R.E. & Hanson, B. (1984). Vegetation-albedo feedbacks. In *Climate Processes and Climate Sensitivity: Geophysical Monograph 29*, ed. J.E. Hansen & T. Takahashi, pp. 180–186. Washington, DC: American Geophysical Union.

Dorman, J.L. & Sellers, P.J. (1989). A global climatology of albedo, roughness length and stomatal resistance for atmospheric general

194   M.J. BARNSLEY, S.L. BARR AND T. TSANG

circulation models as represented by the Simple Biosphere model (SiB). *Journal of Applied Meteorology*, 28, 833–855.

Forshaw, M.R.B., Haskell, A., Miller, P.F., Stanley, D.J. & Townshend, J.R.G. (1983). Spatial resolution of remotely sensed imagery, a review paper. *International Journal of Remote Sensing*, 4, 497–520.

Frank, A.U. & Mark, D.M. (1991). Language issues for GIS. In *Geographical Information Systems: Principals and Applications*, ed. D.J. Maguire, M.F. Goodchild & D.W. Rhind, pp. 147–163. London: Longman.

Freeman, J. (1975). The modelling of spatial relations. *Computer Graphics and Image Processing*, 4, 156–171.

Gardner, R.H., Milne, B.T., Turner, M.G. & O'Neill, R.V. (1987). Neutral models for the analysis of broad-scale landscape patterns. *Landscape Ecology*, 1, 19–28.

Goffredo, S., Wilkinson, G.G. & Fisher, P.F. (1993). Spatial generalisation of thematic maps derived from satellite imagery for operational land cover mapping. In *19th Annual Conference of the Remote Sensing Society*, Nottingham: Remote Sensing Society.

Gong, P. & Howarth, D. (1992). Frequency-based contextual classification and grey-level vector reduction for land-use identification. *Photogrammetric Engineering and Remote Sensing*, 58, 423–437.

Gonzalez, R.C. & Wintz, P. (1987). *Digital Image Processing*. New York: Addison-Wesley.

Goward, S.N., Tucker, C.J. & Dye, D.G. (1987). North American vegetation patterns observed with the NOAA-7 Advanced Very High Resolution Radiometer. *Vegetatio*, 64, 3–14.

Greig-Smith, P. (1979). Pattern in vegetation. *Journal of Ecology*, 67, 755–779.

Gurney, C.M. (1981). The use of contextual information to improve land cover classification of digital remotely sensed data. *International Journal of Remote Sensing*, 2, 379–388.

Gustafson, E.J. & Parker, G.R. (1992). Relationships between land cover proportion and indices of landscape spatial pattern. *Landscape Ecology*, 7, 101–110.

Irons, J.R., Markham, B.L., Nelson, R.F., Toll, D.L. & Williams, D.L. (1985). The effects of spatial resolution on the classification of Thematic Mapper data. *International Journal of Remote Sensing*, 6, 1385–1403.

Iverson, L.R., Cook, E.A. & Graham, R.L. (1989). A technique for extrapolating and validating forest cover across large regions: calibrating AVHRR data with TM data. *International Journal of Remote Sensing*, 10, 1805–1812.

Justice, C.O., Townshend, J.R.G., Holben, B.N. & Tucker, C.J. (1985). Analysis of the phenology of global vegetation using meteorological satellite data. *International Journal of Remote Sensing*, 6, 1271–1318.

Justice, C.O., Markham, B.L., Townshend, J.R.G. & Kennard, R. (1989). Spatial degradation of satellite data. *International Journal of Remote Sensing*, 10, 1539–1561.

Justice, C.O., Townshend, J.R.G. & Kalb, V.L. (1991). Representation of vegetation by continental data sets derived from NOAA–AVHRR data. *International Journal of Remote Sensing*, 12, 999–1021.

LaGro, J. (1991). Assessing patch shape in landscape mosaics. *Photogrammetric Engineering and Remote Sensing*, 57, 285–293.

Lam, N. S.-M. (1990). Description and measurement of Landsat TM images using fractals. *Photogrammetric Engineering and Remote Sensing*, 56, 187–195.

Lam, N. S.-M. & Quattrochi, D.A. (1992). On the issues of scale, resolution, and fractal analysis in the mapping sciences. *Professional Geographer*, 44, 88–98.

Legendre, P. & Fortin, M.-J. (1989). Spatial pattern and ecological analysis. *Vegetatio*, 80, 107–138.

Lloyd, D. (1990). A phenological classification of terrestrial vegetation cover using shortwave vegetation index imagery. *International Journal of Remote Sensing*, 11, 2269–2279.

Loveland, T.R., Merchant, J.W., Ohlen, D.O. & Brown, J.F. (1991). Development of a land-cover characteristics database for the conterminous U.S. *Photogrammetric Engineering and Remote Sensing*, 57, 1453–1463.

Malingreau, J.-P. & Belward, A.S. (1992). Scale considerations in vegetation monitoring using AVHRR data. *International Journal of Remote Sensing*, 13, 2289–2307.

Mark, D.M. (1991). Object modelling and phenomenon-based generalization. In *Map Generalization: Making Rules for Knowledge Representation*, ed. B.P. Buttenfield, & R.B. McMaster, pp. 103–118. Harlow: Longman.

Markham, B.L. & Townshend, J.R.G. (1981). Land cover classification accuracy as a function of sensor spatial resolution. In *Proceedings of the 15th International Symposium on Remote Sensing of Environment*, pp. 1075–1090. Ann Arbor, MI: ERIM.

Matthews, E. (1982). Global vegetation and land use: new high resolution data bases for climate studies. *Journal of Climate and Applied Meteorology*, 22, 474–487.

McMaster, R.B. (1991). Conceptual frameworks for geographical knowledge. In *Map Generalization: Making Rules for Knowledge Representation*, ed. B.P. Buttenfield & R.B. McMaster, pp. 21–39. Harlow: Longman.

Meentemeyer, V. & Box, E.O. (1987). Scale effects in landscape studies. In *Landscape Heterogeneity and Disturbance, Ecological*

Studies 64, ed. M.G. Turner, pp. 15–34. New York: Springer-Verlag.

Mehldau, G. & Schowengerdt, R.A. (1990). A C-extension for rule-based image classification systems. *Photogrammetric Engineering and Remote Sensing*, 56, 887–892.

Milne, B.T. (1991). Lessons from applying fractal models to landscape patterns. In *Quantitative Methods in Landscape Ecology: The Analysis and Interpretation of Landscape Heterogeneity, Ecological Studies 82*, ed. M.G. Turner & R.H. Gardner, pp. 69–84. London: Springer-Verlag.

Moellering, H. & Tobler, W. (1972). Geographical variances. *Geographical Analysis*, 4, 34–64.

Moller-Jenson, L. (1990). Knowledge-based classification of an urban area using texture and context information in Landsat–TM imagery. *Photogrammetric Engineering and Remote Sensing*, 6, 899–904.

Moody, A. & Woodcock, C.E. (1994). Scale-dependent errors in the estimation of land-cover proportions – implications for global land-cover datasets. *Photogrammetric Engineering and Remote Sensing*, 60, 585–594.

Murphy, D.L. (1985). Estimating neighborhood variability with a binary comparison matrix. *Photogrammetric Engineering and Remote Sensing*, 51, 667–674.

Nichol, D.G. (1990). Region adjacency analysis of remotely-sensed imagery. *International Journal of Remote Sensing*, 11, 2089–2101.

Nyerges, T.L. (1991). Representing geographical meaning. In *Map Generalization: Making Rules for Knowledge Representation*, ed. B.P. Buttenfield & R.B. McMaster, pp. 59–85. Harlow: Longman.

Olsen, J.S., Watts, J.A. & Allison, L.J. (1983). *Carbon in Live Vegetation of Major World Ecosystems. Report ORNL-5862*, Oak Ridge, TN: Oak Ridge National Laboratory.

O'Neill, R.V., DeAngelis, D.L., Waide, J.B. & Allen, T.F.H. (1986). *A Hierarchical Concept of Ecosystems*. Princeton, NJ: Princeton University Press.

O'Neill, R.V., Krummel, J.R., Gardner, R.H. *et al.* (1988). Indices of landscape pattern. *Landscale Ecology*, 1, 153–162.

Quarmby, N.A. (1992). Towards continental scale crop area estimation. *International Journal of Remote Sensing*, 13, 981–989.

Quarmby, N.A., Townshend, J.R.G., Settle, J.J. *et al.* (1992). Linear mixture modelling applied to AVHRR data for crop area estimation. *International Journal of Remote Sensing*, 13, 415–425.

Renshaw, E. & Ford, E.D. (1984). The description of spatial pattern using two-dimensional spectral analysis. *Vegetatio*, 56, 75–85.

Ripple, W.J. (1994). Determining coniferous forest cover and forest fragmentation with NOAA-9 Advanced Very High Resolution Radiometer data. *Photogrammetric Engineering and Remote Sensing*, 60, 533–540.

Robinove, C.J. (1986). Spatial diversity index mapping of classes in grid cell maps. *Photogrammetric Engineering and Remote Sensing*, 52, 1171–1173.

Running, S., Justice, C.O., Hall, D. *et al.* (1994). Terrestrial remote sensing science and algorithms planned for EOS/MODIS. *International Journal of Remote Sensing*, 15, 3587–3620.

Schalkoff, R.J. (1989). *Digital Image Processing and Computer Vision*. New York: Wiley.

Sellers, P.J. (1985). Canopy reflectance, photosynthesis and transpiration. *International Journal of Remote Sensing*, 6, 1335–1372.

Smith, M.O., Johnson, P.E. & Adams, J.B. (1985). Quantitative determination of mineral types and abundances from reflectance spectra using principal components analysis. *Journal of Geophysical Research*, 90, 792–804.

Sonka, M., Hlavac, V. & Boyle, R. (1993). *Image Processing, Analysis and Machine Vision*. London: Chapman & Hall.

Strahler, A.H., Woodcock, C.E. & Smith, J.A. (1986). On the nature of models in remote sensing. *Remote Sensing of Environment*, 20, 121–139.

Townshend, J.R.G. (1981). The spatial resolving power of earth resources satellites: a review. *Progress in Physical Geography*, 5, 32–55.

Townshend, J.R.G. (ed.) (1992). *Improved Global Data for Land Applications: A Proposal for a New High Resolution Data Set, IGBP Report No. 20*. Stockholm: International Geosphere-Biosphere Programme.

Townshend, J.R.G. (1992b). Land cover. *International Journal of Remote Sensing*, 13, 1319–1328.

Townshend, J.R.G. (1994). Global data sets for land applications from the Advanced Very High Resolution Radiometer: an introduction. *International Journal of Remote Sensing*, 15, 3319–3332.

Townshend, J.R.G. & Justice, C.O. (1988). Selecting the spatial resolution of satellite sensors for global monitoring of land transformations. *International Journal of Remote Sensing*, 9, 187–236.

Townshend, J.R.G. & Justice, C.O. (1990). The spatial variation of vegetation changes at very coarse scales. *International Journal of Remote Sensing*, 11, 149–157.

Townshend, J.R.G., Goff, T.E. & Tucker, C.J. (1985). Multitemporal dimensionality of images of normalised difference vegetation index at continental scales. *Transactions of the IEEE on Geoscience and Remote Sensing*, 23, 888–895.

Townshend, J.R.G., Justice, C.O. & Kalb, V. (1987). Characterization and classification of South American land cover types using satellite data. *International Journal of Remote Sensing*, 8, 1189–1207.

Townshend, J.R.G., Justice, C.O., Li, W., Gurney, C. & McManus, J. (1991). Global land cover classification by remote sensing: present capabilities and future possibilities. *Remote Sensing of Environment*, 35, 243–255.

Townshend, J.R.G., Justice, C.O., Skole, D. *et al.* (1994). The 1 km resolution global data set: needs of the International Geosphere Biosphere Programme. *International Journal of Remote Sensing*, 15, 3417–3442.

Tucker, C.J. & Sellers, P.J. (1986). Satellite remote sensing of primary production. *International Journal of Remote Sensing*, 7, 1395–1416.

Tucker, C.J., Townshend, J.R.G. & Goff, T.E. (1985). African landcover classification using satellite data. *Science*, 227, 369–375.

Turner, M.G. (1989). Landscape ecology: the effect of pattern on process. *Annual Review Ecology and Systematics*, 20, 171–197.

Turner, M.G. (1990). Spatial and temporal analysis of landscape patterns. *Landscape Ecology*, 4, 21–30.

Turner, M.G. & Ruscher, C.L. (1988). Changes in landscape patterns in Georgia, USA. *Landscape Ecology*, 1, 241–251.

Turner, M.G., Dale, V.H. & Gardner, R.H. (1989a). Predicting across scales: theory development and testing. *Landscape Ecology*, 3, 245–252.

Turner, M.G., O'Neill, R.V., Gardner, R.H. & Milne, B.T. (1989b). Effects of changing spatial scale on the analysis of landscape pattern. *Landscape Ecology*, 3, 153–162.

Turner, S., O'Neill, R.V., Conley, W., Conley, M. & Humphries, H. (1991). Pattern and scale: statistics for landscape ecology. In *Quantitative Methods in Landscape Ecology: The Analysis and Interpretation of Landscape Heterogeneity*, ed. M.G. Turner & R.H. Gardner, pp. 17–49. New York: Springer-Verlag.

Weiler, R.A. & Stow, D.A. (1991). Spatial analysis of land cover patterns and corresponding remotely sensed image brightness. *International Journal of Remote Sensing*, 12, 2237–2257.

Wharton, S.W. (1982). A contextual classification method for recognizing land use patterns in high resolution remotely sensed data. *Pattern Recognition*, 15, 317–324.

Whitehouse, S. (1990). A spatial land use classification of an urban environment using high resolution multispectral satellite data. In *Proceedings of the 16th Annual Conference of the Remote Sensing Society: Remote Sensing and Global Change*, pp. 433–437. Nottingham: Remote Sensing Society.

Wiens, J.A. (1989). Spatial scaling in ecology. *Functional Ecology*, 3, 385–397.

Wilson, M. & Henderson-Sellers, A. (1985). A global archive of land cover and soils data for use in general circulation models. *Journal of Climatology*, 5, 119–143.

Woodcock, C.E. & Strahler, A.H. (1987). The factor of scale in remote sensing. *Remote Sensing of Environment*, 21, 311–332.

Woodcock, C.E., Strahler, A.H. & Jupp, D.L.B. (1988a). The use of variograms in remote sensing: I. Scene models and simulated images. *Remote Sensing of Environment*, 25, 323–343.

Woodcock, C.E., Strahler, A.H. & Jupp, D.L.B. (1988b). The use of variograms in remote sensing: II. Real digital images. *Remote Sensing of Environment*, 25, 349–379.

Zhu, Z. & Evans, D.L. (1994). U.S. forest types and predicted percent forest cover from AVHRR data. *Photogrammetric Engineering and Remote Sensing*, 60, 525–531.

P.J. CURRAN, G.M. FOODY, R.M. LUCAS,
M. HONZÁK and J. GRACE

The carbon balance of tropical forests: from the local to the regional scale

Many small carbon sinks can add up to a single large one
Lugo (1992, p. 4)

Introduction

We do not have an adequate carbon budget for our planet. However, current evidence suggests that tropical forests are acting simultaneously as a major source of atmospheric carbon (mainly as a result of deforestation and land degradation) and a major sink of atmospheric carbon (particularly through regeneration and a fertilisation effect). To quantify the strength of this sink, point estimates of carbon fluxes between the atmosphere and the forest are being scaled-up to the region using Advanced Very High Resolution Radiometer (AVHRR) imagery. This ongoing research illustrates the problems associated with using remote sensing as a tool for scaling-up and emphasises the need to understand both the spectral and spatial characteristics of the imagery. In particular, it was shown that different forest regeneration stages could be detected on AVHRR imagery and these could be aged from finer spatial resolution data, notably Landsat Thematic Mapper (TM) imagery.

The end of the 20th century is approached with only a limited understanding of how carbon cycles between the land, the atmosphere and the oceans (Houghton, Jenkins & Ephramus, 1990; Houghton, Callander & Varney, 1992). The publication in 1973 of the atmospheric CO_2 record from the Mauna Loa Observatory in Hawaii (Moore & Braswell, 1994) resulted in calls for a reduction in the rate at which fossil fuels were being burnt and tropical forests were being cleared (Department of the Environment, 1994). Dealing with carbon sources alone, however, was clearly inadequate as 'the environment cannot speak for itself and we require a clear understanding of its present and future condition to guide its stewardship' (Waldegrave, 1994, p. 16). An understanding of the carbon cycle was sought in carbon budgets (Trabalka, 1985; Jones &

Henderson-Sellers, 1990). Such budgets indicated that the increase in atmospheric CO_2 was not as great as would be expected given the magnitude of the known carbon sinks (e.g. oceanic) and known carbon sources (e.g. fossil fuels) (Lugo, 1992; Lugo & Wisniewski, 1992; Smith *et al.*, 1993; Wisniewski & Sampson, 1993). In other words, some carbon was 'missing' (Edmonds, 1992) and so there must be an unknown and/or underestimated carbon sink.

Subsequent attempts to balance the carbon budget and account for what is estimated to be around 2.2 Pg carbon per year (Pg = petagram = 10^{15} g = Gt) of missing carbon have concentrated on terrestrial ecosystems (Houghton *et al.*, 1992; Sundquist, 1993). These studies were hampered by lack of baseline information as we neither know accurately the distribution of vegetation on our planet (DeFries & Townshend, 1994), nor do we have a global forest carbon inventory for any given year (Dixon *et al.*, 1994). Atmospheric modelling augmented with ground data has suggested both mid-to-high latitudes (Tans, Fung & Takahashi, 1990) and tropical forests (Taylor & Lloyd, 1992) as possible repositories for the missing carbon. Both of these suggestions are credible, but quantification of the mid-to-high latitude sink depends upon the low-altitude sink to set its bounds (Tans *et al.*, 1990) and so research has been concentrated increasingly on the tropics: 'It is clear, therefore, that progress in balancing the global C budget will be made only with improved understanding of the C fluxes from the tropics' (Brown, Iverson & Lugo, 1993, p. 72). Unfortunately, the carbon budget of tropical forests is more uncertain than that of any other terrestrial ecosystem (Table 1). The magnitude of this uncertainty is, however, decreasing. The Intergovernmental Panel on Climate Change (IPCC) estimated a net carbon flux of 1.6 ± 1.0 Pg carbon per year (Houghton *et al.*, 1990) and the estimate for the next IPCC report was to have been 1.7 ± 0.5 Pg carbon per year (Brown *et al.*, 1993) prior to the more recent estimate of 1.65 ± 0.4 Pg carbon per year (Dixon *et al.*, 1994). This degree of uncertainty is both a hindrance to our attempts at managing carbon fluxes and a manifestation of our limited understanding of the global carbon cycle: 'Uncertainties in balancing the C flux from forest landscapes will remain until a coordinated global network of permanent forest inventory plots and the application of remote sensing technology to measure changes in area and conditions of forests, including 'mature' forests, is undertaken' (Dixon *et al.*, 1994, p. 188).

Developing a carbon budget for the tropics

A number of simple budgets have been developed to describe the pools and fluxes of carbon in tropical forests (Brown *et al.*, 1993).

Table 1. *Current and future carbon fluxes in the terrestrial biosphere*

Terrestrial ecosystem	Current carbon flux[a] (Pg carbon per year)	Future carbon flux, if atmospheric CO_2 is doubled[a] (Pg carbon per year)
Tundra, boreal forests	+0.5 to +0.7	−1.0 to −0.5[b]
Temperate forests	+0.2 to +0.5	−2.0 to +2.0
Tropical forests[c]	−2.2 to −1.2	−1.0 to −0.5
Grasslands, savannas and deserts	0 to +0.6	−0.3 to +0.1
Agro-ecosystems	−0.1 to +0.1	0 to +0.1
Wetlands	+0.2	+0.1
Total	−1.4 to +0.9	−4.2 to +1.3

[a]Sink (+); source (−).
[b]During transient (50–100 year) response. In the long term (200–1000 years), it may revert to sink if climate stabilises.
[c]From land-use changes only.
From Sampson *et al.*, 1993.

These outline budgets are useful as they give an indication of the balance between sources and sinks at the regional scale. They also emphasise the temporal component of vegetation–atmosphere interactions. Prior to the industrial revolution, tropical forests were almost in carbon balance; they had a stable carbon pool and a photosynthetic uptake of carbon that was approximately equal to the loss of carbon owing to respiration, decomposition and river export (Brown *et al.*, 1993). By contrast, the current tropical forests are a source of carbon to the atmosphere because the enhanced photosynthetic uptake of carbon (as a result of an increase in atmospheric CO_2) is more than counterbalanced by the loss of carbon through respiration, decomposition, river export and wood removal (Fearnside, 1992) (Fig. 1). Current evidence suggests that by 1990 the total tropical carbon pool had dropped to 428 Pg carbon as a result of this imbalance in carbon flux (Houghton, Lefkowitz & Skole, 1991a; Houghton, Skole & Lefkowitz, 1991b). The rate of loss of carbon to the atmosphere has accelerated since the 1950s, with contributions being made by the tropical forests in Latin America, Africa and Asia (Houghton *et al.*, 1991a,b; Uhling, Hall & Nyo, 1993). If the current imbalance in carbon fluxes between the forest and atmosphere continues, then by

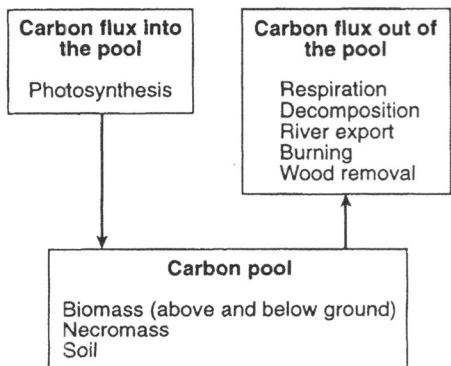

Fig. 1. The flux of carbon into and out of the carbon pool in tropical forests (based on Brown *et al.*, 1993).

2050 the carbon pool will have decreased to between 298 and 334 Pg of carbon (Brown *et al.*, 1993).

These outline budgets make clear what needs to be quantified if the current uncertainty in the estimates of carbon fluxes in the tropics are to be reduced (Fig. 1). However, these outline budgets are aspatial and tend to emphasise the effect of deforestation (burning and wood removal). This is because a spatial component is difficult to incorporate into the budget and because deforestation is an obvious and a readily measurable feature of the tropics (Myers, 1991; Skole & Tucker, 1993; Dale *et al.*, 1994; Turner, Meyer & Skole, 1994). If tropical forests are considered not as a shrinking block of mature forests with burnt edges but as a mosaic of land uses with a changing atmosphere then the importance of *place* becomes clear. From a spatial perspective, the tropical forests at the time of the industrial revolution were a minor source of atmospheric carbon as the conversion of mature forests to shifting cultivation was slightly greater than the conversion of shifting cultivation to mature forests. By the 1950s, tropical forests were being converted to agricultural and subsequently degraded lands at a much faster rate than degraded or agricultural lands were being converted to forests (Whitmore, 1990; Walker & Seffen, 1994). Therefore, the land-use changes that were a carbon source to the atmosphere were greater than the land-use changes that were a sink to the atmosphere (Solomon *et al.*, 1993; Dale, 1994) (Fig. 2). The late 1980s and early 1990s saw a marked reduction in the rate of deforestation (Brown, Lugo & Iverson, 1992); for example, in the Brazilian Amazon the rate of deforestation declined by around 40% between 1988 and 1991 (Dixon *et al.*, 1994).

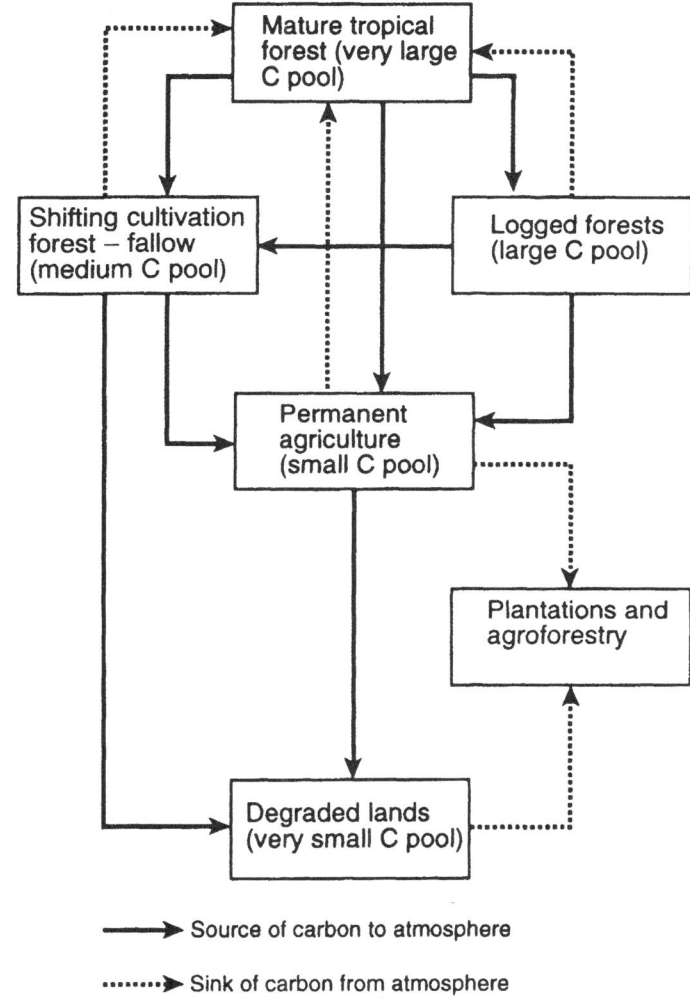

Fig. 2. Major land-use changes in the tropics (based on Brown *et al.*, 1993).

Approximately 217 Mha of the land used for shifting cultivation, permanent agriculture and logged forest is covered with regenerating forest that will, in time, be mature forest, and approximately 134 Mha of land used at some time for agriculture is being used for plantations and agroforestry (Trexler & Haugen, 1993). As a result of these land-use changes, the area of land now acting as a carbon sink is increasing rapidly and it is estimated that regenerating forest alone will account for an increase in carbon sequestration from under 0.1 Pg carbon per year to well over 0.5 Pg carbon per year in the first half of the 21st century (Fig. 3). The early emphasis on the quantification of deforestation and the drawing up of balance sheets is now tempered by a spatial perspective and a drive to quantify changes in all of the land-uses that act as sources of sinks of carbon (Moore & Braswell, 1994; Turner *et al.*, 1994). Interestingly, the workshop that prepared the estimates of global carbon flux in anticipation of the next IPCC assessment reported with guarded astonishment on the significance of land-use change (Wisniewski & Sampson, 1993). They concluded that: 'The most consistent and surprising conclusion to emerge from the deliberations . . . is that land use, not climate change or atmospheric chemistry (e.g. fertilization by CO_2 or N), has been and will continue to be the most import-

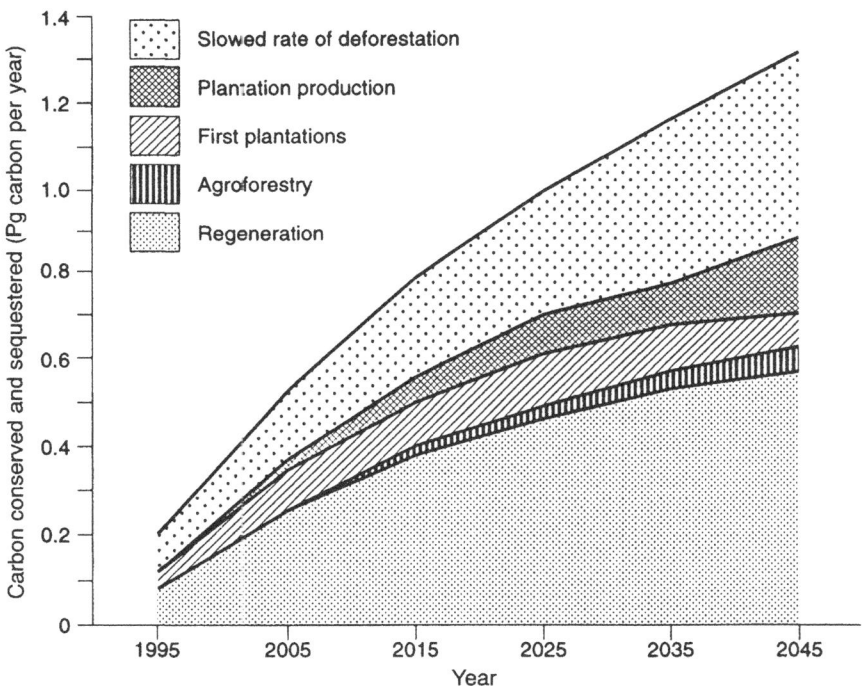

Fig. 3. The potential for conserving and sequestering carbon at low latitudes (based on Trexler & Haugen, 1993).

ant determinant of C storage, uptake and release in all terrestrial ecosystems. The conclusion is a surprise because the participants were focused on the roles of the atmosphere on terrestrial carbon cycling, not on land use' (Sampson *et al.*, 1993, p. 5).

There is a need to obtain information on land-use and land-use change as this is crucial to the understanding of carbon fluxes in tropical forests. 'The most effective way to reduce a large amount of uncertainty in C flux and pool estimates in the tropics is to improve the mapping of area and area-change of land cover/land use using remote sensing technology, including the identification of all classes of disturbed forests (highly disturbed to mature)' (Brown *et al.*, 1993, p. 89).

Quantifying the carbon sink in tropical forests

Many of the areas under the different land-uses outlined in Fig. 2, if allowed, may be restored to mature forest through natural regeneration. In doing so they would increase the size of the carbon pool by sequestering carbon from the atmosphere (Houghton *et al.*, 1992). This sequestration should be greater in the earlier stages than the later stages of forest regeneration owing to rapid growth initially (Brown & Lugo, 1990; Whitmore, 1990). If carbon fluxes were estimated for given stages of regeneration then this could be scaled-up to the region using an accounting model that incorporated information on the spatial extent of each regeneration stage. This is the task before a number of researchers drawn from the Universities of Edinburgh, Wales, Southampton and Salford, the British National Space Centre and the Institute of Terrestrial Ecology as part of the Natural Environment Research Council's (NERC) Terrestrial Initiative in Global Environmental Research (TIGER) programme (Briggs, 1994).

Research is based on Brazil and Cameroon and the methodology being used (Brown *et al.*, 1994) comprises three inter-linked components: first the estimation of carbon fluxes for points using field measurements and a mathematical model (next section); second, maps of regeneration stage derived from remotely sensed data and, third, the production of carbon flux maps using a geographical information system.

Research is focused currently on the first two of these components and these will be outlined here.

Estimation of carbon fluxes

The fluxes of carbon and water may be measured directly by eddy co-variance, using sensors mounted on a tower, exposed 5–15 m above the canopy (Jarvis *et al.*, 1993; Grace *et al.*, 1996). This was done for three vegetation types in Brazil (1992–3) and two in Cameroon (1994). Measurements were made during the wet and dry seasons, in campaigns lasting several weeks.

Carbon dioxide is exchanged between the atmosphere and vegetation by photosynthesis and respiration. The overall carbon balance is obtained by summing the gains and losses over a period of weeks. These measurements must be scaled-up in time to enable statements to be made about fluxes over longer periods. This was achieved using a mathematical model driven by longer-term climatological data (Lloyd *et al.*, 1995). The model incorporates four submodels all of which are specific to the vegetation type: (i) a biochemical model that calculates the uptake of CO_2 inside the leaf by the enzyme ribulose bisphosphate carboxylase oxygenase; (ii) a stomatal model that enables the stomata to respond to light and humidity, thus regulating the supply of CO_2 to the biochemical model; (iii) a respiration model of carbon efflux based on field data of soil and plant respiration; and (iv) a micrometeorological model that calculates the effect of meteorological stability on the supply of CO_2 to the canopy.

When the model was run as a one-year simulation (1992–3), it showed that the undisturbed ecosystem (i.e. mature forest) accumulated carbon and so was a net sink (Grace *et al.*, 1995). The accumulation rate was, however, very sensitive to temperature and the year 1992–3 was perhaps an especially strong accumulation year as a result of lower-than-normal temperatures (which may well have been the result of the Mount Pinatubo eruption). The realisation that inter-annual variation in the weather is significant suggested the need to scale-up for longer periods, perhaps using historical climatological data sets.

Interestingly the carbon flux into the regenerating forest has been less than would be expected from the literature (Brown *et al.*, 1993). This was because the published estimates were derived from the mensuration of forest stands and so take into account only that part of the carbon balance derived from the above-ground biomass (Brown & Lugo, 1990). They neglect the fluxes of carbon from the soil, which come from the decomposition of below-ground carbon, left over from the previous vegetation. In fact, preliminary results from Cameroon show that the early stages of regeneration may have been a net source of carbon, even though the above-ground biomass was accumulating. In the future, more effort will be concentrated on regrowth sites in an attempt to understand the CO_2 fluxes in relation to the carbon content of the soil.

Mapping regeneration stage

The only appropriate operational sensor for the mapping of regeneration stage over large areas of terrain is AVHRR carried on the National Oceanic and Atmospheric Administration (NOAA) satellites. These data have a coarse spatial resolution, 1.1 km at their finest, and are available daily at a low cost. The high temporal frequency of data collection ensures a relatively large probability of acquiring cloud-free imagery (Roller & Colwell, 1986). Several local- to national-scale studies have both demonstrated the value of AVHRR data for the study of tropical forests and discussed the potential of these data for the monitoring of deforestation at regional to global scales. Typically, AVHRR imagery has been used on its own (Malingreau, Tucker & Laporte, 1989; Nelson & Horning, 1993), in a temporal sequence (Achard & Blasco, 1990; Malingreau, 1991) or in unison with finer spatial resolution imagery (Malingreau, Verstraete & Archard, 1992). The NOAA AVHRR sensor was, however, designed for meteorological applications (Hastings & Emery, 1992) and consequently its use in the study of tropical forests poses many problems (Millington *et al.*, 1989; Goward *et al.*, 1991), especially in relation to its coarse spatial resolution, lack of accurate calibration and use of optical wavelengths (Ehrlich, Estes & Singh, 1994). However, tropical forests containing areas of regeneration are distinguishable within AVHRR imagery (Stone *et al.*, 1994) (Fig. 4).

The NOAA AVHRR imagery can be used to estimate regeneration stage over large areas and these stages can be aged at selected test sites. In the absence of accurate ground data, this aging was provided by a time sequence of Landsat TM imagery (Lucas *et al.*,1993). This methodology and the relationships involved are discussed in Curran *et al.* (1994). In this chapter, attention will be drawn to two issues that are vital to this scaling-up that have received relatively little attention. These issues are, first, the indirect relationship between AVHRR-recorded radiation and regeneration stage and, second, the spectral separability of regeneration stages on the Landsat TM imagery used to age the regeneration stages on AVHRR imagery.

The indirect relationship between AVHRR radiance and regeneration stage

The AVHRR records radiation in up to five channels of which three are of relevance here. These are channel 1 (red: 0.58–0.68 μm), channel 2 (near-infrared: 0.72–1.10 μm) and channel 3 (termed thermal infrared: 3.55–3.93 μm). Remotely sensed radiation in these three channels is affected to varying degrees by four inter-related variables that all

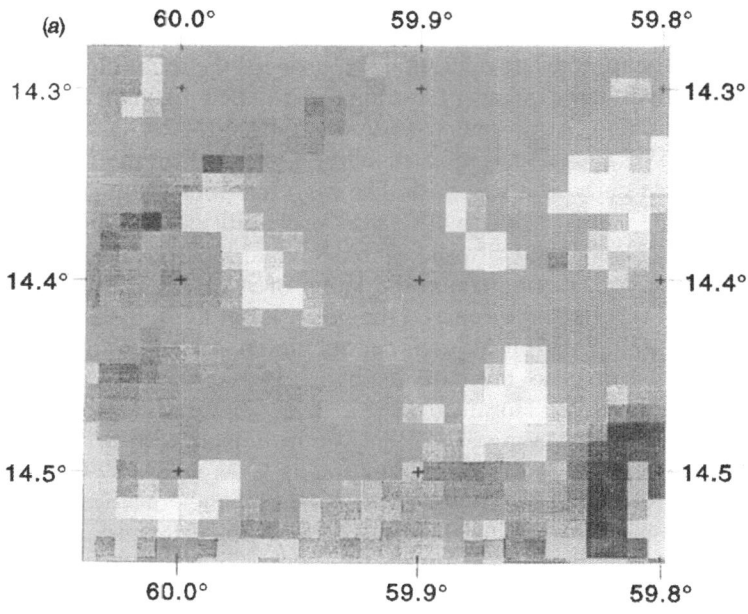

Table 2. *The effect of four canopy variables on the remotely sensed radiation in three wavebands. For example, scattering within leaves increases the amount of reflected near-infrared radiation but has no effect on the amount of reflected red or reflected and emitted thermal infrared radiation*

Waveband	Absorption by chlorophyll	Scattering within leaves	Canopy shadow	Canopy transpiration
Red	Decrease	–	Decrease	–
Near-infrared	–	Increase	Decrease	–
Thermal infrared	–	–	Decrease	Decrease

increase (along with leaf area index and biomass) as the forest regrows (McWilliam *et al.*, 1993). These inter-related variables are absorption by chlorophyll, scattering within leaves, canopy shadow and canopy transpiration (Table 2). However, the effect of shadow is unlikely to be a major determinant of reflectance until a multilayered upper forest canopy starts to form (Curran & Foody, 1994a).

The resultant trends are of a decrease in channel 1 and 3 radiance with age of regeneration and an increase followed by a decrease in channel 2 radiance with age of regeneration. AVHRR imagery of the Capim and Xingu river basins in Para, Brazil and a region north of Manaus in Amazonas, Brazil were used along with information on regeneration stage (Lucas *et al.*, 1993) to quantify these relationships, albeit in relative digital number (DN) (Fig. 5).

To investigate these relationships, a further two sites were selected within the Guapore and Arinos river basins in Northern Mato Grosso, Brazil. The age of regeneration of 138 homogeneous forest blocks was determined using a temporal sequence of Landsat Multispectral Scanning System (MSS) and TM imagery and the procedures outlined in Lucas *et al.* (1993). The Landsat MSS and TM images were co-registered with the AVHRR imagery and digital numbers extracted for each block. The digital numbers in AVHRR channels 1 and 2 were

---

Fig. 4. Areas of regenerating forest on geocorrected imagery of the Guapore area of Northern Mato Grosso, Brazil. The NOAA AVHRR image (*a*) was recorded on 15 May 1992 and the Landsat TM image (*b*) was recorded on 12 July 1992. This figure illustrates that areas of regenerating forest are visible on AVHRR imagery.

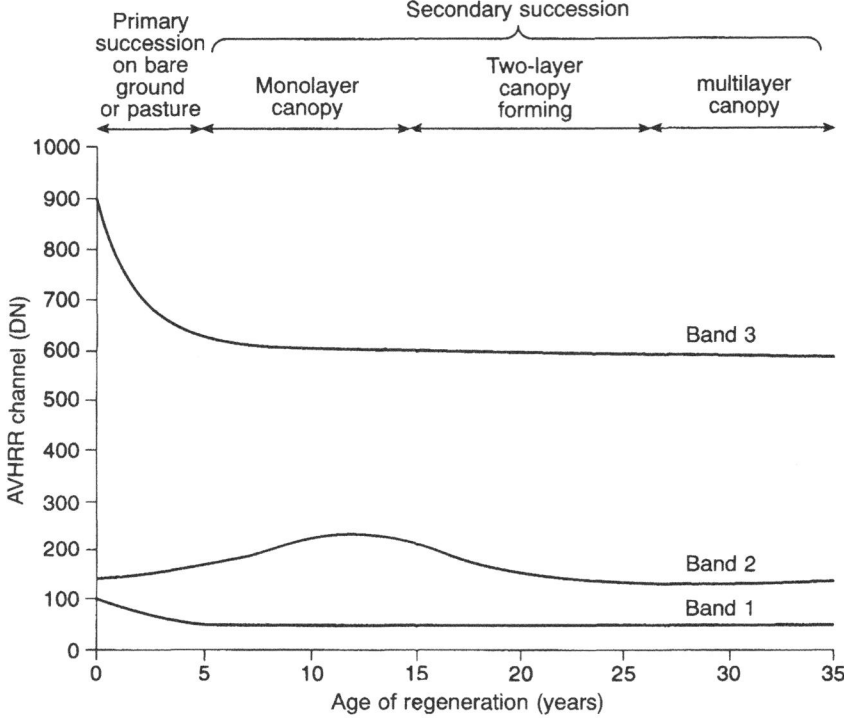

Fig. 5. Hypothesised relationships between relative radiance recorded by AVHRR and age of regeneration (Table 2) with scales derived from AVHRR imagery and ground data from the Capim, Xingu and Manaus areas of Brazil; DN, digital number.

converted to top of atmosphere reflectance (%) and channel 3 to top of atmosphere radiance (mW m^{-2} sr^{-1} cm^{-2}) (D'Souza, 1994). The results in Fig. 6 are in support of the theoretical relationships in Fig. 5. As regeneration ages, so the relative radiance in channels 1 and 3 decreases

Fig. 6. The relationships between age of forest regeneration and relative radiance in AVHRR channels 1–3 for the Guapore and Arinos areas in Northern Mato Grosso, Brazil. (*a*) Channel 3 (thermal infrared); (*b*) Channel 2 (near-infrared); (*c*) Channel 1 (red); (*d*) the NDVI. (*e*) The sample numbers associated with each of the ten regeneration ages.

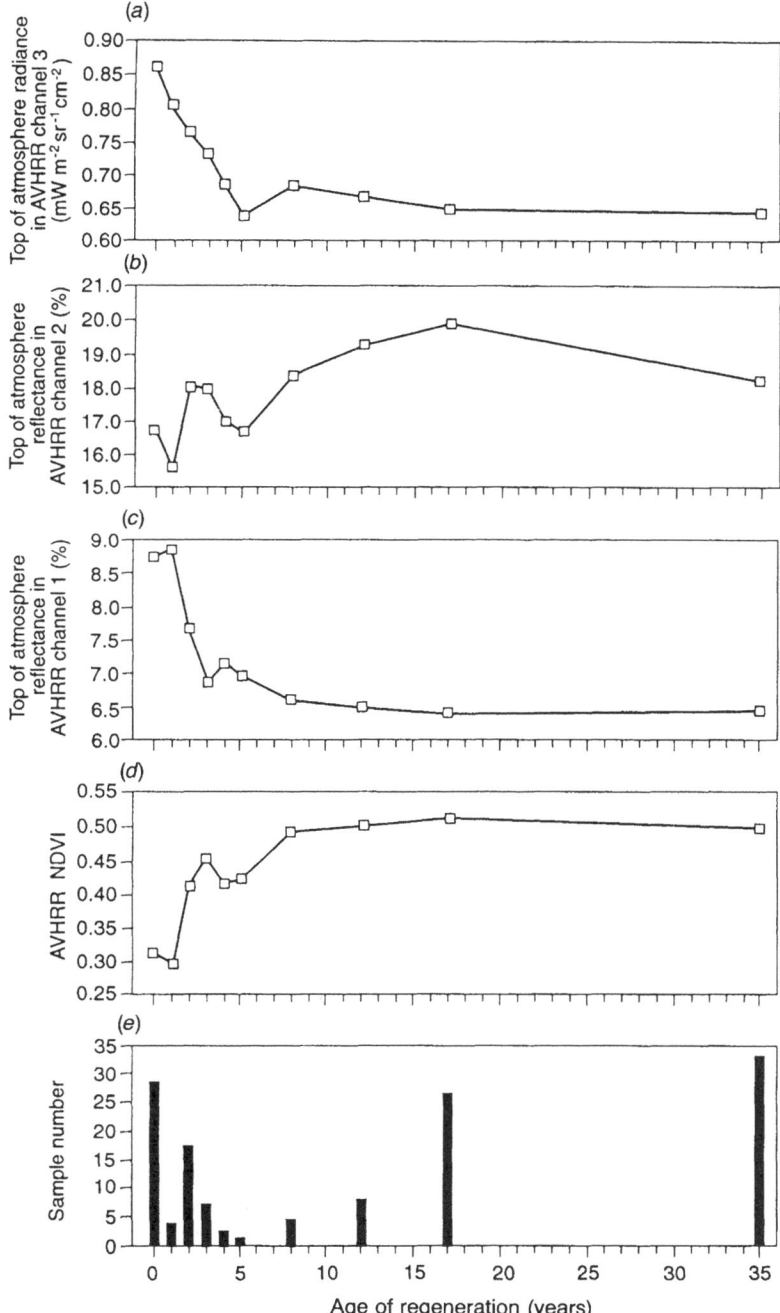

while that in channel 2 increases falteringly and then starts to decrease. Interestingly, the normalised difference vegetation index (NDVI = (near-infrared − red)/(near-infrared + red)) calculated using channels 1 and 2 was relatively insensitive to the age of regeneration after 8 years of growth. Without mixture modelling, correction for the location of plots in the swath and standardisation of the sample number (and thereby error) by age of regeneration, the results presented in Fig. 6 are indicative rather than conclusive. What they do indicate, however, is that more information on regeneration age is contained within channels 1 and 3 than the currently used NDVI (Boyd *et al.*, 1996).

Spectral separability of regeneration stages on Landsat TM imagery

The maps of regeneration stage generated using AVHRR imagery are dependent for their aging on time sequences of Landsat TM imagery. For this to be viable, the six major regeneration stages must be separable on any given Landsat TM image with a reasonably high (over 80%) level of accuracy (Foody *et al.*, 1996).

For the classification discussed here, Landsat TM data acquired in August 1991 for an area north of Manaus in the Brazilian Amazon (2°20′ S, 60°00′ W) were used. Since Landsat TM data are generally three-dimensional in character, with the dimensions relating to reflectance in visible, near- and middle-infrared wavelengths (Townshend *et al.*, 1988) only, three such wavebands were used. Large areas of primary forest were cleared for cattle pastures and plantations in the late 1970s and 1980s, and in 1991 many of these clearances had been abandoned allowing the forest to regenerate. These regenerating forests were identified and aged using a time-series of Landsat MSS and TM data from 1976 and from these data a map depicting forest regenerative age classes was produced (Lucas *et al.*, 1993). This map was refined and verified with the aid of fieldwork performed in July and August 1993 and used as the ground data for this investigation. In addition to a general extensive survey of land cover at the site, the field investigation also involved the collection of data on forest tree species, which allowed the differentiation of two main classes of forest within each regenerative age class and each of these was following a different successional pathway. From detailed inventories of all tress with a diameter of > 3 cm in 15 plots of 100 × 10 m, it was apparent that these two successional pathways differed markedly in species composition (Lucas *et al.*, 1996). One type of forest was dominated by species of the genus *Cecropia* (family: Cecropiaceae) and occurred mainly on pasture that had been abandoned shortly after forest clearance (herein called forest type C).

The other type of forest was dominated by species of the genera *Vismia* (family: Clusiaceae), *Miconia* and *Bellucia* (family: Melastomataceae) and was found typically on sites that had, prior to abandonment, been cleared initially by fire and used for pasture for several years (herein called forest type C/M).

The data were classified with maximum likelihood classifiers. Two approaches were investigated. First, a conventional per-pixel classification of the data and, second, an object-oriented approach to classification was used whereby the image pixels were grouped into objects, typically cross shaped or square blocks of pixels, and each pixel of an object or the whole object was classified (Palubinskas, 1988). Here attention focused on one approach, which is based on the assumption that an image object is a Markov random field (Kalayeh & Landgrebe, 1987) that is represented by a third-order causal autoregressive model. The object was defined as a 3×3 pixel square box. This approach to classification has been found in some investigations to classify data more accurately than the conventional per-pixel maximum likelihood classifier (Mardia, 1984; Palubinskas, 1988).

The performance of the classifications was evaluated by their accuracy. In the absence of a universally acceptable measure of classification accuracy (Congalton, 1991), a number of indices are provided. These are the percentage correct allocation and the kappa coefficient of agreement (Cohen, 1960; Rosenfield & Fitzpatrick-Lins, 1986). Since most of the classes lay along an ordered scale, from recently deforested through regenerating forests to mature forest, these measures may not provide an ideal index of classification performance. For ordinal-level classes, the distribution of error between classes is important, as, for instance, the misallocation of a recently cleared area to the mature forest class is more erroneous than the misallocation of a young regenerating forest to an older regenerating forest class. For such a classification, the use of the percentage correct allocation or kappa coefficient is inappropriate as these indices were derived for application to nominal-level classifications in which all classification errors are of equal magnitude (Jolayemi, 1990; Foody, 1992). For the assessment of the accuracy of an ordinal-level classification, the index of accuracy used should account for the variations in the degree of error that may be associated with a set of class allocations reached by the classification (Jolayemi, 1990). One approach to compensate for the effects of variable degrees of classification error is to use the weighted kappa coefficient (Cohen, 1968). This index of accuracy is an extension of the kappa coefficient in which a pre-defined error weighting is associated with all possible class allocations (i.e. an error weight is associated with each element of the classification confusion matrix). Although determining the weights may be difficult, it does enable the variations in the degree of classification error that may be found in an ordinal-level classification to be accounted for in the assessment of classification accuracy. Furthermore, the weightings may be defined in a fashion that enables the assessment of the accuracy of a classification in which only some of the classes lie on an ordered scale.

To indicate the quality of the classifications for individual classes, the percentage correct allocation for each class from the user's and producer's perspectives (Story & Congalton, 1986) are also provided. These indices of classification accuracy will be supplemented by the classification confusion matrices to allow more detailed study on classification accuracy.

Initial attention focused on the classification of the six classes that were identified from the time series of Landsat sensor data. These classes lay along a forest age and carbon continuum, from pasture (recently deforested/potential forest) to mature forest. Both the per-pixel and object-orientated classifications had a reasonable degree of class separability, with, as expected, a higher accuracy (79.5%) derived from the object-based classification. In both classifications the end-points of the continuum, the pasture and mature forest classes, were classified accurately (an accuracy of > 97%) with much of the confusion limited to between the regenerating forest classes. A large proportion of the error that was observed arose from misallocations between neighbouring classes. Consequently, the weighted kappa coefficient would be expected to provide a more realistic measure of classification accuracy than the percentage correct allocation or the kappa coefficient. The weighted kappa coefficient was calculated for both classifications with weights for the errors determined such that an error weight of 1 was associated with a misallocation to a 'neighbouring' class, a weight of 2 associated with a misallocation to the class adjacent to a neighbouring class and so on, with the highest weight 5 associated with misallocation between the forest and pasture classes. With these weightings, the weighted kappa coefficient was calculated as 0.80 and 0.86 for the per-pixel and object-based classification, respectively.

The confusion matrices, however, revealed that much of the error that was *not* between neighbouring classes was associated with sites that belonged to the youngest forest regenerative age class. This indicated that the youngest age class may be more variable in its composition and spectral response than the older forest classes. This could be a function of a range of successional pathways being followed. Where more than one successional pathway is present, the greatest variability between classes will be apparent in the younger classes rather than in the older classes, as each pathway ends ultimately at a mature forest. At this site, two distinct regenerative pathways were present: one for forest type C and one for forest type C/M. Sites following these different regenerative pathways differed markedly in species composition in the early years of regeneration, but became less distinct with time. Furthermore, the different species associated with these successional pathways may differ in their rate of carbon accumulation and hence biomass (Scatena *et al.*, 1993) and so an ability to discriminate them would not only be beneficial in terms of increasing classification accuracy but also in refining the carbon accounting model.

On the basis of field identification, the regenerating forest classes were split into two groups depending on the successional pathway followed. Three age classes of forest type C and two of forest type C/M were defined in this way (Fig. 7). Together with sites of six other classes also identified in the field the classifications were repeated; the classes identified in Table 3. The results showed a high degree of inter-class separability with accuracies of 78.6% and 86.7% derived from the per-pixel and object-based maximum likelihood classifications (Tables 3 and 4). These accuracies were higher than those derived from the six class classifications, indicating that subdividing the regenerating forest classes into the two successional pathways raised overall inter-class separability.

Whilst the eleven classes did not lie along a simple continuum as in the previous classifications, some classes could be considered related in the sense that misallocation of cases between them was less erroneous than misallocation of other classes. Therefore, errors observed between the forest regenerative classes or the pasture classes may be judged less severe than other misallocations. It was apparent from Tables 3 and 4 that most of the classification error observed was between these related classes. To more appropriately assess the accuracy of the classifications, the weighted kappa coefficient was calculated using weights that attempted to reflect the variations in error severity. For the per-pixel and object-based classifications, the weighted kappa coefficient was calculated as 0.87 and 0.93, respectively. These results indicate that the successional pathways may be identified to a high accuracy. Moreover, the dynamics of the forest succession were to some extent manifest in the remotely sensed data as the pathways from forest to pasture and later regeneration, along one of two routes, were evident in feature space (Fig. 7).

The results indicate that remote sensing may be used to classify accurately regenerating tropical forest classes and, where appropriate,

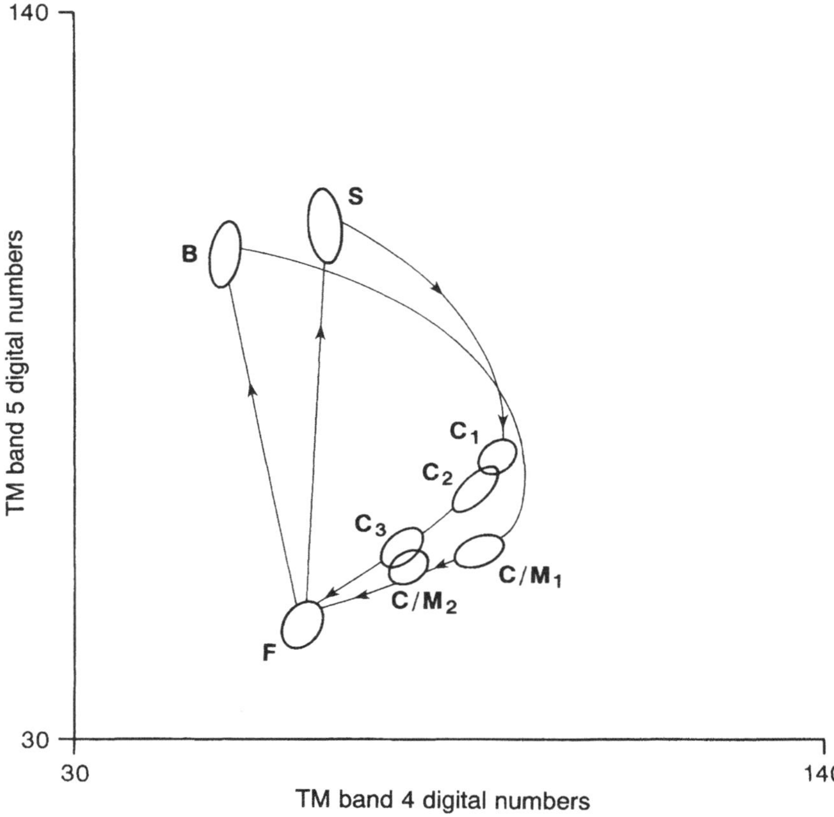

Fig. 7. Location of the forest regenerative classes in a middle infrared (TM band 5) − near-infrared (TM band 4) feature space plot. Note that the two successional pathways that may be followed as the forest regenerates appear spectrally separable; the ellipses plotted represent one standard deviation from the mean value. This feature space plot indicates a potential to observe the dynamics of land cover change from forest to pasture and the succession, along one of two routes, back to forest. F, closed forest, B, remains of burned forest; S, soil and pasture; C_1–C_3, all Cecropiaceae dominant, 4, 7 and 14 years old, respectively; C/M_1 and C/M_2, Clusiaceae/Melastomataceae dominant, 2 and 5 years old, respectively.

Table 3. *Confusion matrix from the per-pixel maximum likelihood classification. The overall percentage correct allocation was 78.6%, the kappa coefficient 0.75 and the weighted kappa coefficient was 0.87*

Actual class[a]	Predicted class[a]											Σ	Producer's accuracy (%)
	B	S	RV	H	CM_1	CM_2	C_1	C_2	C_3	F	Pl		
B	1	161	0	0	0	0	0	0	0	0	4	166	0.6
S	0	75	0	0	0	0	0	0	0	0	0	75	100.0
RV	0	0	25	0	0	0	0	0	0	0	0	25	100.0
H	2	3	0	170	7	0	7	0	0	0	11	193	88.08
CM_1	0	0	0	0	7	9	0	0	0	0	0	16	43.75
CM_2	0	0	5	16	8	166	0	3	9	0	1	192	86.46
C_1	0	0	0	0	0	0	50	2	0	0	4	72	69.44
C_2	0	0	0	0	2	2	77	68	1	0	7	157	43.31
C_3	0	0	1	0	1	2	0	3	68	0	0	75	90.66
F	0	0	1	0	0	0	0	0	0	552	0	553	99.82
Pl	0	0	0	0	0	1	8	9	0	0	141	159	88.68
Σ	3	239	32	186	18	180	142	85	78	552	168	1683	
User's accuracy	33.33	31.38	78.13	91.39	38.88	92.22	35.21	80.0	87.18	100.0	83.93		

[a]Classes: B, burnt pasture; S, soil and pasture; RV, riverine vegetation; H, fallow (herbaceous vegetation); forest type C/M divided into CM_1 (2 year old) and CM_2 (5 year old); forest type C divided into C_1 (4 year old) and C_2 (7 years old) and C_3 (7 years old); F, closed forest, Pl, plantation; Σ, 'indicates summation'.

Table 4. Confusion matrix from the object-based maximum likelihood classification. The overall percentage correct allocation was 86.7%, the kappa coefficient was 0.83 and the weighted kappa coefficient was 0.93

Actual class[a]	Predicted class[a]											Σ	Producer's accuracy (%)
	B	S	RV	H	CM$_1$	CM$_2$	C$_1$	C$_2$	C$_3$	F	Pl		
B	9	81	0	0	0	0	0	0	0	0	0	90	10.0
S	0	25	0	0	0	0	0	0	0	0	0	25	100.0
RV	0	0	9	0	0	0	0	0	0	0	0	9	100.0
H	0	0	0	81	0	0	0	0	0	0	18	99	81.82
CM$_1$	0	0	0	0	1	3	0	0	0	0	0	4	25.0
CM$_2$	0	0	0	0	0	124	0	0	0	0	0	124	100.0
C$_1$	0	0	0	0	0	0	42	0	0	0	0	42	100.0
C$_2$	0	0	0	0	0	0	29	46	0	0	0	75	61.33
C$_3$	0	0	0	0	0	0	0	0	27	0	0	27	100.0
F	0	0	0	0	0	0	0	0	0	397	0	397	100.0
Pl	0	0	0	0	0	0	0	0	0	0	89	89	100.0
Σ	9	106	9	81	1	127	71	46	27	397	107	981	
User's accuracy	100.0	23.58	100.0	100.0	100.0	97.64	59.15	100.0	100.0	100.0	83.18		

[a]Class abbreviations are the same as in Table 3.

identify successional pathways. These results point to the potential of using a classification of forest regeneration stages that have been derived from AVHRR imagery and aged from Landsat TM imagery to scale-up point measurements of the carbon flux associated with forest classes over the whole of the Legal Amazon (Foody *et al.*, 1996).

Discussion and conclusion

Early work in remote sensing concentrated on the local scale and provided information that could also be collected by traditional means, with the emphasis being on convenience, accuracy and cost. Today remote sensing activities are moving from the scale of the field and forest to the landscape and the continent where they can provide direct estimates of environmental phenomena (Curran & Foody, 1994b). More important, remotely sensed data can be used to infer the unmeasurable by extrapolation of point data to areas and the parameterisation and driving of environmental models (Table 5). The research discussed here is illustrative of the third of these four methodologies. Although the research is still in progress, it does illustrate the complexity of scaling-up from point measurements (samples) to areal measurements (population).

In particular this chapter emphasises two practical issues. First, that

Table 5. *The four methods for using remotely sensed data in environmental research at regional to global scales*

Research method	Example
Direct estimation of a physical variable independently of other data	Remotely sensed estimates of sea surface temperature
Inference from the measurement of a related variable	Estimates of methane flux using maps of wetland derived from remotely sensed data
Extrapolation of local-scale measurements to estimates for large areas	Point measurements of evapotranspiration extrapolated to a continent via the correlation between evapotranspiration and remotely sensed estimates of biomass
Modelling environmental processes	Use of remotely sensed data to parameterise and drive environmental models

Modified from Briggs (1991).

of linking of biophysical variables and remotely sensed data at the scale of the plot and, second, that of using fine spatial resolution imagery to provide the accurate classes of land cover needed to classify coarse spatial resolution imagery. This emphasis is set against a wider and all compelling science goal: 'Carbon losses occur in areas of increased human activities but carbon gains also occur on previously disturbed and now abandoned lands. The magnitude of these gains and losses are likely to be different across different geographical regions and what their balance is in time and space at this scale is largely unknown' (Brown *et al.*, 1993, p. 76). The task now is one of using point measurements, remotely sensed data and the techniques of scaling-up to achieve this quantification.

Acknowledgements

The authors acknowledge the financial support provided by NERC through its TIGER programme, award GST/02/604. We are also grateful to the INPE (Brazil) for the provision of some of the data used and the image processing skills of Gintautas Palubinskas of the Department of Data Analysis, Institute of Mathematics and Information, Vilnus, Lithuania, Doreen Boyd of the University of Southampton and Silvana Amaral of INPE Brazil.

References

Achard, F. & Blasco, F. (1990). Analysis of vegetation seasonal evolution and mapping of forest cover in West Africa with the use of NOAA AVHRR HRPT data. *Photogrammetric Engineering and Remote Sensing*, 56, 1359–1365.

Boyd, D.S., Foody, G.M., Curran, P.J., Lucas, R.M. & Honzak, M. (1996). An assessment of Landsat TM middle and thermal infrared wavebands for the detection of tropical forest regeneration. *International Journal of Remote Sensing*, 17, 249–261.

Briggs, S.A. (1991). *Note on the Relevance of Future Earth Observation System Options to Terrestrial Science*. Abbots Ripton, UK: Natural Environment Research Council/British National Space Centre, Remote Sensing Applications Development Unit.

Briggs, S.A. (1994). Remote sensing and terrestrial global environment research – the TIGER programme. In *Environmental Remote Sensing at Regional to Global Scales*, ed. G.M. Foody & P.J. Curran, pp. 8–15. Chichester: Wiley.

Brown, S. & Lugo, A.E. (1990). Tropical secondary forests. *Journal of Tropical Ecology*, 6, 1–32.

Brown, S., Lugo, A.E. & Iverson, L.R. (1992). Processes and lands for sequestering carbon in the tropical forest landscape. *Water, Air and Soil Pollution*, 64, 139–155.

Brown, S., Hall, C.A.S., Knabe, W., Raich, J., Trexter, M.C. & Woomer, P. (1993). Tropical forests: their past, present and potential future role in the terrestrial carbon budget. *Water, Air and Soil Pollution*, 70, 71–94.

Brown, S., Iverson, L.R. & Lugo, A.E. (1994). Land use and biomass of forests in Peninsular Malaysia from 1982 to 1992: a GIS approach. In *Effects of Land Use Change on Atmospheric CO_2 Concentrations*, ed. V. Dale, pp. 117–143. New York: Springer-Verlag.

Cohen, J. (1960). A coefficient of agreement for nominal scales. *Educational and Psychological Measurement*, 20, 37–46.

Cohen, J. (1968). Weighted kappa. *Psychological Bulletin*, 70, 213–220.

Congalton, R.G. (1991). A review of assessing the accuracy of classifications of remotely sensed data. *Remote Sensing of Environment*, 37, 35–46.

Curran, P.J. & Foody, G.M. (1994a). The use of remote sensing to characterise the regenerative states of tropical forests. In *Environmental Remote Sensing from Regional to Global Scales*, ed. G.M. Foody & P.J. Curran, pp. 44–83. Chichester: Wiley.

Curran, P.J. & Foody, G.M. (1994b). Environmental issues at regional to global scales. In *Environmental Remote Sensing from Regional to Global Scales*, ed. G.M. Foody & P.J. Curran, pp. 1–7. Chichester: Wiley.

Curran, P.J., Foody, G.M., Lucas, R.M. & Honzák, M. (1994). A methodology for remotely-sensing the stage of regeneration in tropical forests. In *TERRA-2: Understanding the Terrestrial Environment: Data Systems and Networks*, ed. P.M. Mather, pp. 189–202. Chichester: Wiley.

Dale, V.H. (1994). Terrestrial CO_2 flux: the challenge of interdisciplinary research. In *Effects of Land Use Change on Atmospheric CO_2 Concentrations*, ed. V.H. Dale, pp. 1–14. New York: Springer-Verlag.

Dale, V.H., Brown, S., Flint, E.P. *et al.* (1994). Estimating CO_2 flux from tropical forests. In *Effects of Land Use Change on Atmospheric CO_2 Concentrations*, ed. V.H. Dale, pp. 365–378. New York: Springer-Verlag.

DeFries, R.S. & Townshend, J.R.G. (1994). Global land cover: comparison of ground based data sets to classifications with AVHRR data. In *Environmental Remote Sensing from Regional to Global Scales*, ed. G.M. Foody & P.J. Curran, pp. 84–110. Chichester: Wiley.

Department of the Environment (1994). *Climate Change: The UK Programme*. CM 2427, London: HMSO.

Dixon, R.K., Brown, S., Houghton, R.A., Solomon, A.M., Trexter,

M.C. & Wisniewski, J. (1994). Carbon pools and flux of global forest ecosystems. *Science*, 263, 185–190.

D'Souza, G. (1994). *Geometric and Radiometric Correction of NOAA–LAC HRPT and SHARP Format Data for the TREES Project. Final Report No. 5333-93-06.* Ispra, Italy: Joint Research Centre of the Commission of the European Communities.

Edmonds, J. (1992). Why understanding the natural sinks and sources of CO_2 is important: a policy analysis perspective. *Water, Air and Soil Pollution*, 64, 11–21.

Ehrlich, D., Estes, J.E. & Singh, A. (1994). Applications of NOAA–AVHRR 1 km data for environmental monitoring. *International Journal of Remote Sensing*, 15, 145–161.

Fearnside, P.M. (1992). *Greenhouse Gas Emissions from Deforestation in the Brazilian Amazon.* Washington, DC: Climate Change Division, Environmental Protection Agency.

Foody, G.M. (1992). Classification accuracy: some alternatives to the kappa coefficient for nominal and ordinal level classifications. In *Remote Sensing from Research to Operation*, pp. 529–538. Nottingham: Remote Sensing Society.

Foody, G.M., Palubinskas, G., Lucas, R.M., Curran, P.J. & Honzák, M. (1996). Identifying terrestrial carbon sinks: classification of successional stages in regenerating tropical forest from Landsat TM data. *Remote Sensing of Environment*, 55, 205–216.

Goward, S.N., Markham, B., Dye, D.G., Dulaney, W. & Yang, J. (1991). Normalized difference vegetation index measurements from the Advanced Very High Resolution Radiometer. *Remote Sensing of Environment*, 35, 257–277.

Grace, J., Lloyd, J., McIntyre, J. *et al.* (1995). Net carbon dioxide uptake by an undisturbed tropical rain forest in south-west Amazonia during 1992–1993. *Science*, 270, 778–780.

Grace, J., Lloyd, J., McIntyre, J. *et al.* (1996). Fluxes of carbon dioxide and water vapour over an undisturbed tropical rainforest in south-west Amazonia. *Global Change Biology*, 1, 1–12.

Hastings, D.A. & Emery, W.J. (1992). The Advanced Very High Resolution Radiometer (AVHRR): a brief reference guide. *Photogrammetric Engineering and Remote Sensing*, 58, 1183–1188.

Houghton, J.T., Callander, B.A. & Varney, S.K. (eds.) (1992). *Climate Change 1992 – The Supplementary Report to the IPCC Scientific Assessment.* Intergovernmental Panel on Climate Change. Cambridge: Cambridge University Press.

Houghton, J.T., Jenkins, G.J. & Ephramus, J.J. (eds.) (1990). *Climate Change: The IPCC Scientific Assessment.* Cambridge: Cambridge University Press.

Houghton, R.A., Lefkowitz, D.S. & Skole, D.L. (1991a). Changes in the landscape of Latin America between 1850 and 1985, I.

Progressive loss of forests. *Forest Ecology and Management*, 38, 143–172.

Houghton, R.A., Lefkowitz, D.S. & Skole, D.L. (1991b). Changes in the landscape of Latin America between 1850 and 1985, II. Net release of CO_2 to the atmosphere. *Forest Ecology and Management*, 38, 173–199.

Jarvis, P.G., Moncrieff, J.M., Grace, J., McCracken, P. & Massheder, J. (1993). Carbon dioxide fluxes from vegetation: from boreal to the tropical regions. *Journal of Experimental Botany*, 44 (May Supplement), 14 (Abs).

Jolayemi, E.T. (1990). On the measurement of agreement between two raters. *Biometrical Journal*, 32, 87–93.

Jones, M.D.H. & Henderson-Sellers, A. (1990). History of the greenhouse effects. *Progress in Physical Geography*, 14, 1–18.

Kalayeh, H.M. & Landgrebe, D.A. (1987). Stochastic model utilizing spectral and spatial characteristics. *IEEE Transactions on Pattern Analysis and Machine Intelligence*, 9, 457–461.

Lloyd, J., Grace, J., Miranda, A.C. *et al.* (1995). A simple and calibrated model of Amazon rainforest productivity based on leaf biochemical properties. *Plant, Cell and Environment*, 18, 1129–1145.

Lucas, R.M., Honzák, M., Foody, G.M., Curran, P.J. & Corves, C. (1993). Characterising tropical secondary forests using multitemporal Landsat sensor imagery. *International Journal of Remote Sensing*, 14, 3061–3067.

Lucas, R.M., Curran, P.J., Honzák, M., Foody, G.M., do Amaral, I. & Amaral, S. (1996). Disturbance and recovery of tropical forests: balancing the carbon account. In *Amazonian Deforestation and Climate*, ed. J.H.C. Gash, C.A. Nobre, J.M. Roberts & R.L. Victoria, pp. 383–398. Chichester: Wiley.

Lugo, A.E. (1992). The search for carbon sinks in the tropics. *Water, Air and Soil Pollution*, 64, 3–9.

Lugo, A.E. & Wisniewski, J. (1992). Natural sinks of CO_2: conclusions, key findings and research recommendations from the Palmas del Mar workshop. *Water, Air and Soil Pollution*, 64, 455–459.

Malingreau, J.P. (1991). Remote sensing for tropical forest monitoring: an overview. In *Remote Sensing and Geographical Information Systems for Resource Management in Developing Countries*, ed. A.S. Belward & C.R. Valenzuela, pp. 253–278. Dordrecht: Kluwer Academic.

Malingreau, J.P., Tucker, C.J. & Laporte, N. (1989). AVHRR for monitoring global tropical deforestation. *International Journal of Remote Sensing*, 10, 855–867.

Malingreau, J.P., Verstraete, M.M. & Archard, F. (1992). Monitoring tropical forest deforestation: a challenge for remote sensing. In *TERRA-1: Understanding the Terrestrial Environment – the Role of Earth Observations from Space*, ed. P.M. Mather, pp. 121–131. London: Taylor & Francis.

Mardia, K.V. (1984). Spectral discrimination and classification maps. *Communications in Statistics: Theory and Methods*, 13, 2181–2197.

Millington, A., Townshend, J.R.G., Kennedy, P., Saull, R., Prince, S. & Madams, R. (1989). *Biomass Assessment: Woody Biomass in the SADCC Region*. London: Earthscan Publications.

McWilliam, A.L.C., Roberts, J.M., Cabral, O.M.R. *et al.* (1993). Leaf area indices and above-ground biomass of *terra firme* rain forest and adjacent clearings in Amazonia. *Functional Ecology*, 7, 310–317.

Moore, III, B. & Braswell, B.H. (1994). Planetary metabolism: understanding the carbon cycle. *Ambio*, 23, 4–12.

Myers, N. (1991). Rainforest: the disappearing forests. In *Save the Earth*, ed. J. Porritt, pp. 46–55. London: Dorling Kindersley.

Nelson, R. & Horning, N. (1993). AVHRR–LAC estimates of forest area in Madagascar, 1990. *International Journal of Remote Sensing*, 14, 1463–1475.

Palubinskas, G. (1988). A comparative study of decision making algorithms in images modelled by Gaussian Markov random fields. *International Journal of Pattern Recognition and Artificial Intelligence*, 2, 621–639.

Roller, N.E.G. & Colwell, J.E. (1986). Coarse-resolution satellite data for ecological surveys. *BioScience*, 36, 468–475.

Rosenfield, G.H. & Fitzpatrick-Lins, K. (1986). A coefficient of agreement as a measure of thematic classification accuracy. *Photogrammetric Engineering and Remote Sensing*, 52, 223–227.

Sampson, R.N., Apps, M., Brown, S. *et al.* (1993). Workshop summary statement: terrestrial biospheric carbon fluxes – quantification of sinks and sources of CO_2. *Water, Air and Soil Pollution*, 70,3–15.

Scatena, F.N., Silver, W., Siccama, T., Johnson, A. & Sanchez, M.J. (1993). Biomass and nutrient content of the Bisley Experimental Watersheds, Luquillo Experimental Forest, Puerto Rico, before and after Hurricane Hugo, 1989. *Biotropica*, 25, 15–27.

Skole, D. & Tucker, C. (1993). Tropical deforestation and habitat fragmentation in the Amazon: satellite data from 1978 to 1988. *Science*, 260, 1905–1910.

Smith, T.M., Cramer, W.P., Dixon, R.K., Leemans, R., Neilson, R.P. & Solomon, A.M. (1993). The global terrestrial carbon cycle. *Water, Air and Soil Pollution*, 70, 19–37.

Solomon, A.M., Prentice, I.C., Leemans, R. & Cramer, W.P. (1993). The interaction of climate and land use in future terrestrial carbon storage and release. *Water, Air and Soil Pollution*, 70, 595–614.

Stone, T.A., Schlesinger, P., Houghton, R.A. & Woodwell, G.M. (1994). A map of the vegetation of South America based on satellite imagery. *Photogrammetric Engineering and Remote Sensing*, 60, 541–551.

Story, M. & Congalton, R.G. (1986). Accuracy assessment: a user's perspective. *Photogrammetric Engineering and Remote Sensing*, 52, 397–399.

Sundquist, E.T. (1993). The global carbon dioxide budget. *Science*, 259, 934–941.

Tans, P.P., Fung, I.Y. & Takahashi, T. (1990). Observational constraints on the global CO_2 budget. *Science*, 247, 1431–1438.

Taylor, J.A. & Lloyd, J. (1992). Sources and sinks of atmospheric CO_2. *Australian Journal of Botany*, 40, 407–418.

Townshend, J.R.G., Cushnie, J., Hardy, J.R. & Wilson, A. (1988). *Thematic Mapper Data: Characteristics and Use*. Swindon: Natural Environment Research Council.

Trabalka, J.R. (ed.) (1985). *Atmospheric Carbon Dioxide and the Global Carbon Cycle*. DOE/ER-0239. Washington, DC: US Department of Energy.

Trexler, M.C. & Haugen, C. (1993). *Keeping it Green: Evaluating Tropical Forestry Strategies to Mitigate Global Warming*. Washington, DG: World Resources Institute.

Turner, II, B.C., Mayer, W.B. & Skole, D.C. (1994). Global land-use/land-cover change: towards an integrated study. *Ambio*, 23, 91–95.

Uhling, J., Hall, C.A.S. & Nyo, T. (1993). Changing patterns of shifting cultivation in selected countries of Southeast Asia and their effect on the global carbon cycle. In *Effects of Land Use Change on Atmospheric CO_2 Concentrations*, ed. V.H. Dale, pp. 145–200. New York: Springer-Verlag.

Waldegrave, W. (1994). Response to a report on Environmental Research Programmes prepared by the Advisory Council on Science and Technology. In *Biodiversity: The UK Action Plan*, ed. R. Sharp, p. 16. CM 2428. London: HMSO.

Walker, B. & Seffen, W. (1994). Land use change in the humid tropics. *Global Change Newsletter*, 17, 6–7.

Whitmore, T.C. (1990). *An Introduction to Tropical Rain Forests*. Oxford: Clarendon Press.

Wisniewski, J. & Sampson, R.N. (eds.) (1993). *Terrestrial Biospheric Carbon Fluxes: Quantification of Sinks and Sources of CO_2*. Dordrecht: Kluwer Academic.

R.J. HARDING, E.M. BLYTH
and C.M. TAYLOR

Issues in the aggregation of surface fluxes from a heterogeneous landscape: from sparse canopies up to the GCM grid scale

Within general circulation models (GCMs), and in the new generation of large-scale ecological and hydrological models (Vorosmarty *et al.*, 1989; Dumenil & Todini, 1992), the turbulent transports of energy and mass from the surface play a central role. Better descriptions of these transports will significantly increase the ability of these models to simulate the real world (e.g. Betts *et al.*, 1996) and hence the effects of anthropomorphic changes on a future climate. GCMs currently have a grid cell of a few hundred kilometres across. This spatial resolution is unlikely to be increased by more than a factor of 10 in the foreseeable future. At present the surface fluxes of energy and mass are parameterised in GCMs using comparatively simple equations with little attempt to describe the variability which occurs within a single grid cell. This chapter explores the errors that are likely to arise from this simplification and introduces possible schemes to include subgrid-scale landscape variability into the flux calculations.

The surface fluxes of heat, momentum, water vapour and, when included, CO_2 are modelled using simple equations describing flux–gradient relationships with parameters that are found by calibration against observations. These equations, along with equations describing the surface radiation exchange and the soil processes, are collectively known as the SVAT (soil–vegetation–atmosphere transfer) scheme. A typical SVAT scheme contains between 10 and 50 parameters that describe the vegetation and soil characteristics (for example: albedo, roughness length, soil hydraulic conductivity) (Berry *et al.*, this volume). These parameters are specified for a number (typically 20) of vegetation types (Table 1). Within present GCMs, each cell is assigned one, or at most two, cover types. The SVAT parameters appropriate to that square are then looked up from a table. When there is more than one cover type within a single cell, an average of the parameters is made to give an 'effective parameter'.

Table 1. *Vegetation types used in UK Meteorological Office GCM with specification of three of the SVAT parameters*

Vegetation type	z_0 (m)	r_s (s m^{-1})	Albedo
Ice	1×10^{-4}	0	0.75
Inland lake	3×10^{-4}	0	0.06
Evergreen needleleaf tree	1.00	85	0.14
Evergreen broadleaf tree	1.20	130	0.12
Deciduous needleleaf	1.00	85	0.13
Deciduous broadleaf tree	1.00	100	0.13
Tropical broadleaf tree	1.20	130	0.13
Drought deciduous tree	1.00	100	0.13
Evergreen broadleaf shrub	0.40	80	0.17
Deciduous shrub	0.40	80	0.16
Thorn shrub	0.40	80	0.16
Short grass	0.01	60	0.19
Long grass	0.04	80	0.20
Arable	0.04	60	0.20
Rice	0.01	40	0.12
Sugar	0.08	40	0.17
Maize	0.08	40	0.19
Cotton	0.10	35	0.19
Irrigated crop	0.04	40	0.25
Urban	1.50	200	0.18
Tundra	0.01	40	0.15
Swamp	0.10	40	0.12
Soil	3×10^{-4}	100	0.20

r_s, surface resistance; z_0, roughness length.

Current GCMs use SVAT schemes to calculate the fluxes of heat, moisture and momentum as a lower boundary condition to the main body of the atmosphere. The description of the hydrological response of the surface is, therefore, limited to those processes that strongly affect the partition of available energy into latent and sensible heat flux. As more demands are put on GCMs (for example to estimate the effects of doubled CO_2 on river flows), SVAT schemes are becoming more complex and it is becoming increasingly important to represent all the significant hydrological processes. There are many challenging problems in the representation of hydrological processes, which often occur at small scale (such as macropores and hill slopes), compared with the GCM scale. The representation of variable hydrological fields is still in its infancy and although a number of researchers have tackled this question (notably: Entekhabi & Eagleson, 1989; Wood, Lettenmaier & Zartarian, 1992; Quinn, Beven & Culf, 1994), a consensus on methodology has not yet been reached.

Measurements of surface fluxes and the development of equations to describe them have generally been on the plant or patch scale (see Fig. 1). This chapter will focus on the problems of aggregating flux–gradient relationships developed on the patch scale up to the GCM grid scale. Some of these problems arise from the non-linearity of the SVAT equations and these are evident at all scales. A second problem arises from the interaction between one patch and another. Also, problems may arise from the use of grid box-averaged boundary layer variables in calculating surface fluxes, and the possibility that large areas of strongly contrasting surface types may trigger sea breeze-type circulations unresolved by the GCM.

Non-linearity of SVATs

Variation of surface resistance or surface conductance

The latent heat flux from a surface can be calculated using the Penman–Monteith equation:

$$\lambda E = \frac{\Delta A + \rho c_p \delta q / r_a}{\Delta + c_p / \lambda (1 + r_s / r_a)} \tag{1}$$

where λE is the latent heat flux, Δ is the slope of the saturation humidity versus temperature curve, A is the available energy, ρ is the density of air, c_p is the specific heat capacity of air, δq is the humidity deficit, r_a is the aerodynamic resistance and r_s is the surface resistance.

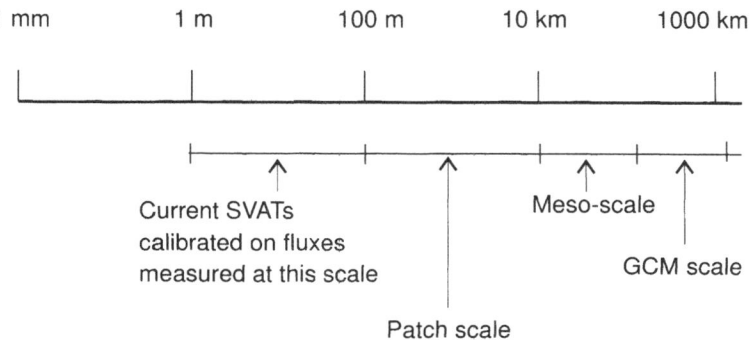

Fig. 1. Length scales of heterogeneity.

The greatest cause of heterogeneity in evaporation is heterogeneity in r_s. By examining the dependence of evaporation on r_s and its reciprocal, the surface conductance (g_s), with fixed values of r_a, A and δq, the non-linear relationship becomes apparent. Figure 2 shows λE plotted against r_s and g_s for $A = 500$ W m^{-2}, $\delta q = 0.005$ kg kg^{-1}, $\Delta = 0.0008$ kg kg^{-1} K^{-1} and $r_a = 10$ s m^{-1}. For a surface made up of equal areas of two values of r_s, e.g. 10 and 150 s m^{-1}, the area-average evaporation is obtained by bisecting the straight line between the two points on the r_s curve given by these two resistances. The effective surface resistance is the value of r_s that would give the correct area-average evaporation flux (found simply in Fig. 2 by drawing a horizontal line from the point of bisection to the curve). By using an area-average value of r_s, the evaporation rate obtained is underpredicted and given by the point on the curve at the average value of r_s (found simply in Fig. 2 by drawing a vertical line from the point of bisection to the curve). By comparison, the g_s curve is arched the other way and using area-average values of g_s overpredicts the evaporation. Hence, it is the non-linear relationship between

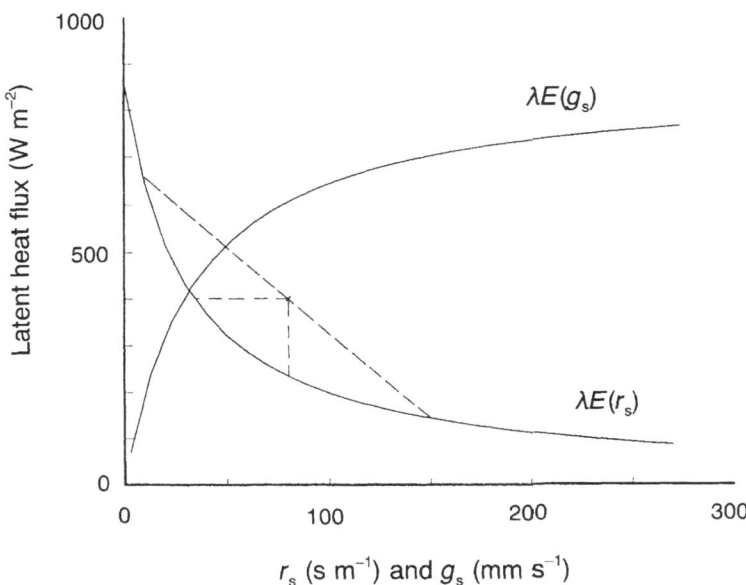

Fig. 2. Evaporation plotted as a function of surface resistance and surface conductance as predicted by the Penman–Monteith equation, using constant values of aerodynamic resistance and available energy.

evaporation and r_s or g_s that prevents the use of simple area-average values of these parameters.

If the range of r_s or g_s values is small, the curve can be considered linear. For instance, the curve is almost linear between r_s of 50 s m^{-1} or g_s of 20 mm s^{-1} (well-watered grass) and r_s of 100 s m^{-1} or g_s of 10 mm s^{-1} (well-watered trees). This result was confirmed by the HAPEX–MOBILHY experiment based in south-west France, an area consisting of forest and crops (Noilhan & Lacarrere, 1995). By modelling the area with a meso-scale model and comparing the results with a GCM scale model using average parameters, Noilhan and Lacarrere found that area-average values of surface conductance correctly predicted the area-average evaporation rates over this region.

By studying the curves in Fig. 2, it might be tempting to conclude that, since the curve is roughly linear at $r_s > 100$ s m^{-1}, area-average values of r_s can be used with minimal error in dry conditions. However, one might equally conclude from the g_s curve that, since the curve is roughly linear at $g_s > 100$ mm s^{-1}, area-average values of g_s can be used when the surface is wet or moist (Dolman, 1992). The problem with studying the problem of heterogeneity with the non-linear equations alone is that it gives no indication of the likely variation in r_s or g_s values over a terrain. However, there are few data on this at present. Variations in hydrological conditions can produce large variations in r_s. For instance, large differences in the value of r_s occur after a patchy rainfall event, with values of r_s then ranging from 0 to 100 s m^{-1}. Large variations can also occur between vegetated and non-vegetated areas, with r_s ranging from 50 to 1500 s m^{-1}, or between irrigated and non-irrigated, with r_s ranging from 30 to 300 s m^{-1}. In these cases, the area-average parameters would be significantly different from the effective parameters. It is interesting to note that for all these problem situations, it is the source of water that is varying.

Figure 2 shows that, for a constant r_a, the average r_s always underpredicts the evaporation while an average g_s always overpredicts it. This result is usually true when variations in aerodynamic resistance are also considered (Blyth, Dolman & Wood, 1993). The only time when the effective surface resistance strays significantly outside the limits given by area-average r_s ($\langle r_s \rangle$) and area-average g_s ($\langle g_s \rangle$) is when the limits are close and the range of r_s values is low. In this case, the effective surface resistance is close to the average resistance anyway. Therefore, it can be assumed, without great risk, that the effective r_s is bounded by two limits; $\langle r_s \rangle$ and $1/\langle g_s \rangle$. For two values of r_s, the maximum error in assuming that the effective resistance is equal to the average resistance is, therefore, given by:

$$Err = 1 - 4 \left[\frac{r_{s,1} \, r_{s,2}}{(r_{s,1} + r_{s,2})^2} \right] \tag{2}$$

This equation can be used to quantify the likely errors in using area-average surface resistances.

Variation of aerodynamic resistance or conductance

A factor that is missing from the graphs shown in Fig. 2 is the effect of a variation in the roughness of the surface. The problem was explored by Blyth *et al.* (1993). An example of how the value of r_a changes the energy balance of a heterogeneous surface is given below. Consider a landscape made up of two surface types: one rough and one smooth. The first case is when a rainfall event covers only the rough surface and the second case is when the rain falls only on the smooth surface. Figure 3 shows the evaporation rates predicted by a numerical model in these two cases. The overall evaporation rate from the first case (Fig. 3*a*) is 460 W m^{-2} and the second case (Fig. 3*b*) is 310 W m^{-2}. Using area-average values of surface and aerodynamic resistance, there is no distinction between the two cases.

Raupach (1993) tackled the problem of the effect of the combination of r_s and r_a on the evaporation flux by re-writing the Penman–Monteith equation as follows:

$$\frac{\varepsilon H - E}{\rho \lambda} = \frac{\varepsilon}{\varepsilon + 1} \frac{A}{\rho \lambda} \left[\frac{r_s}{r_a + \dfrac{r_s}{\varepsilon + 1}} \right] - \delta q \left[\frac{1}{r_a + \dfrac{r_s}{\varepsilon + 1}} \right] \tag{3}$$

where ε is $\Delta c_p / \lambda$ and H is the sensible heat flux.

If an average of the combination of resistances found in the first and second terms on the right-hand side of the equation is used in combination with area-average values of δq and A, then the correct average value of H and λE will be obtained.

In a similar vein, Lhomme (1992) re-wrote the equation of the surface temperature (T_s) as a linear sum as follows:

$$T_s = T_s + \frac{A}{\rho c_p} \frac{1}{\varepsilon + 1} \left[\frac{r_a (r_a + r_s)}{r_a + \dfrac{r_s}{\varepsilon + 1}} \right] - \frac{\delta q}{\gamma (\varepsilon + 1)} \left[\frac{r_a}{r_a + \dfrac{r_s}{\varepsilon + 1}} \right] \tag{4}$$

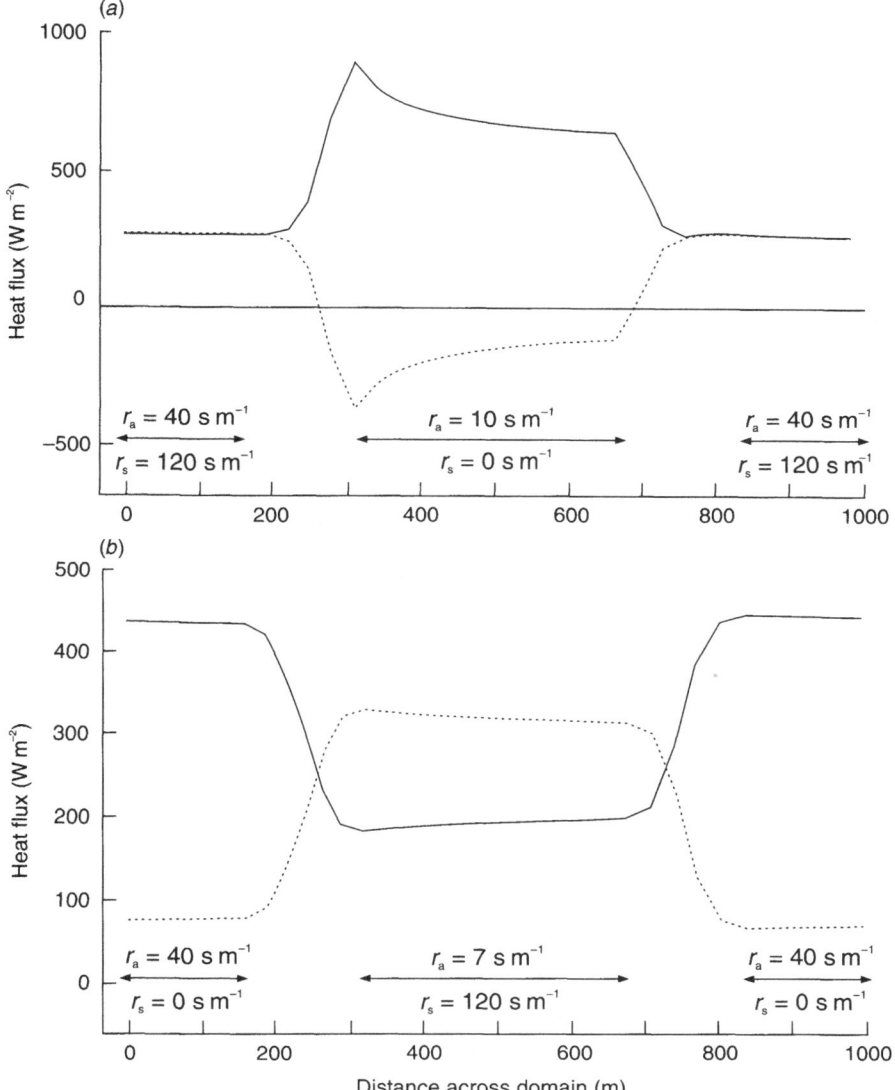

Fig. 3. Evaporation modelled over two extreme cases of heterogeneity. (*a*) When a rainfall event only covers the rough surface, and (*b*) when it only covers the smooth surface. Latent heat flux (—) and sensible heat flux (- - - -).

Again, if average values of the resistances in the groupings in the second and third terms on the right-hand side of this equation are used, then the correct average surface temperatures can be found.

When applied to the cases displayed in Fig. 3, Raupach's method estimated the evaporation more accurately than Lhomme's method, while Lhomme's method calculated the surface temperature more accurately. Clearly, effective parameters can be found to give a specified area-average property, even with extreme heterogeneity, but there is always a compromise. Only one property can be correct: either the evaporation or the surface temperature, not both. The other compromise with effective resistances is that an aggregate description cannot distinguish between sources of water. For instance, an aggregate description of vegetation and a soil surface must take its water from the same reservoir when in reality the soil evaporation will come from the surface and the transpiration will be extracted from the root zone. As discussed earlier, complex effective parameters are required when there is a distinct hydrological difference between the surfaces. In these cases, the source of water is important and effective parameters should not be used. Also, surfaces that are similar hydrologically, such as unstressed trees and unstressed grass, might become hydrologically dissimilar in another season or in a future climate.

Tile methods

In areas where effective parameters cannot be used, the energy and water balance can be calculated over each surface type and then the fluxes can be 'area averaged'. This procedure, the so-called tile method, automatically overcomes the non-linearity problem and the other compromises associated with using effective resistances. The method has been used by several researchers. Avissar & Pielke (1989) used tiles in a meso-scale model, while Taylor (1995), who studied the modelling of interception in GCMs, used a 'wet' and 'dry' tile to improve significantly the evaporation prediction. Taylor demonstrated the strong dependence of the aggregate evaporation on the boundary layer temperature. A comprehensive way of modelling the heterogeneity is to calculate the heat fluxes from a statistical distribution of surface parameters. Avissar (1992), Entekhabi & Eagleson (1989) and Bonan, Pollard & Thompson (1993) all adopted this approach. The question of how many classes in the distribution are needed has not yet been addressed. It is possible that the non-linearity can be summed up with just two representative surface types. However, each separate water source (e.g. intercepted rain, soil, vegetation) will need a separate calculation of the water and energy balances.

Patch scale aggregation (100 m to 10 km)

It can be seen clearly in Fig. 3 that evaporation is enhanced at a dry–wet transition but not equally depressed at a wet–dry transition. The enhanced evaporation is proportionally more important as the length scale of the variation is reduced. Or, to put it another way, the more edges there are per unit area, the more evaporation there will be. The same is true for momentum flux. A transition from a smooth to a rough surface sees a transitory increase in momentum flux that is not matched in a rough to smooth transition. This advection effect on surface fluxes is not accounted for in either the effective parameter methods or the tile methods described in the last section.

Using effective parameters or the tile method, it is assumed that the evaporation is independent of the height used for the aerodynamic resistance and the humidity deficit in Equation (1). This approximation is valid for homogeneous terrain as the fluxes are almost constant over the first 100 m of the atmosphere. It is not, however, the case for heterogeneous terrain. The average of the evaporation fluxes calculated over two different surfaces is dependent on the height at which average values of δq are obtained. As this height tends to zero, the relative influence of the rougher surfaces on the evaporation flux increases. The process is also true for momentum flux, so that the lower the height taken for area-average wind velocity, the higher the momentum flux.

Mason (1988) argued that this dependence of surface flux on reference height is linked to the advection effects. By comparing the length scale associated with the advection process, which mixes the atmospheric properties, to the length scale associated with the diffusion process, which describes how the surface interacts with the overlying meteorology, he found the 'blending height', l_b:

$$l_b \left(\ln \frac{l_b}{z_0} \right)^2 \approx 2k^2 L \tag{5}$$

where L is the Monin–Obukov length, k the von Karman constant and z_0 the roughness length for momentum. If the wind speed at the blending height is used to calculate the momentum fluxes over a mixture of surfaces with different roughnesses, then the advection effects would be accounted for and the correct area-average momentum flux would be obtained.

Wood & Mason (1991) then extended their length scale analysis to heat and other scalar fluxes. They found another, smaller blending height, l_{bh}:

$$l_{bh}\left(\ln \frac{l_{bh}}{z_0}\right)\left(\ln \frac{l_{bh}}{z_{0h}}\right) \approx 2k^2 L \tag{6}$$

where z_{0h} is the roughness length for heat. Claussen (1991) independently developed this idea of using average properties at an intermediate height. By semiempirical methods of comparing analytical solutions with numerical model output, he found that the ideal height was given by the diffusion height, l_d:

$$l_d\left(\ln \frac{l_d}{z_0}\right) = 2kL \tag{7}$$

Blyth (1995) showed that the method of solving the energy balance over two separate surfaces proposed by Shuttleworth & Wallace (1985) for modelling sparse canopies could be combined with the blending height concept to provide a comprehensive scale-dependent SVAT scheme. Figure 4 shows the layout of such a SVAT, where

$$\frac{r_a^3}{r_a^1 + r_a^3} = \frac{\ln(z_{ref}) - \ln(l_{bh})}{\ln(z_{ref}) - \ln(z_{0h})} \tag{8}$$

Equation (8) was shown in Blyth (1995) to predict results from several numerical model simulations at different length scales and different environmental forcings.

The numerical model and the blending height theory were developed for flat terrain. For small patch scales, the height of the vegetation itself plays a role in the size of the advection effect. For instance, Blyth & Harding (1995) calculated values of the resistances required for the SVAT scheme illustrated in Fig. 4 to describe the heat flux and surface temperature of the tiger bush in the Sahel. This distinctive surface cover consists of 10–30 m wide strips of vegetation separated by 50–100 m strips of bare soil. Use of the blending height given by Equation (6) underestimated the value of r_a^3. There was more interaction between the two surface types than could be accounted for by this method.

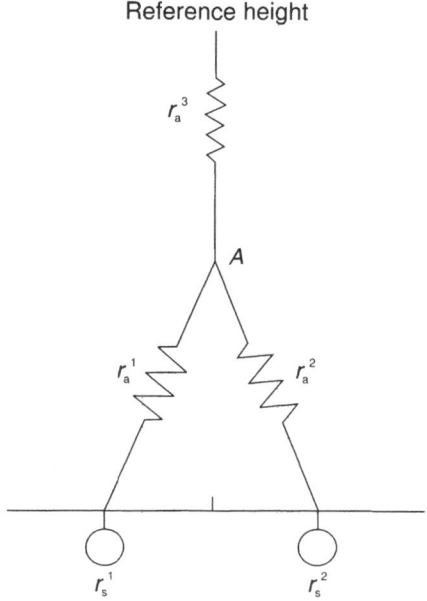

Fig. 4. Layout of the scale-dependent SVAT scheme.

Meso-scale heterogeneity

Introduction

In this section, we address issues of scaling-up of fluxes from surfaces that exhibit variability on scales of greater than about 10 km. On this scale, the influence of a locally homogeneous surface extends throughout the planetary boundary layer, up to 1 or 2 km (Raupach, 1993). Atmospheric interactions become possible between heterogeneous boundary layers on this scale. In particular, strong horizontal gradients in surface sensible heat flux can generate organised meso-scale circulations in the boundary layer (Pielke & Segal, 1986). These meso-scale flows are most commonly associated wtih contrasts in surface heating at coastlines (sea breezes), and in mountainous terrain (valley flows).

Variability in land cover and surface hydrology on the meso-scale may also generate boundary layer circulations. For example, contrasts in near-surface soil moisture have been shown to initiate sea breeze-type circulations in modelling studies (Ookouchi *et al.*, 1984), and in observations around irrigated areas in semiarid climates (Segal *et al.*, 1989; Mahrt, MacPherson & Desjardins, 1994). In this situation, there is low-level flow from the moist cool areas to the drier warmer areas, accompanied by ascent over the dry areas and descent over the wet areas, with a return horizontal flow aloft. Similar circulations have been generated in numerical models for contrasts in vegetation type (e.g. Mahfouf, Richard & Mascart, 1987; Segal *et al.*, 1988) and observations of a 'forest breeze' were made during the HAPEX–MOBILHY experiment (Bougeault *et al.*, 1991).

In the boundary layer, these meso-scale circulations can transport large quantities of humidity, heat and momentum, both horizontally and vertically (Pielke *et al.*, 1991; Avissar & Chen, 1993). To calculate the evaporation from a GCM grid box with strong meso-scale heterogeneity, the effect of these meso-scale fluxes has to be quantified. The effect is dependent on the particular situation, with the magnitude of the circulation itself being sensitive to the length scale and the strength of the heterogeneity, orographic features and the synoptic flow (Ookouchi *et al.*, 1984; Pielke *et al.*, 1991).

Numerical simulations that resolve the meso-scale surface heterogeneities can be used to draw some general conclusions to this problem. Here we present an example of a hypothetical heterogeneous soil moisture distribution in the semiarid environment of the Sahel. The effect on area-averaged evaporation by this surface heterogeneity is examined, along with the effects of the boundary layer feedbacks. The representation of such a surface in a large-scale model is discussed.

Numerical simulations of the boundary layer above strong heterogeneity

The meso-scale version of the UK Meteorological Office Unified Model (Cullen, 1993) was used in this experiment. This model has a horizontal resolution of 16.8 km and a domain covering about 1500 km × 1500 km, which, for this experiment, was centred around Niamey, Niger (2° E, 13.5° N). The aim of the experiment was to apply meso-scale land surface heterogeneity to an atmosphere that initially only exhibited synoptic scale variability. Meso-scale features developing during the 24-hour simulation can then be identified with land surface heterogeneities.

The initial atmospheric analysis is derived from the UK Meteorological Office global forecast model, an operational high-resolution GCM, at 00:00 on 8 October 1992. Conditions are characterised by strong coupling between the water-stressed sparse vegetation and a boundary layer of high vapour pressure deficit. The winds are south-westerly in the boundary layer and north-easterly in the very dry free troposphere, with cloud-free skies.

The land surface parameters are interpolated onto the meso-scale domain from the global forecast model. Two simulations are performed, with differences in the initialisation of the soil moisture field. In the 'homogeneous' case, the soil moisture is treated the same as the other surface parameters (Fig. 5a) and simply interpolated. The heterogeneous simulation differs only in that the soil moisture from one GCM grid box is re-distributed to create one unstressed wet patch of approximately 70 km × 100 km, with a completely dry area adjacent (Fig 5b). The effect of this extreme meso-scale heterogeneity can be studied by examining the differences between the two simulations.

Area-averaged evaporation from the two simulations is plotted in Fig. 6. For much of the day, evaporation from the heterogeneous surface is about 60% of evaporation from the homogeneous surface. In addition, the boundary layers develop differently in the two simulations. These differences can be summarised by the following three factors.

1. As outlined previously, a heterogeneous and homogeneous surface with the same area-averaged surface parameters and coupled to the same boundary layer variables can give very different area-averaged fluxes. This is particularly so with strongly contrasting surfaces and high boundary layer temperatures. In this case, both surfaces have the same average soil moisture, and the same effective resistance to evaporation averaged over one GCM square, yet the evaporation from the heterogeneous surface is reduced because of the non-linearity of the SVAT scheme.

2. Over the wet patch, sensible heat fluxes are low and the boundary layer fails to develop beyond about 1 km. This is in contrast to maximum inversion heights of about 2 km over the homogeneous surface and values of over 3 km over the dry area adjacent to the wet patch. The low boundary layer moistens appreciably over the wet patch (Fig. 7) and remains relatively cool. This tends to inhibit evaporation.

3. A meso-scale circulation develops above the heterogeneous surface (Fig. 7), with drier air descending over the wet patch and horizontal advection of moist air from the wet patch to the dry area. This acts to increase evaporation.

The effects of processes (1) and (2) can be quantified by considering the differences in boundary layer properties between the two

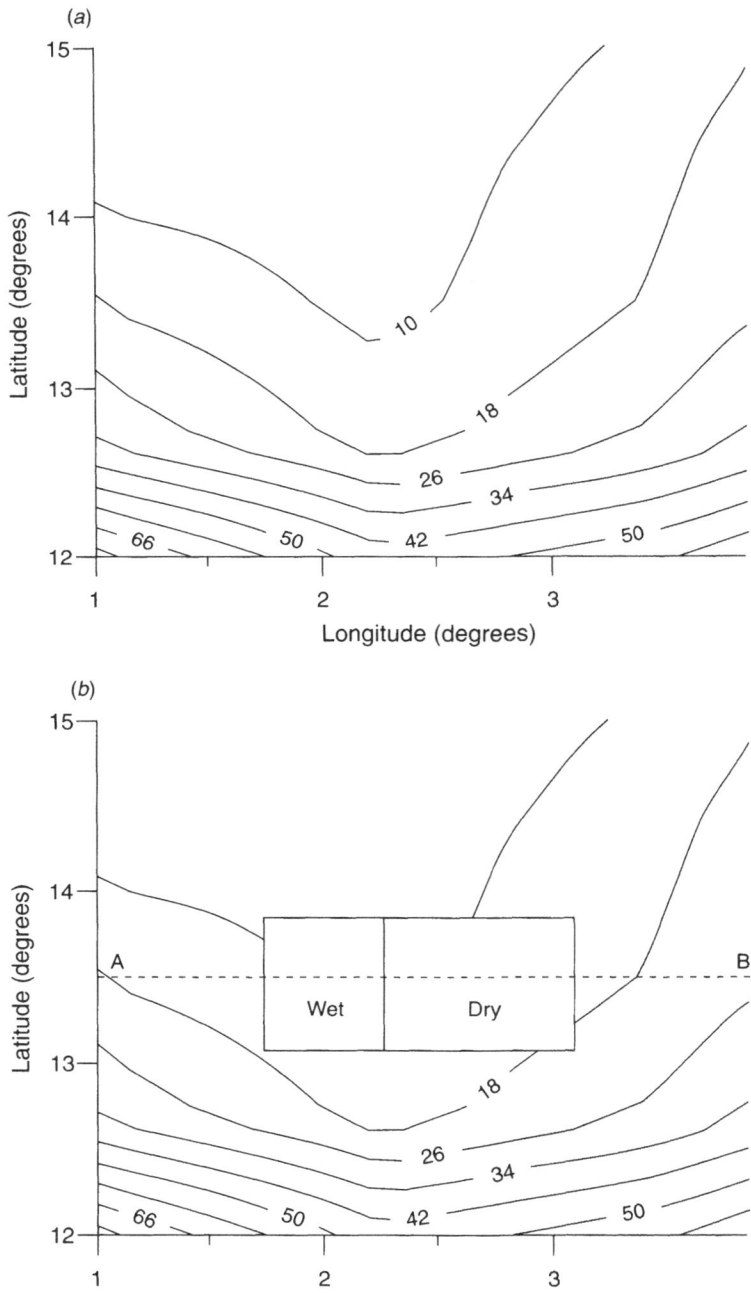

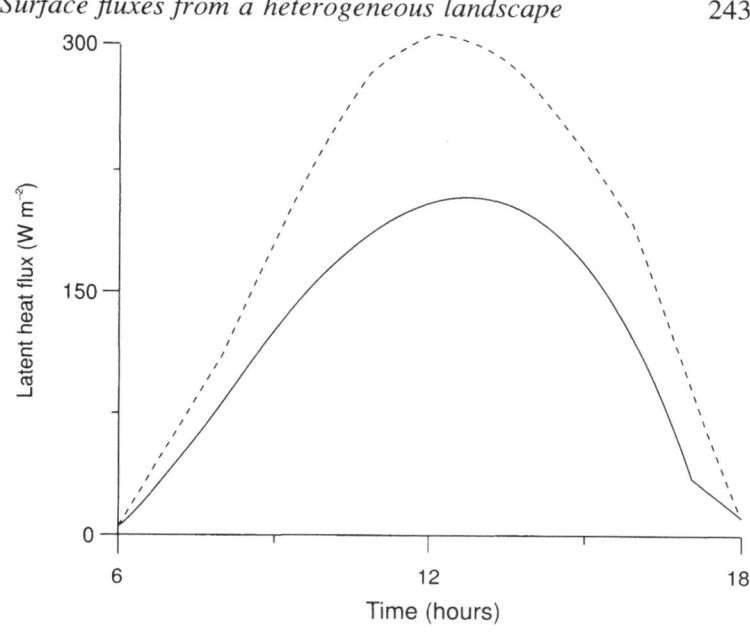

Fig. 6. Evaporation from the homogeneous (- - -) and heterogeneous (—) simulations averaged over the GCM grid box through the day.

simulations. One can partition an atmospheric variable over the heterogeneous surface X into perturbations from the homogeneous simulation X_{hom}, owing to resolved and parameterised processes.

$$X = X_{\mathrm{hom}} + X_{\mathrm{r}} + X_{\mathrm{p}} \qquad (9)$$

Here X_{r} represents the difference between the two experiments in the vertical and horizontal advection of X by the resolved flow. The term X_{p} represents changes in X owing to differences between the parameterised processes in the two simulations, notably subgrid-scale vertical fluxes. The temperature and humidity at the first model level can be partitioned in this way and used in the Penman–Monteith equation for a wet tile of

Fig. 5. Distributions of soil moisture (mm) over the centre of the meso-scale domain: (*a*) 'homogeneous' distribution on the meso-scale; (*b*) heterogeneous simulation with a wet and dry area covering one GCM grid box. The line AB corresponds to the cross-section shown in Fig. 7.

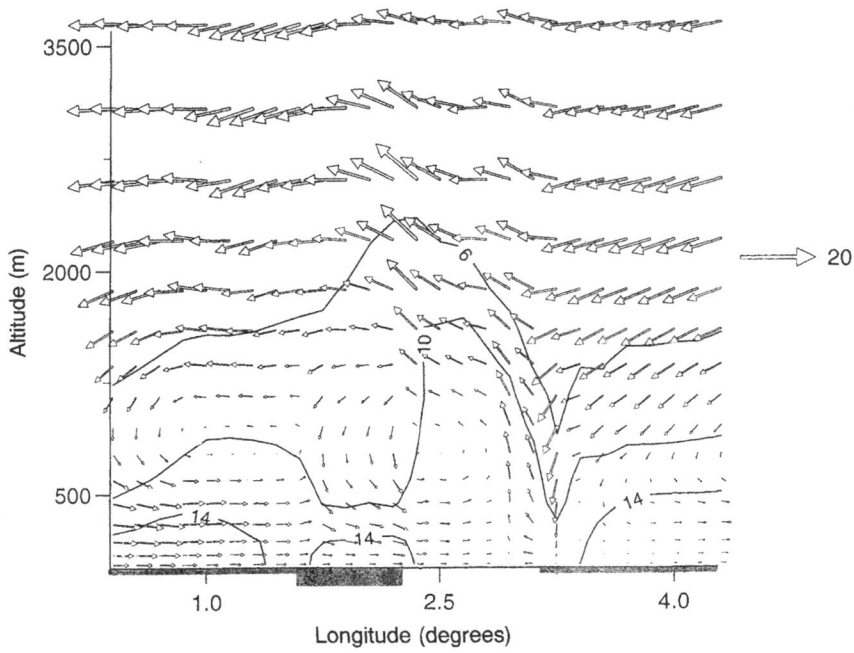

Fig. 7. Vertical slice through boundary layer simulation at 14:00 showing zonal (m s^{-1}) and vertical (cm s^{-1}) wind components above soil moisture heterogeneity (shaded areas). Contours of specific humidity (g kg^{-1}) in the boundary layer are also shown.

fractional area α. The equations defining the tiled surface evaporation with no feedbacks (E_{tile}) and the evaporation into a boundary layer modified by parameterised (E_{p}) and resolved (E_{r}) processes are therefore:

$$\lambda E_{\text{tile}} = \alpha \frac{A\Delta(T_{\text{hom}}) + (\rho c_{\text{p}}/r_{\text{a}})(q_{\text{sat}}(T_{\text{hom}}) - q_{\text{hom}})}{\Delta(T_{\text{hom}}) + (\lambda/c_{\text{p}})(1 + r_{\text{s}}/r_{\text{a}})} \tag{10}$$

$$\lambda E_{\text{p}} = \alpha \frac{A\Delta(T_{\text{hom}} + T_{\text{p}}) + (\rho c_{\text{p}}/r_{\text{a}})(q_{\text{sat}}(T_{\text{hom}} + T_{\text{p}}) - q_{\text{hom}} - q_{\text{p}})}{\Delta(T_{\text{hom}} + T_{\text{p}}) + (\lambda/c_{\text{p}})(1 + r_{\text{s}}/r_{\text{a}})} \tag{11}$$

$$\lambda E_{\text{r}} = \alpha \frac{A\Delta(T_{\text{hom}} + T_{\text{r}}) + (\rho c_{\text{p}}/r_{\text{a}})(q_{\text{sat}}(T_{\text{hom}} + T_{\text{r}}) - q_{\text{hom}} - q_{\text{r}})}{\Delta(T_{\text{hom}} + T_{\text{r}}) + (\lambda/c_{\text{p}})(1 + r_{\text{s}}/r_{\text{a}})} \tag{12}$$

The temperature and humidity at the first model level ($T_{\text{hom}} + T_{\text{r}}$) and

($q_{hom} + q_r$) are defined by Equation (9). The differences between these evaporation rates can be interpreted as perturbations to the evaporation from the tiled surface caused by boundary layer feedbacks. The term ($E_{tile} - E_p$) represents the effect of different subgrid vertical fluxes of heat and moisture on evaporation (process (2)). Similarly, ($E_{tile} - E_r$) denotes the perturbation to the evaporation E_{tile} owing to advection of heat and moisture by the resolved flow, principally the meso-scale circulation (process (3)).

The temperature and humidity perturbations in Equations (10)–(12) can be found by calculating the differences between the relevant flux divergences in the two simulations. These terms, averaged over the wet patch, are shown in Fig. 8. The parameterised turbulent flux terms T_p and q_p tend to moisten and cool the boundary layer during the morning and late afternoon above the wet patch relative to the homogeneous simulation. In the afternoon, the meso-scale advection of warmer dry air from above the inversion tends to heat and dry the boundary layer over the wet area.

The area-averaged evaporation rates E_{tile}, E_r and E_p are plotted in Fig. 9 through the day as calculated using Equations (10)–(12) and taking area-averaged values of available energy and aerodynamic resistance. Also plotted is the evaporation E_{ave}, found by solving only one surface energy balance with average surface resistance, coupled to the boundary layer variables T_{hom} and q_{hom}. This represents the evaporation from the simulation with the homogeneous surface.

The reduction in area-averaged evaporation from the heterogeneous surface apparent in Fig. 6 may now be explained using Fig. 9. The large difference between E_{ave} and E_{tile} suggests that the majority of the variation in Fig. 6 stems from the non-linearity of the SVAT scheme. The relative moistening and cooling of the boundary layer owing to parameterised mixing above the wet patch (Fig. 8) inhibits evaporation during the day. This can further reduce evaporation by up to 20%. However, the drying and warming of the boundary layer because of the meso-scale circulation occurs too late in the day in this particular simulation to have much impact on evaporation.

In the calculation of evaporation from this particular heterogeneous surface in a GCM, the use of area-averaged parameters is clearly a poor approximation. Coupling a wet and dry tile to area-averaged boundary layer variables would capture much of the reduction in evaporaion owing to the heterogeneity. A further improvement to the evaporation could be made by calculating two sets of boundary layer variables with no interaction between them. This would account for much of the difference between E_{tile} and E_p in Fig. 8. One might expect the relative

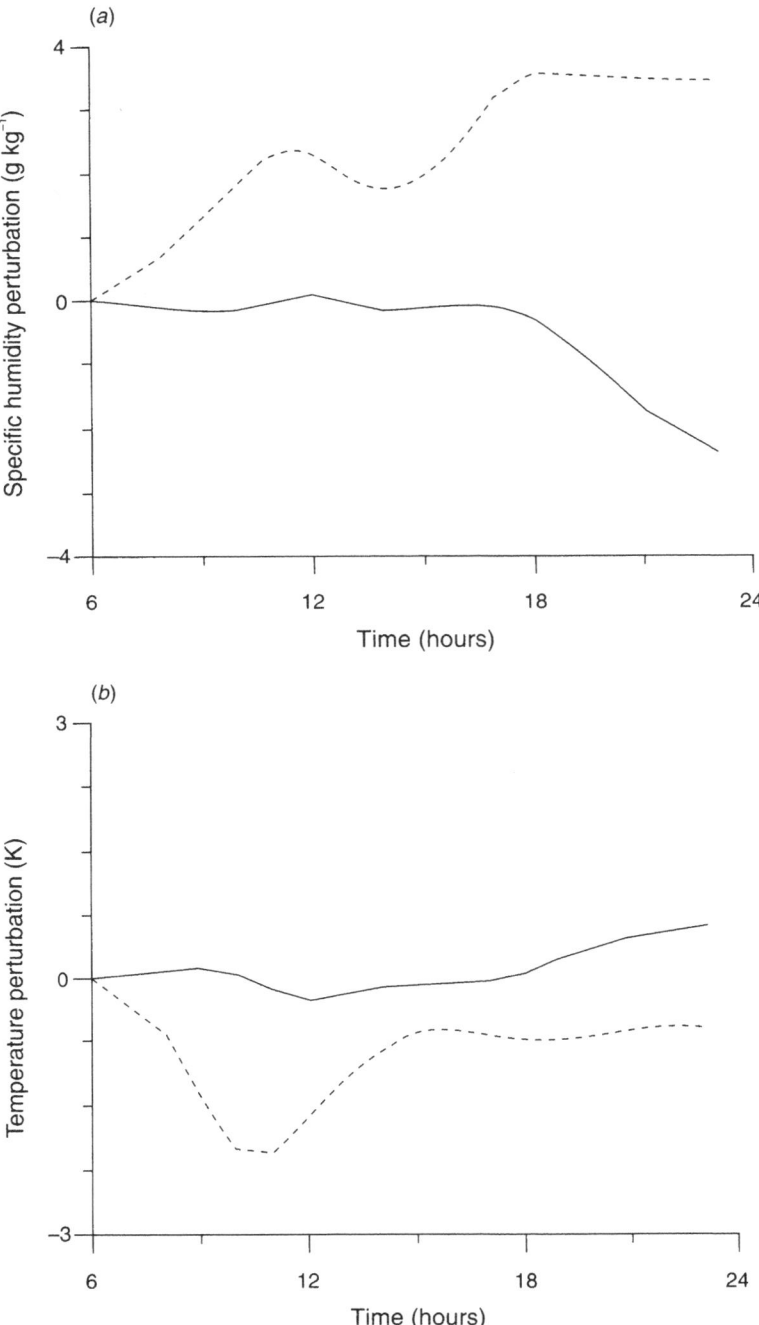

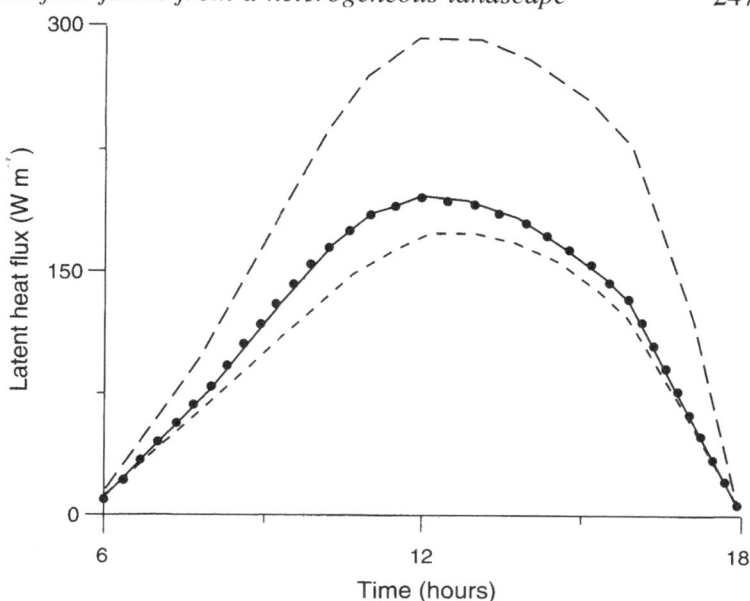

Fig. 9. Evaporation rates calculated through the day: E_{ave} (– –), E_{tile} (—), E_p (- - -) and E_r (●).

importance of these two effects to reverse in the wet season, when the temperature and vapour pressure deficits are lower and boundary layers are more shallow. In this case, increased vertical flux divergence and a lower surface temperature would provide stronger feedbacks on evaporation.

No parameterisation of the meso-scale circulation and its interaction with the synoptic and subgrid fluxes currently exists (Avissar & Chen, 1993). Such a parameterisation may not be necessary for calculating evaporation from the particular GCM grid square considered here. However, as is evident from the moist plumes shown in Fig. 7, the meso-scale circulation is very important in transporting humidity to higher levels in the atmosphere. Indeed, one would expect that in conditions of higher relative humidity such a circulation would generate convective

Fig. 8. Perturbations to homogeneous first model level variables above the wet patch owing to resolved (—) and parameterised (- - -) processes. (*a*) Specific humidity (g kg⁻¹); (*b*) temperature (K).

cloud and possibly rainfall, with large indirect feedbacks on the surface energy and water balances.

Conclusions

This chapter has considered some of the problems that arise when calculating area-average surface fluxes for use in numerical climate, hydrological and ecological models. The biggest errors associated with representing heterogeneous terrain come from using linear average parameters in the non-linear SVAT equations. This problem occurs at all spatial scales.

Analyses show that if the range of surface parameters (such as surface resistance) is small the error caused by making a weighted average of the parameters from the individual landscape components is also small. Therefore, if the vegetation is unstressed, a typical range of surface resistance is 60 to 120 s m^{-1} and the error in an effective parameter, calculated as a simple weighted average, is about 10%. The error in the calculated evaporation is less than this value. However, if there is a large range of surface resistance, with, for example, a proportion of the surface suffering water stress or, conversely, wetted through rainfall, the use of weighted mean parameters can cause errors in the evaporation in excess of 50%. In these more extreme cases, a tile model, in which the fluxes are calculated for each surface individually, should be used.

The horizontal advection of heat and water vapour from one surface to another depends on scale and meteorological conditions. Significant positive feedback on evaporation only arises when there are large contrasts in the surfaces. These effects are largest when the size of the vegetation patches is small. For the same mixture of vegetation type, the evaporation can increase by 50% as the scale decreases from 10 km to 10 m. Blyth (1995) has developed a scale-dependent SVAT to take account of this advection effect.

At larger scales, there is a negative feedback on evaporation through the atmosphere. Each surface develops its own boundary layer. The high evaporation rates from the wet surface moisten and cool the relatively small enclosed planetary boundary layer, which acts to suppress further evaporation. This effect could be incorporated into GCMs by having a separate boundary layer scheme for each surface type.

Finally, when strong surface heterogeneity generates a meso-scale circulation, the heat and water transported horizontally and vertically by the flow may have limited direct feedback on the surface energy balance. This is because of the long timescale of the circulation, of the order of a day.

This chapter has concentrated on the calculation of surface fluxes and has ignored many other facets of the aggregation problem, which are certainly important. It has ignored, for example, soil and hydrological processes. Runoff is dominated by inhomogeneities in the soil and landscape. As large-scale hydrological models become more widely used, these issues will have to be addressed. Equally, it has ignored some extreme landscape contrasts, such as a patchy snow-covered landscape, where it may be expected that non-linear and advective effects would dominate the energy and mass exchanges. Finally, it has ignored the very small scale, the aggregation from the leaf to the canopy and the detailed processes at a boundary between two vegetation types.

Acknowledgements

This chapter summarises a number of studies undertaken under the TIGER-funded project 'Understanding SVATs for global modelling using mesoscale meteorological and hydrological models'.

References

Avissar, R. (1992). Conceptual aspects of a statistical–dynamical approach to represent landscape subgrid-scale heterogeneities in atmospheric models. *Journal of Geophysical Research*, 97, 2729–2742.

Avissar, R. & Chen, F. (1993). Development and analysis of prognostic equations for mesoscale kinetic energy and mesoscale (subgrid scale) fluxes for large-scale atmospheric models. *Journal of Atmospheric Science*, 50, 3751–3774.

Avissar, R. & Pielke, R.A. (1989). A parameterisation of heterogeneous land surfaces for atmospheric numerical models and its impact on regional meteorology. *Monthly Weather Review*, 117, 2113–2136.

Betts, A.K., Ball, J.H., Beljaars, A.C.M., Miller, M.J. & Viterbo, P. (1996). The land-surface–atmosphere interaction: a review based on observational and global modelling perspectives. *Journal of Geophysical Research*, 101, 7209–7226.

Blyth, E.M. (1995). Using a simple SVAT scheme to describe the effect of scale on aggregation. *Boundary Layer Meteorology*, 72, 267–285.

Blyth, E.M. & Harding, R.J. (1995). Application of aggregation models to surface heat flux from the Sahelian tiger bush. *Agricultural and Forest Meteorology*, 72, 213–236.

Blyth, E.M., Dolman, A.J. & Wood, N. (1993). Effective resistance to sensible and latent heat flux in heterogeneous terrain. *Quarterly Journal of the Royal Meteorological Society*, 119, 423–442.

Bonan, G.B., Pollard, D. & Thompson, S.L. (1993). Influence of sub-grid scale heterogeneity in leaf area index, stomatal resistance, and soil moisture on grid-scale land–atmosphere interactions. *Journal of Climate*, 6, 1882–1897.

Bourgealt, P., Bret, B., Lacarrere, P. & Noilhan, J. (1991). An experiment with an advanced surface parameterisation in a mesobeta-scale model. Part II: the 16 June 1986 simulation. *Monthly Weather Review*, 119, 2374–2392.

Claussen, M. (1991). Estimation of areally-averaged surface fluxes. *Boundary Layer Meteorology*, 54, 387–410.

Cullen, M.J.P. (1993). The Unified Forecast/Climate model. *The Meteorological Magazine*, 122, 81–94.

Dolman, A.J. (1992). A note on areally-averaged evaporation and the value of the effective surface conductance. *Journal of Hydrology*, 138, 583–589.

Dumenil, L. & Todini, E. (1992). A rainfall-runoff scheme for use in the Hamburg climate model. In *Advances in Theoretical Hydrology, A Tribute to James Dooge*, ed. J.P. O'Kane, pp. 129–157. Amsterdam: Elsevier.

Entekhabi, D. & Eagleson, P.S. (1989). Land surface hydrology parameterization for atmospheric general circulation models including subgrid scale spatial variability. *Journal of Climate*, 2, 816–831.

Lhomme, J.-P. (1992). Energy balance of heterogeneous terrain, averaging the controling parameters. *Agricultural Forest Meteorology*, 61, 11–21.

Mahfouf, J.F., Richard, E. & Mascart, P. (1987). The influence of soil and vegetation on the development of mesoscale circulation. *Journal of Climate and Applied Meteorology*, 26, 1483–1495.

Mahrt, L., MacPherson, J.I. & Desjardins, R. (1994). Observations of fluxes over heterogeneous surfaces. *Boundary Layer Meteorology*, 67, 345–367.

Mason, P.J. (1988). The formation of areally-averaged roughness lengths. *Quarterly Journal of the Royal Meteorological Society*, 114, 399–420.

Noilhan, J. & Lacarrere, P. (1995). GCM gridscale evaporation from mesoscale modelling. *Journal of Climate*, 8, 206–223.

Ookouchi, T., Segal, M., Kessler, R.C. & Pielke, R.A. (1984). Evaluation of soil moisture effects on generation and modification of mesoscale circulations. *Monthly Weather Review*, 11, 2281–2292.

Pielke, R.A. & Segal, M. (1986). Mesoscale circulations forced by differential terrain heating. In *Mesoscale Meteorology and Forecasting*, ed. P.S. Ray, pp. 516–548. Boston, MA: American Meteorology Society.

Pielke, R.A., Dalu, G.A., Snook, J.S., Lee, T.J. & Kittel, T.G.F. (1991). Nonlinear influence of mesoscale land use on weather and climate. *Journal of Climate*, 4, 1053–1069.

Quinn, P., Beven, K. & Culf, A. (1994). The introduction of macrocale hydrological complexity into land surface–atmosphere transfer models and the effect on the planetary boundary layer development. *Journal of Hydrology*, 166, 421–444.

Raupach, M.R. (1993). The averaging of surface flux densities in heterogeneous landscapes. In *Proceedings of the Yokohama Symposium, July 1993*. International Association of Hydrological Societies Publication No. 212, pp. 343–355.

Segal, M., Avissar, R., McCumber, M.C. & Pielke, R.A. (1988). Evolution of vegetation effects on the generation and modification of mesoscale circulations. *Journal of Atmospheric Science*, 45, 2268–2292.

Segal, M., Schreiber, W.E., Kallos, G. *et al.* (1989). The impact of crop areas in northeast Colorado on midsummer mesoscale thermal circulations. *Monthly Weather Review*, 117, 809–825.

Shuttleworth, J.S. & Wallace, J.W. (1985). Evaporation from sparse crops – an energy combination theory. *Quarterly Journal of the Royal Meteorological Society*, 111, 839–855.

Taylor, C.M. (1995). Aggregation of wet and dry land surfaces in interception schemes for general circulation models. *Journal of Climate*, 8, 441–448.

Vorosmarty, C.J., Moore, B., Grace, A.L. *et al.* (1989). Continental scale models of water balance and fluvial transport: an application to South America. *Global Biogeochemical Cycles*, 3, 241–265.

Wood, E.F., Lettenmaier, D.P. & Zartarian, V.G. (1992). A land surface hydrology parameterization with sub-grid variability for general circulation models. *Journal of Geophyical Research*, 97, 2717–2728.

Wood, N. & Mason, P.J. (1991). The influence of static stability on the effective roughness lengths for momentum and temperature. *Quartery Journal of the Royal Meteorological Society*, 117, 1025–1056.

B. MARSHALL, J.W. CRAWFORD
and J.R. PORTER

Variability and scaling: matching methods and phenomena

Introduction

What image does the word 'scale' conjure up in your mind? What does changing scale involve? Do we need to concern ourselves with scaling-up or scaling-down. Is it a simple and prudent task to tag on to the end of a research programme to provide answers at scales appropriate for policy-makers and implementers? Or are the laws of scaling intrinsic to the biological systems we are trying to understand, in many cases still waiting to be discovered?

We intend to show that it is the latter. Until these laws are discovered, caution should be exercised in interpreting observations and averaging across and extrapolating to different scales. The measurements made, and the boundaries drawn, can impose critical constraints on the understanding that can be obtained. The sampling procedure must, therefore, match the phenomenon under investigation. Coupling and heterogeneity are ubiquitous characteristics of biological systems. Behaviour on one scale influences behaviour on another – to elucidate these mechanisms requires measurements on a range of scales. We will illustrate these points with a series of case studies, for example weather variability and plant growth, denitrification and water flow in soils.

Variability, scale and scaling

Since all natural objects are structured, possessing characters that are non-constant in space and time, variability is both pervasive and intrinsic to natural phenomena. The significance of this variability to the scientific process depends on a number of factors. The first is the magnitude of the variability compared with the precision required by the study or imposed by measurement. The second is the origin of the variability, and in particular whether significant variability is externally imposed or intrinsic to the system under study. Finally, the manner in which the variability changes with the scale of measurement is central to the problem of extrapolating knowledge based on measurements at one scale to different scales, i.e. scaling. While the first factor can generally be dealt with, the second and third are more difficult and, as will be shown, are related. It is this relationship between the phenomenon and scaling that presents the central challenge to understanding systems which are manifest across a broad range of scales, particularly where there are a number of cross-scale interacting components. It may also harbour the key for unlocking the nature of the scaling laws that operate. Before discussing these factors in detail, it is important to establish the underlying assumptions regarding variability associated with the terms *homogeneity* and *heterogeneity*.

Homogeneity implies a quantity that is invariant in space and hence a homogeneous system is structureless. The opening sentence of this section suggests that such a situation does not exist in nature, and so the term implies an approximation. In practice, it is usually used in the sense of quantities that vary in space according to some frequency distribution which is sufficiently narrow so as to be unimportant to the precision of the study. Implicit in this definition must be the scale at which the quantities are measured, since a system may appear homogeneous at one scale and heterogeneous at another.

A simple illustration is the analysis of pattern in the spatial distribution of vegetation. Suppose we sampled a patch of vegetation by laying quadrats and counting the number of individuals of a given species in each of the quadrats. Hypothetically, we might arrive at the result depicted in Fig. 1, where every other quadrat contained three individuals. The question is: does this species show a clumped spatial distribution or not?

Pattern analysis (Greig-Smith, 1964), a form of nested block analysis of variance, compares the mean square number of individuals between adjacent quadrats, between adjacent pairs of quadrats, and so on with the variance expected for samples drawn from a random distribution. For counts of individuals, the expected distribution is Poisson with the mean number per quadrat equal to its variance. If the ratio mean square: mean number per quadrat is plotted against block size, clustering is indicated by a peak at the block size nearest the size of the patch – with the implication that this does not have to be an integer value. In our example, the variance ratio will clearly have a value greater than unity (= a random distribution) at block size 1; the variance between blocks is greater than the mean number per block. At block size 2, a different conclusion is reached. Here, variation between blocks of two is zero, the variance ratio is less than one and we have a regular spatial pattern. Moving to block size 3 again reverses our perception to that of a clumped distribution, and so on. Whether clumping exists and the form

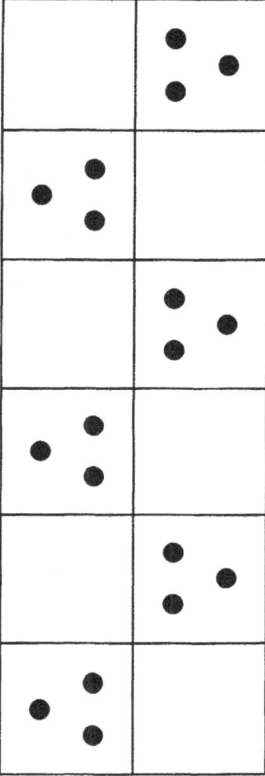

Fig. 1. A hypothetical spatial distribution of individuals of a plant species.

of any pattern is at least partly dependent on the scale at which we choose to view it.

For subsequent discussion, we define the smallest interval of measure at which a quantity appears homogeneous as the characteristic scale. A somewhat more subtle assumption is made when this link between measurement and scale is lost in defining a spatial average. Spatial averages imply the simplest form of scaling law, whereby the quantity is linearly proportional to length, area or volume. However, while a system may be homogeneous at a particular scale according to the above definition, this does not guarantee that a spatial average is scale independent. This is because the above definition is based on making a number of independent measurements at different locations and takes no account of spatial correlation. This is crucial, as will be discussed below, and is central to the problem of scaling.

When the system displays structure at scales of significance to the scientist, the quantity of interest is defined to be heterogeneous. This is most commonly the case, and arguably always so, in the natural sciences because the systems of interest comprise processes occurring across a wide range of scales, and variability at each of these scales plays a key role in viability and Darwinian evolution. Heterogeneous systems present the greatest challenge to scaling, and the tractability of the problem depends on the strength and nature of the couplings, or correlations, in the structure of the system. An operational distinction may be made between the instances where the function describing the spatial autocorrelation can be best charracterised by an exponential or a power law decrease in length scale. In the former case, the rapid drop-off in autocorrelation with length scale means that, for practical purposes, the spatial average is independent of scale above the characteristic length scale where the correlation drops to $1/e \approx 0.368$. In the latter case, there is no such rapid drop-off in autocorrelation, spatial averages cannot be uniquely defined and scaling is a non-trivial task.

It is interesting that variation in time is usually considered distinct from spatial variability. The distinction is artificial, as will be shown, since the same considerations to those in relation to spatial scaling apply. Hence, scaling is trivial (i.e. linear) only for systems that are homogeneous in space and time, i.e. unchanging. Complications potentially arise in all systems that have the capacity for change and, therefore, in most systems of scientific interest.

The conclusion of this work will be that there are no general recipes for scaling, but that the nature of the phenomenon itself largely dictates the scaling properties of the system, and the way in which it should be observed. For this reason, we continue with a study of specific phenomena that illustrate the different issues raised above. Common considerations that will be identified are the necessity for data collection across scales and strategies for determining the nature of spatial and temporal correlations, the importance of understanding the mechanisms underlying a phenomenon for extrapolating its manifestation at different scales and the significance of the definition of boundaries and boundary conditions for scaling behaviour. It is hoped that these general considerations will lead both to improved understanding of the mechanisms of scaling and to a more informed appreciation of variability and its relation to the phenomenon under study.

Phenomena that are scaling in time

As already stated, processes that are linear in time can be thought of as temporally homogeneous. This could be a system that is static in time, or one that changes at a constant rate. In both instances, scaling is trivial since the rates associated with the processes are constant and cumulative quantities are linearly proportional to the time interval of observation. Phenomena of more general scientific relevance are heterogeneous in time, with associated complex temporal behaviour. It is usually assumed that there exists a characteristic timescale longer than which meaningful definitions may be employed. The heart rate of an individual at rest, while being heterogeneous on time scales close to and below the interval between beats, is clearly homogeneous to a high level of approximation on time scales of around 1 min, as witnessed by the utility of the measure in determining health irregularities across the spectrum of the population (although see Goldbeter & West, 1987). Therefore, a good approximation to the daily total number of heartbeats in an individual at rest can be made from a few minutes' observation and a linear scale factor. However, not all processes have an associated characteristic timescale. A very common type of temporal scaling in which no characteristic scale exists occurs where the magnitude of the variability is inversely proportional to its frequency. Most of the variability is in components that change only slowly with time. This type of behaviour is an example of Hurst noise, a more general class of phenomenon exhibiting a power law relationship between magnitude of variability and its frequency. Such 'broad-band' variability is widespread in physical systems (Schroeder, 1990) but is also increasingly being recognised in the social, ecological and biological forums (West, 1985; Pimm & Redfearn, 1988; Hastings & Sugihara, 1993). The origin of the behaviour is still uncertain, but it arises in situations where there are many parallel processes whose capacity for change ranges across a broad spectrum of time scales or in certain types of non-linear system (West, 1985; Schroeder, 1990). An important consequence of this type of behaviour is '*persistence*', manifested as the average growth, over the time interval t_1 to t_2, in the magnitude of the increment

$$\Delta X(t) = |X(t_1) - X(t_2)| \propto |t_1 - t_2|^H = \Delta t^H$$

$$0 < H < 1$$

There are trends on all time scales, t, and the largest deviations are at the longest time increments. Persistence implies that the *rate* cannot be defined independently of timescale since

$$\text{rate} = \Delta X(t)/\Delta t \propto \Delta t^{H-1}$$

Therefore, extrapolation from one timescale to another requires knowledge of the underlying temporal correlations in the data. An illustration of this point is provided in Fig. 2*a*,*b* where data collected over 400 hours suggests an apparent increasing trend that is reversed in a longer data train. Extrapolation to time scales longer than those intrinsic to the measurement requires knowledge of the details of the underlying scaling law characteristic of the specific phenomenon. The lack of a characteristic timescale in the variability leads directly to a simple scaling law of the form

$$X(t) = \lambda^{-H} X(\lambda t)$$

relating the time series on a timescale t to that on a timescale λt. This is illustrated in Fig. 2*c* where a time series of length 1000 hours has been generated by re-scaling the first 200 hours according to the above scaling law. The similarity of the form of the curve with that of the actual 1000 hours time series is evident.

In practice, the power law relation between amplitude and frequency of the variations cannot hold over an infinite range of frequency. Figure 2*a* was constructed with an upper cut-off in frequency of $2\pi/5000$ meaning that $t = 5000$ is a characteristic timescale above which the fluctuations average out. Characteristic time scales mark the range of validity of the scaling laws, and their determination may necessitate collection of large data sets. Time series of ecological relevance (abundances of several species of forest fly) are known where the characteristic timescale exceeds 20 years (Hanski, 1990). Therefore, care must be taken in interpreting trends or population viability in data sets collected over time scales that are too short. Until the underlying mechanisms are identified and understood, the appropriate range of time scales over which data should be collected is not known. An iterative and close interaction between theory and experiment is required to maximise the chances of collecting data sets with appropriate time scales. This also implies that for competent prediction, interpretation and experimental design the underlying origin of persistence must also be understood if the nature of the components comprising the system under study are likely to change with time.

Power law relations, while pervasive, by no means constitute a universal form of scaling. Exponential decay, or more complex relationships between amplitude of variability and frequency, exists. Chaotic

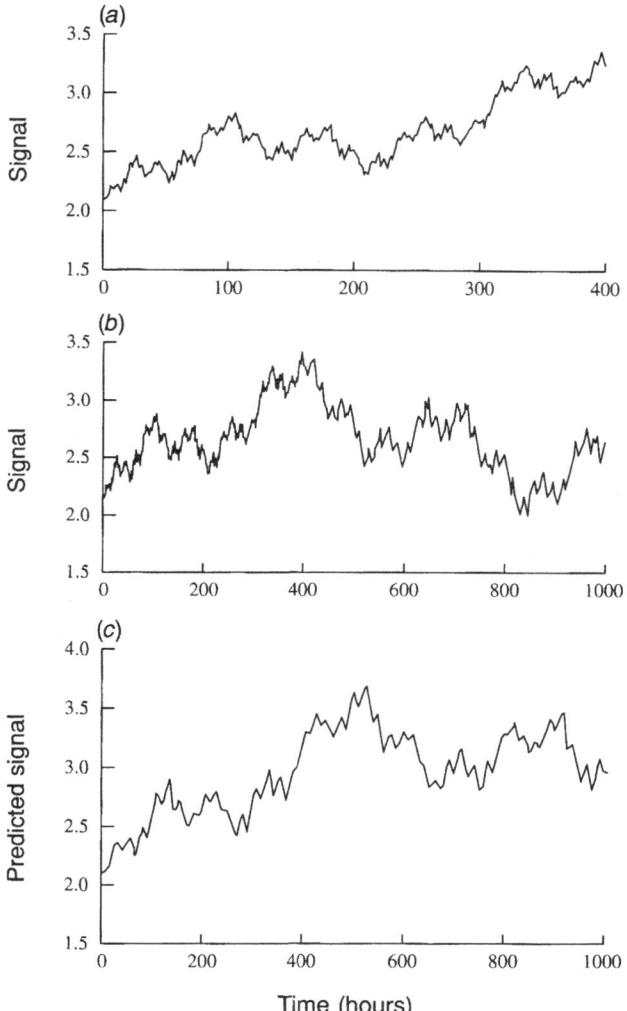

Fig. 2. Hypothetical time series data (*a*) first 400 hours' observations; (*b*) observations over 1000 hours; (*c*) time series reconstructed using correlations derived from data in (*a*) to extrapolate to 1000 hours.

phenomena can be understood as locally persistent and long-term random (West, 1990). In general, however, these narrow-band forms of variability present less of a problem for scaling because they are less persistent than power law forms. Provided the scales of concern fall outwith the correspondingly narrower range between lower and upper characteristic time scales, the assumption of homogeneity introduces little error in prediction. However, the nature of variability contains most of the important information about the mechanisms intrinsic to a system. Ignoring it precludes the possibility of reaching detailed understanding of the phenomenon, and hence it will not be possible to predict confidently the behaviour of the system under any change of circumstance. Because of their importance, a range of techniques exist for analysing time series, the most popular of which are based on the analysis of the correlations between measurements made at successive time increments as outlined above (Sugihara, Grenfell & May, 1990; Hastings & Sugihara, 1993).

So far, the examples we have discussed are concerned with cases where the mechanisms generating variability are intrinsic to the system as defined. Complications to scaling also arise from extrinsic sources when a non-linear system is forced by some external agent that varies with time. A simple but common example is the interaction between weather and crop growth.

Plant growth and development are connected to the environment via a combination of linear and non-linear responses (Porter & Delécolle, 1988; Campbell & Norman, 1989). Crop simulation models (Penning de Vries & van Laar, 1982; Porter, 1993) reflect the above mixture of responses and, in the broadest sense, transpose a distribution of weather sequences to a distribution of total dry matter and, in the case of crop plants, harvestable yield.

We present two examples where ignoring the non-linearity of 'climate impact' models gave diverse predictions of the effect of climate change on crop yields. The first arises from the method used to synthesise predictions of climate change from individual global circulation models (GCMs). Although based on the same physical equations, the predictions of different GCMs of possible climate change are very different. For example, the Geophysical Fluid Dynamics Laboratory (GFDL) model predicts an increase in global mean temperature of about 4 K while the UK Meteorological Office Low Resolution (UKMO-LO) model predicts an increase of 5.2 K for the same equilibrium $2 \times CO_2$ scenario (Barrow, 1993). Differences between models in their regional estimations of climate change are similarly large. To reduce inter-model variation, we can either average information from individual GCMs to construct a 'composite' input scenario (Santer *et al.*, 1990) for a crop model or we can preserve the diversity of individual GCM scenarios and aggregate the output from the crop models (Fig. 3). With a linear system, the output is independent of the order in which these procedures

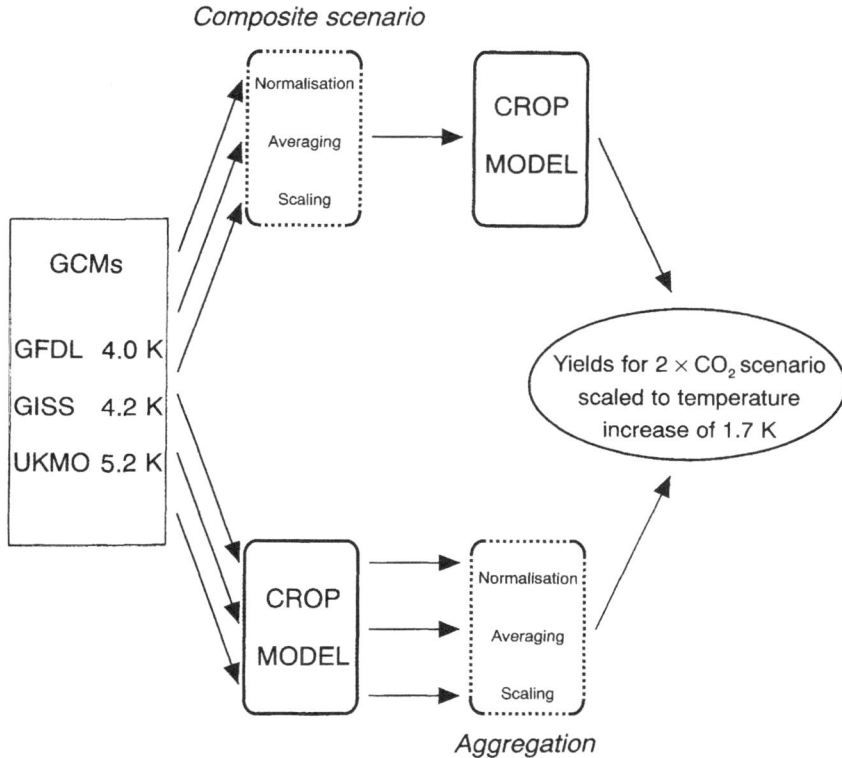

Fig. 3. Two ways to treat information on the effects of climate change on crop production using different GCMs (GFDL, GISS and UKMO-LO (UKMO)) and a crop–climate model. The increase in global mean temperature by each GCM at equilibrium under $2 \times CO_2$ conditions is shown in the GCM box. The value of 1.7 K is the best estimate of the increase in global mean temperature according to IPCC A 2050 scenario (Houghton, Callander & Barney, 1992).

are performed but this is not what we found. Using a crop model (AFRCWHEAT2) to simulate changes in wheat grain yields at four sites in the UK with three individual GCMs produced divergent results (Fig. 4). In comparison with the 30 years (1951–80) baseline climate at each site, the composite approach predicted average yields to increase by about 10% whereas the aggregation route led to an average yield decrease of about 2%. The non-linearity of the response reverses the sign of the predicted change in grain yield.

Output variables from GCMs are for atmospheric circulation at

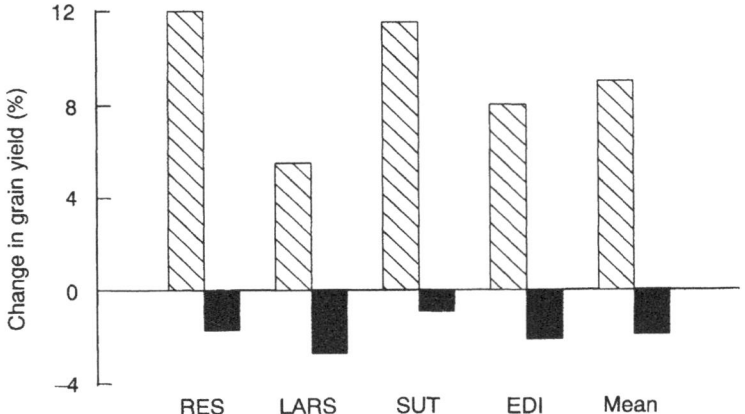

Fig. 4. Relative changes (%) in grain yield of winter-wheat simulated by the AFRCWHEAT2 crop model at four sites in the UK (RES, Rothamsted; LARS, Long Ashton; SUT, Sutton Bonington; EDI, Edinburgh) dependent on whether data from GCMs was averaged (composite scenario; ▧) before passing to the crop–climate model or aggregated after emerging from the crop–climate model (■). Climate change scenarios were those given for equilibrium $2 \times CO_2$ conditions by the GFDL, GISS and UKMO-LO GCMs and re-scaled to 1.7 K.

coarse regional scales, typically grid boxes of 3–5° of latitude and longitude, encompassing perhaps 10^5 km². These have to be downscaled to a finer resolution by a factor of about 10^3 to be usable by crop models, which also need daily weather data. There are several ways of doing this. The simplest is to append differences in average conditions between control and $2 \times CO_2$ GCM runs to historical weather data (Giorgi & Mearns, 1991). The current variability of weather sequences is maintained while the averages reflect those of the climatically changed conditions. However, the frequency of occurrence of extreme weather events, such as drought, is better correlated with the variability of climatic variables than with their mean values (Katz & Brown, 1992), and GCMs have not been validated to project changes in climatic variability (Rind, Goldberg & Ruedy, 1989; Mearns et al., 1990). Incorporation of changes in variability into climatic scenarios can be made via a stochastic local weather generator in conjunction with a crop simulation model. In climatic change studies, a stochastic weather generator can provide flexibility in the construction of weather scenarios and allows the effects of changes in local mean conditions to be explored separately from changes in their variance (Wilks, 1992). We designed

AFRCWHEAT3S to do this. It comprises the wheat simulation model AFRCWHEAT2 (Porter, 1993) and a stochastic weather generator, LARS-WG (Semenov & Porter, 1994). AFRCWHEAT3S preserves the statistical properties of generated weather sequences including their means and variances and the correlation between meteorological variables and their distributions. When AFRCWHEAT3 was run for a period of 30 years of increased mean temperature, predicted yield was reduced but there was little effect on its coefficient of variation (Fig. 5); solely doubling the standard deviation of temperature reduced the predicted mean yield similarly but substantially increased its coefficient of variation (Fig. 5). The scenario of simultaneous increases in the mean and variance of temperature decreased mean yield further. The overall effect on mean yield and its variance is not predictable on the basis of the simple arithmetic sum of the components. Analysis of changes in

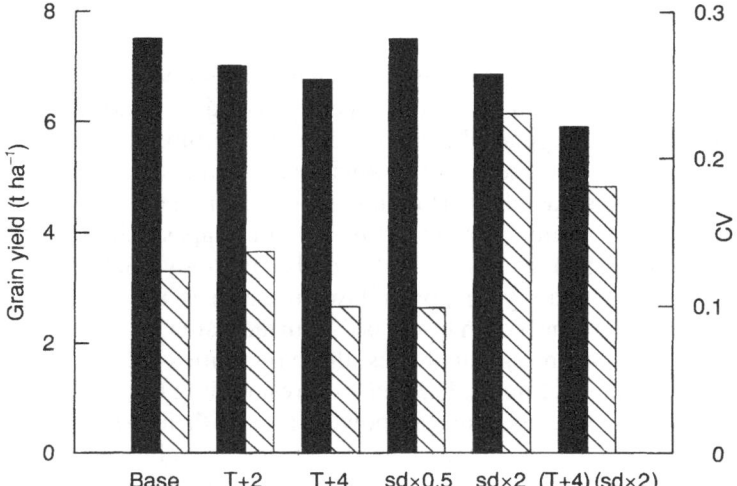

Fig. 5. The importance of temperature variability in the impact assessment of climate change using AFRCWHEAT3S. Simulated average grain yield (■) and its coefficient of variation (CV; ◌) for winter-wheat at IACR-Rothamsted for different changes in temperature: base, baseline weather conditions simulated by LARS-WG; T + 2 and T + 4, increase in mean daily temperature of 2 K and 4 K without a change in its variability; sd × 0.5 and sd × 2, halving and doubling of the standard deviation of temperature without a change in its mean value; (T + 4)sd × 2, combined 4 K increase in the daily mean temperature and a doubling of its standard deviation.

the daily variability of temperature from the UKMO-HI GCM shows that our assumptions of the scale of possible climate change in our sensitivity experiments are close to those derived from the model.

Processes that are scaling in space

Regular, smooth shapes pervade classical geometry yet are manifestly absent in nature. Therefore, many of the available mathematical tools are inappropriate for application to biological phenomena. When spatial variability in nature is uncorrelated, a spatial average can be defined that is statistically independent of location and spatial scale. However, many biological structures are highly correlated in space, in the same manner as the corelations in time discussed above, and this introduces similar complications in scaling. Analogously, the power law relationship between the magnitude and spatial scale of variability is the origin of the lack of a characteristic scale at which variability is smoothed out and averages can be defined.

Power law spatial correlations are at the heart of fractal geometry, a subject that appeared in pure mathematics around the end of the 19th century, but which has more recently found a home in applied fields. Fractals are irregular objects possessing structure on all scales that have been found to be useful approximations to many spatially complex natural forms such as plants (Prusinkiewicz & Lindenmayer, 1990), physiological structures (West, 1990), landscapes (Mandelbrot, 1982; Peitgen & Saupe, 1988) and clouds (Lovejoy, Schertzer & Tsonis, 1987). The underlying power law structure is reflected in the scaling behaviour of the volume ('mass'), surface area or length of the structure. Whereas in regular shapes, these quantities depend on spatial scale as power laws of index 3, 2 and 1 corresponding to the respective spatial dimension, fractals scale as power laws with non-integer indices, i.e. fractal dimensions (in direct analogy, the 'roughness' of a time series can be associated with a fractal dimension related to the power law index describing variability). As an example, the solid matrix of soil can be approximated by a fractal (Young & Crawford, 1991) and this implies that density (ρ) is scale dependent according to

$$\rho \propto r^{D-3}$$

where r is the spatial scale and D is the fractal dimension. The fractal nature of the structure of soil has implications for the processes that take place within it. Water and gaseous movement will be discussed below, but the observation that pore walls are also well described by a fractal surface (Young & Crawford, 1991; Kampichler & Hauser, 1993) has consequences for microbial dynamics. By virtue of their ability to exploit the smaller convolutions in the fractal pore wall, a greater pore surface area is available to small microorganisms compared with large. Therefore, approximately half of the pore surface available to bacteria ($r = 5$ μm) provides a refuge from the larger predatory protozoa ($r = 30$ μm) (Crawford, Ritz & Young, 1993). The consequences of ignoring the scaling properties of the pore surface for understanding microbial dynamics are clearly considerable. Since biological processes in soil occur across scales ranging from microbial ($r = 1$ μm) to plant roots ($r = 1$ m) and above, it is crucial that appropriate scaling laws are employed to study the phenomena. Similar comments apply to the effect of above-ground heterogeneity on habitat space (Morse *et al.*, 1985; Marquet, Navarrete & Castilla, 1990). As before, it is necessary to understand the origin of the scaling laws.

Fractal structure can arise either deterministically via some bifurcation process such as branching (West, Bhargava & Goldberger, 1986; Crawford & Young, 1990) or from correlations with an entirely random origin. The growth and development of biological organisms through a repetition of structural units across scale exemplifies fractal scaling. The relation between metabolism and body mass leads to another example of fractal scaling derived from deterministic causes. In contrast, diffusion limited aggregation (DLA) (Witten & Sander, 1981) is a stochastic process that generates a complex structure which is highly ordered, and which has been used to model a diverse range of phenomena such as electrical discharge and viscous fingering. Individual particles diffusing towards a central seed and sticking on impact slowly generate a branched pattern that is fractal (as well as having five-fold symmetry). The implied spatial correlations arise because the largest-scale structures in the pattern prevent particles from reaching the interior parts to generate small-scale structure. A process perhaps more relevant to soil is cluster–cluster aggregation, where individual diffusing particles stick together and the resulting aggregates can interact and unite (Meakin, 1988). For reasons similar to those in DLA, the structures are fractal with the resulting dimension depending on the nature of the sticking rules. Where fragmentation is included with a probability increasing with aggregate size, the possibility of dynamic equilibrium arises (J. W. Crawford, S. R. Verrall & I. M. Young, unpublished results). At equilibrium, a range of particle sizes exists up to a maximum that represents a characteristic scale for the collective. Above this scale, the mass of the aggregate ensemble scales linearly with volume, and intrinsic variability is generated by heterogeneity in the parameters in the nearest-neighbour interactions at the smallest scales. The occurrence of upper and lower characteristic scales that bound the region of fractal scaling will be a general property of all systems in which fractal scaling is observed. They mark the points at which the scaling laws change and are crucial in studies of multiscale phenomena.

In both DLA and cluster–cluster aggregation, short-range interactions give rise, somewhat paradoxically, to structures that are correlated across a wide range of scales. These are examples of more general phenomena where space and time interact to produce structures that are both spatially and temporally variable. Scaling of such phenomena presents the most challenging problem.

Phenomena that are scaling in space and time

Diffusion in soil is an example of a process that is temporally persistent (Brownian motion is characterised by a power law relation between magnitude of variability and frequency) confined to a domain that is a spatial fractal (Peitgen & Saupe, 1988). Traditionally, this fact has been ignored by assuming either that the soil is homogeneous, or that it comprises a few (usually two) pore size classes. These assumptions lead to incorrect scaling models; a more realistic treatment reveals the sensitivity of the process to the intrinsic variability (Crawford *et al.*, 1993). Under the assumption of homogeneity, where a constant diffusion coefficient is implied, the mean-square displacement, $\langle r^2 \rangle$, of the diffusing particles scales linearly with time. However, in a fractal soil the relationship is no longer linear and a different law is implied, of the form

$$\langle r^2 \rangle \propto t^{d_s/d_p}$$

where d_p is the fractal dimension of the pore space and d_s is a measurable scaling exponent relating to pore connectivity (Crawford *et al.*, 1993). Diffusion in soil is consequently characterised by a non-constant diffusion coefficient, $D(r)$, which decreases with r as a power law where the index is a function of the parameters that characterise the spatial variability. As a result, diffusion rates at scales smaller than those of the cores on which the measurements are made will be larger than predicted under the assumption of homogeneity.

Water movement in soil, although analogous to gas movement in some respects, scales according to a fundamentally different scaling law. By virtue of its surface tension, water adheres to the surface of the solid rather than moving freely throughout the pore space. This is reflected in the fact that the scaling law for hydraulic conductivity, again a power law, is a function of fractal and connectivity exponents corresponding to the solid matrix (Crawford, 1994). However, the form of the scaling law is also sensitively dependent on the structure, changing discontinuously from a power law to linear scaling as the fractal dimension increases continuously below a critical value dependent on porosity (Crawford, 1994). Another critical value exists, above which the matrix ceases to conduct. Both of these cases illustrate the sensitivity of a scaling law to the details of the underlying phenomenon and serve to reiterate the point that the origin of scaling behaviour must be properly understood. Empirical derivation of the scaling law for water movement obtained from a sample of soil would not provide sufficient information for general application.

Since microbial activity is governed by oxygen and water it should not be a surprise, following the above discussion, to find that it scales in a complex manner both in space and time. Despite this, measures of activity, such as denitrification, are usually made at a single scale and linearly interpolated. Figure 6 shows the calculated distribution of oxygen in a fractal soil structure, randomly seeded with microorganisms (J. W. Crawford, C. Rappoldt & I. M. Young, unpublished). Steep gradients of oxygen concentration occur over short distances, even in the proximity of large, well-ventilated pores. The oxygen is rapidly depleted in the vicinity of the respiring microorganisms, but diffusion is limited by the tortuous and heterogeneous structure of the pores. Consequently, anaerobic zones, which are clearly heterogeneous, quickly form and the resulting rates of denitrification at the scale of the anaerobic zones will be far higher than bulk measurements made from cores. Because the microorganisms are confined to the fractal soil matrix, measured denitrification rates will probably decrease as the size of the soil samples under investigation is reduced, although the complicating interaction between structure and gas flow may modify the form of the scaling law.

Conclusions

Variability can be generated by extrinsic or intrinsic mechanisms, while the consequences for scaling are intrinsic to the details of the biological systems we are trying to understand. The distinction between mechanisms that are extrinsic or intrinsic to the system relates to a terminology based on the definition of the boundary to the system. An intrinsic source can become extrinsic if the boundary is re-defined and vice versa. These facts illustrate the point that both variability and scaling are always intrinsic to the system, and also that when trying to approach an understanding we are not free to

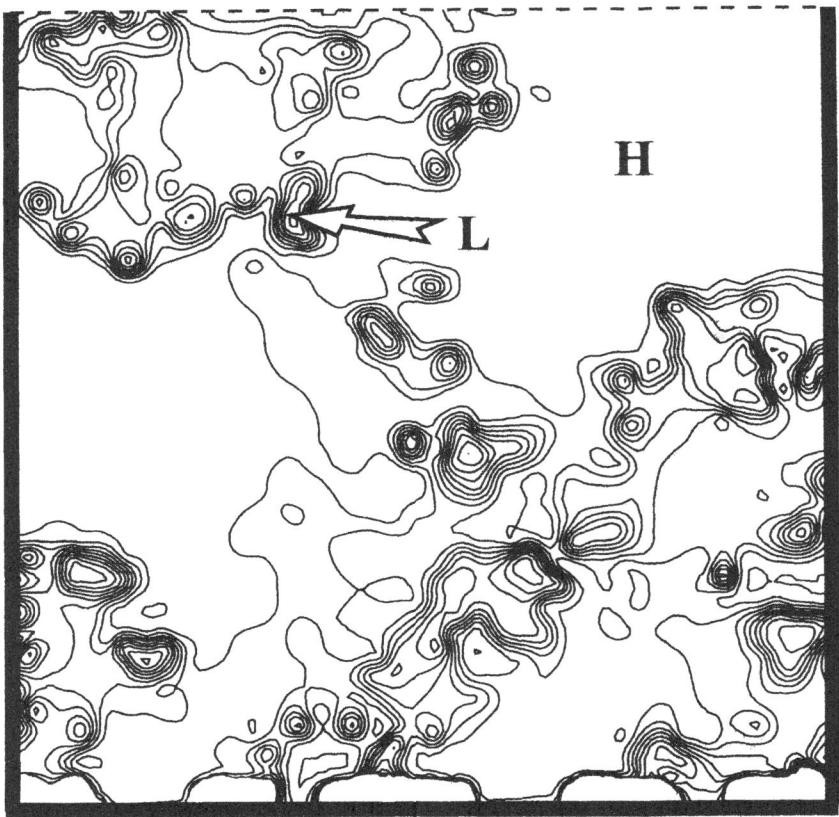

Fig. 6. Simulated oxygen concentration distribution in a structured soil, randomly seeded with microorganisms. The top of the soil core is open to the atmosphere (constant oxygen concentration) and the sides and bottom are sealed. The starting concentration was uniform. The contours are lines of equal concentration. An example of a region of high oxygen concentration (H) in a large pore well connected to the atmosphere above the soil and an anaerobic region (L) close by are indicated. The number of contours and the short distance over which they occur indicate the steepness of gradients that can be created in heterogeneous systems.

define our boundaries. In situations where logistics impose a boundary, it is, therefore, clear that attention should be focused on the couplings across the boundary, i.e. the boundary conditions, since these can have the most important bearing on the behaviour of the system. Also important is the scale of the boundary in relation to the characteristic scales. When Sugihara *et al.* (1990) analysed measles data bulked to the scale of the UK, they missed the chaotic signal that characterised the dynamics of the disease at the scale of individual cities. Boundaries and boundary conditions are as significant for scaling behaviour as the mechanisms underlying the system.

The distribution in magnitude of variability across spatial and temporal scales defines the correlations that are the basis of scaling laws. It also determines the characteristic scales where the scaling laws change. Experiments that are designed to look only at a single scale cannot lead to an understanding of the behaviour of the system at any other scale. To reach this understanding, it is essential to collect multiresolutional data and to analyse spatial and temporal correlations. Furthermore, it is necessary to develop a theory for the origin of the scaling behaviour in order to understand how scaling laws may change in circumstances different to those of the experiments, in particular when the boundaries change. In turn, this development of a detailed understanding leads to more informed sampling strategies through identification of key processes and scales involved in the scaling behaviour. Theory and experiment must, therefore, develop in tandem, not in sequence.

The rewards for this effort are considerable. Identification of the key processes involved in defining the characteristic scales and scaling laws, as well as the influence of boundary conditions, leads to informed prediction and estimation of precision. Identification of the scale at which these processes act also leads to informed management of the system, since it is at these scales that behaviour at all other scales is influenced. At a time where there is an increasing awareness of the growing strength of the coupling across the boundary between humans and their environment, it is particularly important to take these issues on board. By ignoring or attempting to average out variabiity in data, we throw away the most valuable key to understanding.

Acknowledgements

The authors would like to thank I. M. Young and N. Matsui for their work on soil structure, C. Rappoldt for discussions on denitrification and M. A. Semenov for his work in relation to crop modelling and climate change. This research is supported by both the Scottish Office Agriculture Environment and Fisheries Department and the Biotechnology and Biological Science Research Council (formerly AFRC).

References

Barrow, E.M. (1993). Scenarios for climate change for the European Community. *European Journal of Agronomy*, 2, 247–260.

Campbell, G.S. & Norman, J.M. (1989). The description and measurement of plant canopy structure. In *Plant Canopies: Their Growth, Form and Function*, ed. G. Russell, B. Marshall & P.G. Jarvis, pp. 1–19. Cambridge: Cambridge University Press.

Crawford, J.W. (1994). The relationship between structure and the hydraulic properties of soil. *European Journal of Soil Science*, 45, 493–502.

Crawford, J.W. & Young, I.M. (1990). A multiple scaled fractal tree. *Journal of Theoretical Biology*, 145, 199–206.

Crawford, J.W., Ritz, K. & Young, I.M. (1993). Quantification of fungal morphology, gaseous transport and microbial dynamics in soil: an integrated framework utilising fractal geometry. *Geoderma*, 53, 157–172.

Giorgi, F. & Mearns, L.O. (1991). Approaches to the simulation of regional climate change: a review. *Reviews of Geophysics*, 29, 191–216.

Goldbeter, A.L. & West, B.J. (1987). Chaos in physiology: health or disease? In *Chaos in Biological Systems*, ed. W. Degn, A.V. Holden & L.F. Olsen, pp. 1–4. New York: Plenum.

Greig-Smith, P. (1964). *Qualitative Plant Ecology*, 2nd edn. London: Butterworths.

Hanski, I. (1990). Density dependence, regulation and variability in animal populations. *Philosophical Transactions of the Royal Society of London*, 330, 141–150.

Hastings, W. & Sugihara, G. (1993). *Fractals: A User's Guide for the Natural Sciences*. Oxford: Oxford University Press.

Houghton, J.T., Callander, B.A. & Barney, S.K. (1992). *Climate Change 1992. The Supplementary Report to the IPCC Scientific Assessment*. Cambridge: Cambridge University Press.

Kampichler, C. & Hauser, M. (1993). Roughness of soil pore surface and its effect on available habitat space of microarthropods. *Geoderma*, 53, 223–232.

Katz, R.W. & Brown, B.G. (1992). Extreme events in a changing climate: variability is more important than averages. *Climatic Change*, 21, 289–302.

Lovejoy, S., Schertzer, D. & Tsonis, A.A. (1987). Functional box-counting and multiple elliptical dimensions in rain. *Science*, 235, 1036–1038.

Mandelbrot, B.B. (1982). *The Fractal Geometry of Nature.* New York: Freeman.

Marquet, P.A., Navarrete, S.A. & Castilla, J.C. (1990). Scaling population density to body size in rocky intertidal communities. *Science,* 250, 1125–1127.

Meakin, P. (1988). Models for colloidal aggregation. *Annual Reviews of Physical Chemistry,* 39, 237–267.

Mearns, L.O., Schneider, S.H., Thompson, S.L. & McDaniel, L.R. (1990). Analysis of climate variability in general-circulation models – comparison with observation and changes in variability in $2 \times CO_2$ experiments. *Journal of Geophysical Research,* 95, 20 469–20 490.

Morse, D.R., Lawton J.H., Dodson, M.M. & Williamson, M.M. (1985). Fractal dimension of vegetation and the distribution of arthropod body lengths. *Nature,* 314, 731–732.

Peitgen, H. & Saupe, D. (1988). *The Science of Fractal Images.* Berlin: Springer.

Penning de Vries, F.W.T. & van Laar, H.H. (1982). *Simulation of Plant Growth and Crop Production.* Wageningen: Pudoc.

Pimm, D.L. & Redfearn, A. (1988). The variability of population densities. *Nature,* 334, 613–614.

Porter, J.R. (1993). AFRCWHEAT2: a model of the growth and development of wheat incorporating responses to water and nitrogen. *European Journal of Agronomy,* 2, 69–82.

Porter, J.R. & Delécolle, R. (1988). The interaction between temperature and other environmental factors in controlling the development of plants. In *Plants and Temperature,* ed. S.P. Long & F.I. Woodward, pp. 133–156. Cambridge: The Company of Biologists.

Prusinkiewicz, P. & Lindenmayer, A. (1990). *The Algorithmic Beauty of Plants.* New York: Springer-Verlag.

Rind, D., Goldberg, R. & Ruedy, R. (1989). Change in climate variability in the 21st century. *Climatic Change,* 14, 5–37.

Santer, B.D., Wigley, T.M.L., Schesinger, M.E. & Mitchell, J.F.B. (1990). *Developing Climate Scenarios from Equilibrium GCM Results, Report No. 47.* Hamburg: Max-Plank-Institut für Meteorologie.

Schroeder, M.R. (1990). *Fractals, Chaos and Power Laws: Minutes from an Infinite Paradise.* New York: Freeman.

Semenov, M.A. & Porter, J.R. (1994). The implications and importance of non-linear responses in modelling of growth and development of wheat. In *Predictability and Non-linear Modelling in Natural Sciences and Economics,* ed. J. Grasman & G. van Straten, pp. 157–171. Dordrecht: Kluwer.

Sugihara, G., Grenfell, B. & May, R.M. (1990). Distinguishing error from chaos in ecological time series. *Philosophical Transactions of the Royal Society of London,* 330, 235–251.

West, B.J. (1985). An essay on the importance of being non-linear. *Lecture Notes in Biomathematics*, Vol. 62. Berlin: Springer-Verlag.

West, B.J. (1990). *Fractal Physiology and Chaos in Medicine*. Singapore: World Scientific.

West, B.J., Bhargava, V. & Goldberger, A.L. (1986). Beyond the principle of similitude: renormalisation in the bronchial tree. *Journal of Applied Physiology*, 60, 1089–1097.

Wilks, D.S. (1992). Adapting stochastic weather generation algorithms for climate change studies. *Climate Change*, 22, 67–84.

Witten, T.A. & Sander, L.M. (1981). Diffusion limited aggregation, a kinetic critical phenomenon. *Physical Review Letters*, 47, 1400–1403.

Young, I.M. & Crawford, J.W. (1991). The fractal structure of soil aggregates: its measurement and interpretation. *Journal of Soil Science*, 42, 187–192.

G. RUSSELL and P.R. VAN GARDINGEN

Problems with using models to predict regional crop production

Introduction

Since the 1960s, our understanding of crop yield has been revolution-ised by the development of process-based crop models such as AFRC WHEAT (Weir *et al.*, 1984) and WOFOST (van Diepen *et al.*, 1989). Models like these, which were initially developed, parameterised and validated at a field plot scale, are now being used not only as research tools at this scale but also as potential aids for decision making at a regional level (e.g. NSCGP, 1992; van Lanen *et al.*, 1992; Wolf, 1993). In this context, regions are taken to be the administrative regions for which agricultural statistics are collected; these are generally more than 1000 km^2 in area. The aim of this chapter is to review the factors that affect the accuracy of this type of activity.

Two main problems arise when a crop model is scaled-up from the field to the regional scale. The aims of the modeller who created the model may not match the requirements of the user and it is often diffi-cult to derive the inputs needed to run the model at the new scale. Moreover, since soil characteristics vary significantly within areas of less than 1 ha (Beckett & Webster, 1971), modellers inevitably have to make compromises between the incorporation of environmental varia-bility on the one hand and data availability and computing efficiency on the other.

Biologists and agronomists deal with variability by taking random or stratified random samples and applying standard statistical techniques to the results to calculate a measure of central tendency, normally the arithmetic mean, and its associated confidence limits. Crop modellers have not, in general, adopted this approach, although the major within-region sources of variation can be taken account of by stratifying each region into a series of homogeneous land evaluation units (LEUs) and running the model for each in turn. This procedure reduces but does not remove the problems of scale. Estimates of confidence limits for regional yield predictions are rare in the modelling literature.

In this chapter, a discussion of spatial variability in crop yields and phenology will be followed by an examination of sources of error in model predictions.

Measured variability of state and output variables

Within-field variability

For most of the current generation of crop models, the size of the basic area, or experimental unit, for which predictions are made must be similar to the size of the field trials used for parameterisation, namely a few hundred square metres. Field trials used for this purpose commonly consist of a series of replicated plots, each of which may be 40 m^2 in area, in apparently homogeneous fields. In such trials, the coefficient of variation (CV) for the yield of well-grown crops is rarely less than 5% and is often much more. In many parts of Europe, the individual parcels of land used for crop growing are typically two orders of magnitude larger and are less homogeneous. There are few published data on crop variability within real farmers' fields. However, the recent introduction of combine harvesters equipped with grain-flow meters and Global Positioning System sensors (Searcy et al., 1989; Blackmore, 1995) has allowed yields to be mapped and their variability assessed. Figure 1 shows the distribution of yields in a typical arable field monitored in this way. The mean yield was 6.6 t ha^{-1} and the coefficient of variation 10%. Although the data were negatively skewed, the median was only about 3% more than the mean and any errors introduced by assuming normality would consequently be small.

Within-region variability

In spite of decades of collection of agricultural statistics, few estimates of intra-regional variation in yield have been published. Clearly, the variability depends on the heterogeneity of the regions both in terms of the physical environment and the types of farming system used. Hay, Galashan & Russell (1986) divided Scotland into three regions and estimated the within-region coefficient of variation for mean farm cereal yields to be about 20%.

Published phenological data are even more scarce. Observations of development stage have, however, been routinely made on a sample of farmers' fields in south-east Scotland (about 15 000 km^2) to help advise farmers about crop protection (M. Farquhar, personal communication). Five years of data from commonly grown winter-wheat cultivars were analysed by plotting the frequency distribution of the dates on which

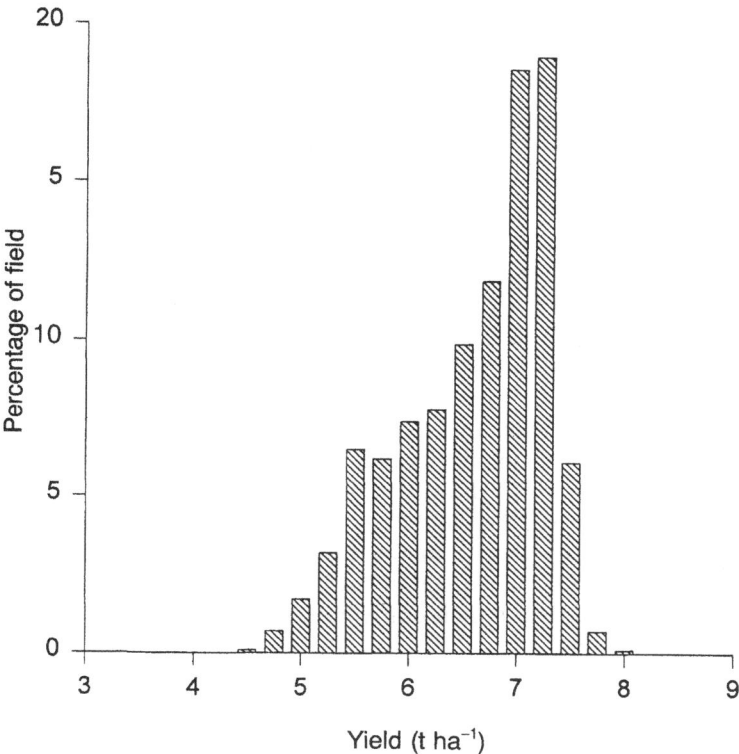

Fig. 1. Yield variation in a field of winter-wheat. Data courtesy of C. Dawson and Massey Ferguson.

the crops achieved particular growth stages (GS). Key development stages in winter-wheat are GS 30 (Zadoks, Chang & Konzak, 1974), which marks the onset of stem extension, and GS 55, which marks the mid-point of ear emergence. In practice, field observations of GS 31 (one node) and GS 51 (the start of ear emergence) are more reliable.

Individual crops achieve GS 30 after a certain thermal time (accumulated temperature) has passed, the exact total required depending on cultivar and photoperiod, and, therefore, on sowing date and year. In a maritime climate like that of south-east Scotland, temperatures increase slowly and erratically in spring time and relatively small differences in thermal time can be associated with large differences in calendar time. In each year of the study, 90% of the monitored crops achieved growth stage 31 (one node) within a four-week period

(Fig. 2) and the distribution of these dates was essentially constant from year to year even though the median date varied by 16 days.

The predicted yield of many wheat models is sensitive to the date of anthesis, which occurs about 12 days after GS 51. Modellers take comfort from the narrow range of dates found in single-site trials incorporating a wide range of year-sowing date combinations. On a regional scale, however, the variability is rather larger (Fig. 3). GS 51 was achieved by 90% of the crops within a two-week period and the median date varied by 17 days. For a typical above-ground growth rate of 15 g m^{-2} per day and a harvest index of 0.45, a two-week delay in the date of anthesis is equivalent to 0.9 t ha^{-1} increase in grain dry weight. In practice, the advantage of delayed anthesis would be less. Low temperatures delay not only the rate of development but also canopy expansion and are often correlated with lower radiation receipts. The topographic heterogeneity of this region and its proximity to the sea suggest that most other arable areas would show less phenological variability. Therefore, in most circumstances, the yield of a crop in the middle of the phenological range should be close to the average. This assumption will be least tenable when conditions worsen in the latter part of the growing season. For crops like maize in which cultivars differ markedly in their time to

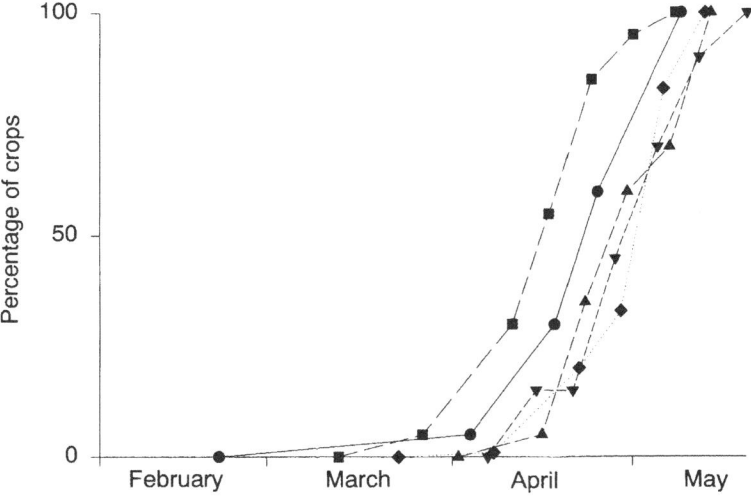

Fig. 2. The percentage of sampled winter-wheat crops in south-east Scotland at or beyond GS 31 in harvest years 1989 (●), 1990 (■), 1991 (▲), 1992 (▼) and 1993 (◆). Data courtesy of M. Farquhar and the Scottish Agricultural College.

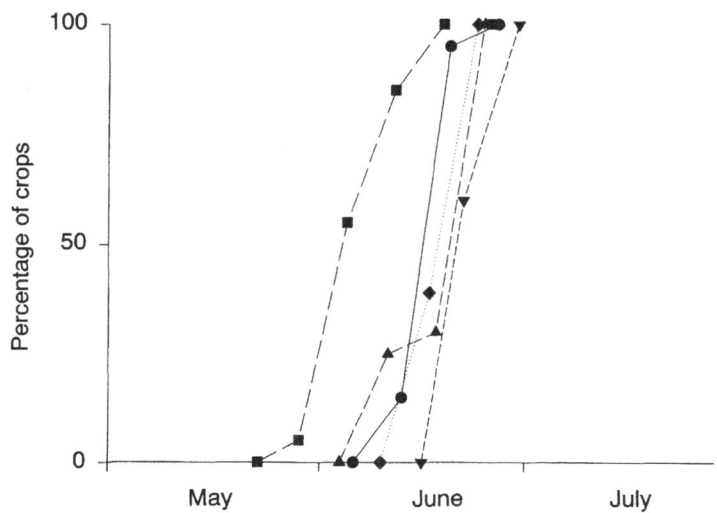

Fig. 3. The percentage of sampled winter-wheat crops in south-east Scotland at or beyond GS 51 in harvest years 1989 (●), 1990 (■), 1991 (▲), 1992 (▼) and 1993 (◆). Data courtesy of M. Farquhar and the Scottish Agricultural College.

maturity, it may be necessary to make separate predictions for the different maturity groups.

Sources of error

Four main sources of error are introduced when existing crop models are used to predict regional yields:

1. The model structure
2. The parameter values
3. The input values
4. The weather data.

Model structure

Model structure in the present chapter refers to the components or compartments in a model and the equations linking them. Models that have been satisfactorily validated may give inaccurate predictions in other environments because they omit or inadequately represent processes that are not limiting in the region of origin. Goudriaan *et al.* (1996) compared the outputs of several wheat models using meteorological data from two contrasting sites. They found that, although each model was considered well validated, there were large differences in the predicted yields. The reason for these discrepancies was not obvious but may be because an approximation that is sufficiently good in the home region is inappropriate elsewhere. For example, models developed in a region in which yield variation is dominated by a single limiting factor such as water shortage are likely to emphasise processes influenced by that factor. When such models are applied to regions where yield can be limited by other factors, either singly or in combination, their accuracy is likely to decline.

Physiological processes

Slafer & Savin (1994) have provided evidence that wheat grain yield can be either sink limited or co-limited by both source and sink. The sources of assimilate for grain growth are current photosynthesis and stem reserves. The reserves act as a buffer against unfavourable conditions during grain-fill and their importance, therefore, varies from year to year and place to place. Figure 4 shows the time course of water-soluble carbohydrate (WSC) content of wheat stems in 1993 at six sites

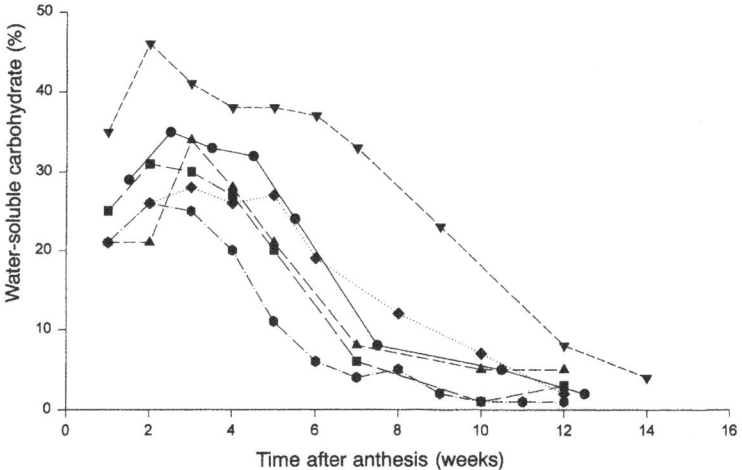

Fig. 4. Postanthesis stem reserves (expressed as percentage dry matter) in winter-wheat for harvest year 1993. Data from Edinburgh (▼) and five other sites in the UK. The data are from a Home-Grown Cereals Authority-sponsored trials series. Data courtesy of R. Sylvester-Bradley.

in the UK. Five sites, in central or southern England, conformed to one pattern while the site in Scotland showed consistently higher values. Models developed using data from places where the WSC level is normally low may well underestimate the degree of sink limitation. Therefore, models used to predict regional yield should be tested against the results of field trials from throughout the areas of interest.

Most crop models have been developed to predict potential yield, defined as the yield that would be attained if only limited by meteorological, and sometimes soil, factors directly affecting growth. Meteorological inputs to the models are usually obtained from a station representative of the region. It should be noted that there is not an inevitable correspondence between conditions at the reference station and the effect of local weather conditions on the crop at the level of an individual field or farm. Most models are essentially deterministic and ignore stochastic processes such as extreme weather events, which are nevertheless important at the farm scale. As part of a larger project (Russell & Wilson, 1994), the reasons for low wheat yields in particular years were established for countries of the European Union by questioning staff of National Statistical Offices. Table 1 shows a small part of that data set. Note that the factors responsible for most between-year variation in yield may not be the same as those responsible for variation between regions and indeed it is conceivable that they are not the major yield-limiting factors. The effects of drought and possibly of temperature extremes would be included in most crop models. Although average temperatures can be extrapolated satisfactorily, the duration of temperatures above or below a threshold is much more difficult to establish, particularly in the case of frost, where local topography plays a dominant role. The incidence of ear diseases and the effects of harvest weather are not normally included in models since their effect depends not only on external conditions but also on the response of the farmer.

Table 1. *Some causes of low wheat yield at a country scale*

Country	Year	Cause
Belgium	1980	Ear diseases
UK	1987	Weather at harvest
France	1987	Late frost
The Netherlands	1990	Drought
France	1992	High temperature

From Russell & Wilson (1994).

Although these factors appear to operate stochastically, it should be possible to develop algorithms to predict the *proportion* of crops affected in a region.

Factors reducing yields below the potential

Inherent in many models is the assumption that yield is limited only by the direct effects of weather and soil. For the well-managed, high-input crops typical of much of western Europe, yield is indeed largely a function of temperature, solar radiation and rainfall. Elsewhere, inputs may be suboptimal (for maximum yield) for a range of reasons related to farm size, risk averseness of farmers, distance from markets, availability of capital and legislation. Yield can fail to reach its potential because of suboptimal nutrition, poor timing of farming operations or losses caused by competition from weeds, pests and diseases. These factors are extremely difficult to quantify effectively (see later).

A more satisfactory alternative to modelling input usage and management skill is to correct the forecast yield using the ratio of actual to modelled yields found in previous years. This correction factor (CF) should approach but not reach 1.00, since there are harvest losses and low yielding edges even in well-managed crops. This approach was used by de Koning & van Diepen (1993) to investigate the productivity of regions of the European Union. They used the model WOFOST in its water-limited mode and found that although the mean CF was the same for wheat as for potatoes (0.61) the variation in CF was greater for wheat (0.14–0.95 for individual regions) than for potatoes (0.30–0.83). The reason for the lower maximum for the potato crop is presumably because of higher losses during harvesting. We might expect the CF for wheat to be related to input level and that for potatoes to be related to other factors such as the weather at harvest (water shortage is already included in the model).

Greece had a low CF for wheat and a high CF for potatoes (low inputs for wheat, good conditions for potato harvest) whereas Scotland and Ireland showed the opposite trend (high inputs for wheat but poor conditions for harvesting potatoes). These observations are consistent with the above hypothesis (Table 2). A key question is whether these ratios are constant from year to year. Input levels, rotations and management skill as well as systematic errors caused by the use of unrepresentative input data are unlikely to lead to large year-to-year variations in the CFs. Any trend caused by changes in farming practice can be analysed and corrected for. However, unfavourable weather conditions not included in the model and uncontrolled outbreaks of pests or diseases will cause inter-year variability.

Table 2. *The ratio between actual and modelled yield (CF)*
averaged over country groups

Country	Wheat (CF)	Potato (CF)
England	0.77	0.74
Benelux	0.75	0.66
Denmark and Germany	0.71	0.57
Ireland, Scotland and Wales	0.68	0.53
France	0.64	0.61
Spain and Portugal	0.48	0.54
Italy	0.42	0.60
Greece	0.38	0.64

Data from de Koning & van Diepen (1993).

Moen, Kaiser & Riho (1994) examined the relationship between modelled and measured maize yields in the state of Illinois in the USA. The model included some but not all of the factors responsible for reducing yield below the potential. The authors found a mean CF of 0.95 but a discouraging inter-year CV of 28%. It is not always the model that is wrong, however, and estimates of agricultural production have associated errors, and reliability varies between years.

Parameters

Model parameters are variables that are constant for an entire model run. They fall into two categories: those needed as inputs to make the model work, e.g. the date of sowing, and those used to describe the exact shape of the relationships specified by the equations that make up the model. Experimental errors associated with estimating parameter values from measurements made on the crops are not considered in this chapter. Parameters depend on the size of measurement unit and need to be adjusted to take account of the frequency distribution of the independent variables. Parameter values are usually obtained from field trials based in a small number of representative locations. For a trial to be 'representative' it must respond to environmental conditions in the same way as the cropping area of interest as a whole. The trial sites (Fig. 5) used as the basis for the developmental part of AFRCWHEAT2 (Porter *et al.*, 1987) covered virtually the entire range of combinations of thermal regime and photoperiod encountered in the wheat-growing parts of north-west Europe. The field plots were located on research stations and experimental farms where management is of a high

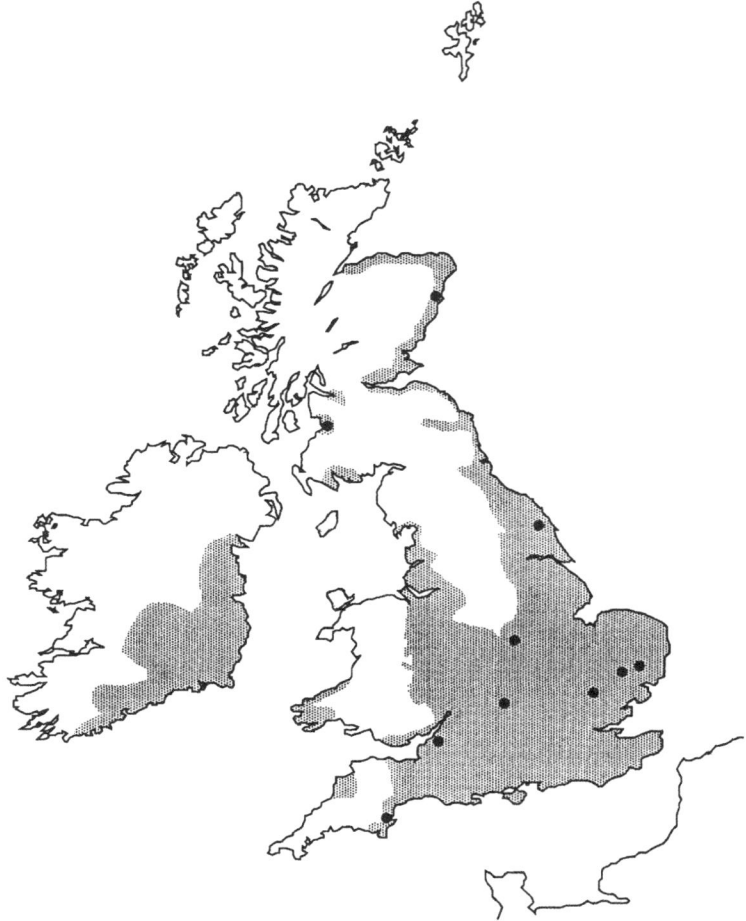

Fig. 5. The sites used by Porter *et al.* (1987) for their research on wheat development. The wheat-growing areas of the British Isles are stippled.

standard and commercially impractical treatments are often used to remove the effect of limiting factors such as low levels of weed competition. The subject of interest of these trials was the phenology of the crop. If the trials had been established to investigate the processes involved in crop growth and yield, the selected sites would have been less appropriate since soil–climate combinations are not represented in proportion to their contribution to the wheat-growing areas of the

British Isles (the stippled areas in Fig. 5). In particular, the less fertile and climatically less favoured regions are poorly represented. This problem of the non-representative nature of trial sites used for parameterisation and validation is widespread. It is a sound principle of ecological research that measurements should be made in extreme conditions as well as in typical ones so that hypotheses can be adequately tested. This principle should also be applied to agricultural research.

It has long been known that even the best managed field experiments and trials can show an unwelcome degree of variability in attributes such as plant population density and components of yield (e.g. Anon, 1926). Visual inspection of fields often shows variation in crop cover and, therefore, green area index (GAI), which is a key measure of canopy development used to compute canopy photosynthesis. GAI is a similar measure of leafiness to LAI (leaf area index) but excludes dead or senescent tissue. The time course of GAI is modelled as a function of factors, which may include supply of assimilate, thermal time from crop emergence, soil water status and nitrogen availability. It should, therefore, be interesting to assess the within-field variation of GAI over a growing season. In one winter-wheat trial near Edinburgh, Scotland, GAI was monitored at regular intervals during the 1993 field season using an LAI-2000 canopy analyser (LI-COR, Lincoln, Nebraska). On each sampling date, the GAI was approximately normally distributed within a relatively narrow range (Fig. 6). However, in order to assess scaling-up errors, it is the distribution of canopy photosynthesis, and, therefore, crop growth rate (CGR), that is of more importance. Crop growth rate can be estimated from the data in Fig. 6 for the period before anthesis using a simple Beer's law radiation absorption model and a constant dry matter:radiation quotient (DMRQ) (Russell, Jarvis & Monteith, 1989). When this is done using typical values for the UK (light extinction coefficient 0.50; mean PAR irradiation 7 MJ m^{-2} per day; DMRQ 3 g MJ^{-1}), the absolute variability is found to be greatest in the early part of the season when PAR receipts are low and least during the major period of crop growth. Using these parameters and the measured values of GAI shown in Fig. 6 produces estimates of the coefficient of variation of growth rate of 40% at the first observation, dropping to 32% at the second and to between 1 and 2% at maturity.

There are two ways of calculating the PAR (photosynthetically active radiation flux) absorbed by a crop. The GAI estimates from each plot can be averaged before computing a single value of absorbed PAR (APAR) or the calculated values of APAR for each plot can be averaged. In the former case, the mean APAR is overestimated in the early part of the year because the relationship between APAR and GAI is

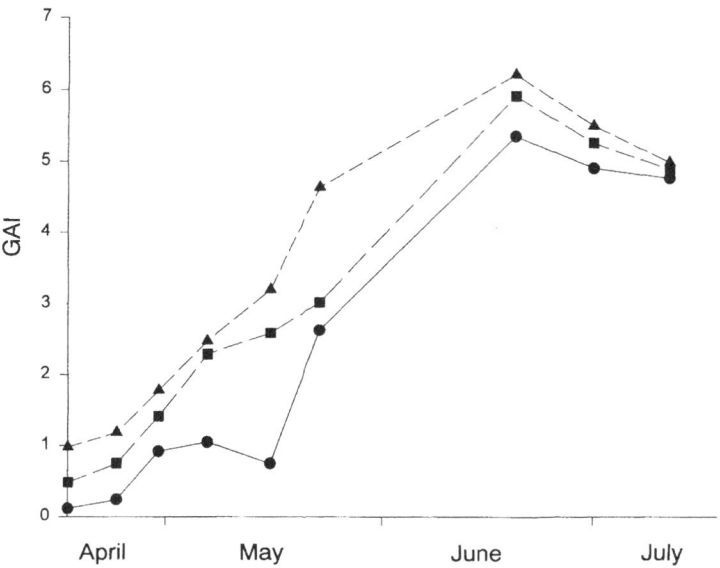

Fig. 6. The time course of GAI of a winter-wheat crop grown in south-east Scotland in harvest year 1993. The crop was sown on 7 October 1992 and the data span the period from 17 April to 12 July. The lines mark the minimum (●), the median (■) and the maximum (▲) of the replicate values.

negatively exponential. In the data presented in Fig. 6, the difference in accumulated APAR at anthesis between the leafiest and least leafy parts of the crop was less than 10%. Consequently, simple averaging should normally suffice for estimating the parameters specifying the time course of GAI. The design of an effective sampling programme should, however, be an important component of the process of obtaining appropriate parameter values.

Input data

In homogeneous regions, it should be relatively easy to identify a single set of weather and other input data that can be used to drive a crop model so that the response of the model to weather is the same as the average of all the fields of the crop in the region. However, there will be many situations in which the mean regional yield calculated in this way differs from the true mean. The main reason for this discrepancy is that many plant processes, such as the PAR absorption mentioned earlier, are curvilinearly related to environmental variables. The consequence of this curvilinearity varies with pedo–climatic zone (soil–climatic zone). For example, only small errors in calculated growth rate are introduced by assuming that the daily net canopy photosynthesis of crops growing in north-west Europe is a linear function of daily irradiation because these crops remain on the linear portion of the response curve for most of the growing season. The extreme instance of curvilinearity is when there is a conditional switch with a threshold above, or below, which a response does not take place. For example, the rate of grain-fill of wheat is reduced when the maximum air temperature exceeds 28 °C on three successive days (Russell & Wilson, 1994). If the averaged regional weather data suggests no limitation and a significant part of the region actually exceeds the threshold, the regional mean yield will be overestimated. Conditional switches have already been identified as potential sources of error in field-scale models (e.g. Kocabas *et al.*, 1993). In deterministic models, they can result in unrealistic step changes in the predicted response of the crop to environmental variables.

Soil
Soils data can be even more difficult to deal with than weather. Administrative regions are often bounded by major topographic features such as rivers or mountain ranges with the result that they include a range of topographic land units with contrasting soils. All that is often available are tabulations of data for typical profiles from broad classes of soil type. Information about some attributes that influence crop growth may, therefore, have to be derived from mapped data using pedo-transfer functions (King, 1990). This problem may explain why many crop models specify soil only in terms of available water capacity and possibly rooting depth, although in reality other factors such as air–soil temperature differences and trafficability can cause significant between-site variation in yields. Not surprisingly, the problems of spatial variability in soils, which have been well recognised by soil scientists, have been largely ignored by crop modellers. Ragg & Henderson (1980) investigated the homogeneity of four Scottish arable soil series derived from glacial drift and mapped at a scale of 1:25 000. Even at this scale, they found that the mapping purity (i.e. the probability that a random site within the mapping unit matches the taxonomic unit) was less than 66%, although the reliability for topsoil texture was much higher (84%). The authors explained the variability in terms of the presence of small and unmappable inclusions of other series. Unfortunately, for modellers, the only digital soils map that completely covers the

countries of the European Union is at a nominal scale of 1:1 000 000 (King, 1990). Areas that are smaller than 1 km² cannot be distinguished at this resolution. The actual number of mapping units within a particular region depends on the resolution (Table 3) and on the criteria used for differentiating soil types.

Not all distinctions on a soil map are significant for crop growth and many units could be amalgamated for modelling purposes, although phases (e.g. stony) are often important. For each mapping unit, it is possible to estimate the mean value of attributes of the dominant soil type. However, some parts of otherwise suitable units will not be in arable cultivation because of limitations set by slope, soil depth or drainage status. Therefore, the mean value of key attributes for the cropped area could differ from, and the variation in soil attributes be less than, the unit as a whole. Farmers have tried to reduce the effects of suboptimal soil characteristics with varying degrees of success. Consequently, the fertility of a soil cannot be deduced from the map unit alone, because of the effect of crop rotation and fertiliser policy. Other important attributes of soil, such as available water capacity, are not amenable to manipulation in this way.

Crop type and farming system

Crop yield also depends on the skill of the individual farmer and the level and type of input applied. Therefore, although farms often appear similar across relatively large pedo–climatic zones in terms of their component enterprises, economic surveys show considerable variation in their profitability. Farmers attempt to reduce the effect of a spatially variable environment by making strategic decisions of cultivar, rotation and sowing date and tactical decisions of rate and timing of fertilisers and pesticides in order to compensate for suboptimal conditions. Few current models incorporate information about important aspects of the farming system, such as the rotation employed and the intensity of input usage: EPIC (Jones *et al.*, 1991) is a noteworthy exception. It is also

Table 3. *The number of soil mapping units, including phases, in a 100 km² grid square on the southern outskirts of Edinburgh, Scotland; the 1:10 000 map covered only 15% of the area*

Map scale	1:1 000 000	1:250 000	1:63 360	1:10 000
Mapping units	4	12	27	>42

not easy to see how to obtain the specification for a typical farm in a region. The rate of nitrogen fertiliser usage seems at first to be a reasonable candidate for an index of the intensity of farming. A regional-average application rate could, for example, be calculated for the previous three years and used as an input to a model. These data exist for at least some countries. For example, geo-referenced data are collected annually on this and other aspects of fertiliser usage in the UK and published in the *British Survey of Fertiliser Practice* (e.g. Burnhill, Chalmers & Fairgrieve, 1994). The data of Fig. 7, which show a mean nitrogen application rate of 185 kg ha^{-1} and a median that is slightly higher, suggest that nitrogen is not a limiting factor in this region. However, further investigation would be needed to ascertain whether the

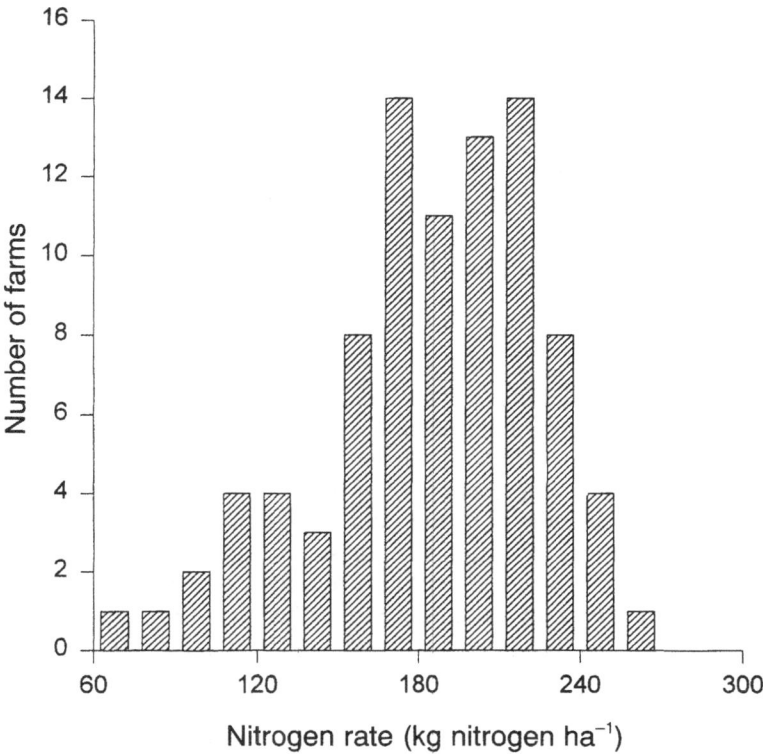

Fig. 7. Frequency distribution of the rate of application of nitrogen fertiliser to fields of winter-wheat in Scotland in 1993. Data from the *British Survey of Fertiliser Practice* (Burnhill *et al.*, 1994).

farms with low application rates operated low input systems or whether the rates were low because of the crop rotation adopted. Moreover, since weather can have a strong effect on leaching, denitrification and mineralisation, the timing of applications can affect their efficacy and farmers may have adjusted the rate to take account of losses of nitrogen earlier in the season. Uncertain data like these need to be used with caution (Bouman, 1994).

Some crop models assume a typical generic cultivar and others require genetic coefficients of the cultivar to be specified in the input data. Brisson, Bona & Bouniols (1989) carried out a sensitivity analysis on the result of modifying the genetic parameters of SOYGRO to establish the required accuracy of these parameters for soybean crops. The parameters specifying the phenology of the cultivar seem to be particularly important. There is debate as to how far genetic parameters, as currently defined, can be considered truly independent of environment. Although farmers choose cultivars that are appropriate for the expected conditions, the pressures of the market may encourage the growing of lower yielding but higher value types. As a result, it is normal for a range of cultivars to be grown in each region, with early maturing types being preferred in climatically marginal parts. Thus, there is a certain element of confounding of cultivar with geographical location and the error in yield estimates caused by ignoring cultivar will tend to be systematic rather than random.

Weather

Effective modelling of regional crop yield depends on estimating regional weather values accurately from point data collected at meteorological stations. Modellers usually either assume that the data from a single station is representative of the whole area of interest or else apply statistical procedures to the data to interpolate local weather variables from a network of stations in, or surrounding, the region. The problem of variation in weather within a region is analogous to the problem of using weekly, rather than daily, weather data to run a model. Nonhebel (1994) found that potential yield was overestimated by 5–15% if ten-day averages rather than daily weather data were used to drive a spring-wheat model, but that water-limited potential yields were underestimated in dry conditions. The two responses can be explained in terms of the non-linear responses of crop growth rate to weather variables. The rate of change of crop growth rate with solar radiation declines with solar radiation while the rate of change of crop growth rate with soil water deficit increases as the deficit increases.

Modellers concerned with regional production generally divide the countries of interest into agro-climatic zones and assume homogeneity of weather within these zones. There are three main problems with this approach. First, the agricultural portions of the agro-climatic zones may not be homogeneous in terms of some important weather variables. Consequently, it may be difficult to choose a representative meteorological station. Second, existing agro-climatic maps may be inappropriate for some crops. For example, the maps in Thran & Broekhuizen (1965) were drawn up for cereals. Conditions after cereal harvest influence the growth of crops such as sugar beet and potatoes but will not have been taken account of in the classification. Finally, the mean values for weather variables will be biased whenever they are statistically interpolated from observations made, as is often the case, at meteorological stations in coastal, urban and mountain locations that are unrepresentative of the cropped area. Little use has yet been made of co-variates such as altitude in order to improve interpolation.

Some modellers chose a single representative meteorological station to characterise the weather experienced in each region. The representative station should ideally be located where the mean crop yield is the same as the regional average. In practice, there will be a limited choice of stations and one will be chosen near the middle of the agricultural part of the region or where the yields are subjectively assessed as being typical of the region. Although the 'typical', i.e. modal, and mean yields will only be the same where the frequency distribution of yields is normal, in practice any difference will probably be small. A second source of error worth noting is that one meteorological station may not be best for all crops in a region. The distribution of forage and cereal crops, for example, may be quite diferent because their climatic requirements are not the same.

In a few parts of Europe, the network of meteorological stations is dense enough to allow an analysis of intra-regional variation in weather. De Koning & van Diepen (1993) chose one representative meteorological station to characterise the weather of each of the 109 agro-climatic zones (average area of each zone is $22\,000\,\mathrm{km}^2$, approximately $150 \times 150\,\mathrm{km}$) into which they divided the countries of the European Union. The region covering eastern England, which is topographically similar to many of the agricultural regions of Europe and which contains 57 official meteorological stations, was represented by the station at Waddington in Lincolnshire. Weather data from 1991 were analysed as an example in order to assess the variation in thermal and solar radiation environments across the region and to establish the difference between the regional mean and the figures for Waddington. Clearly, a fuller analysis would be necessary in order to develop rules for estimating the maximum error introduced by using data from a representative station rather than a regional average.

Thermal time (base temperature 0 °C) was chosen as a measure of temperature because it drives the phenological modules of many crop models. Temperature data from 1 March to 30 June were analysed and the average thermal time for the region was calculated after weighting the data to take account of the uneven distribution of stations. The mean thermal time for Waddington was 1164 °Cd, compared with the regional average of 1159 °Cd, suggesting that it was indeed representative of the region as a whole. If the rate of development of a crop depends on temperature, the duration of a phase in days will depend on the inverse of temperature. Therefore, if mean temperature is distributed normally, the durations will be positively skewed. Although the data from Waddington were positively skewed, the deviation from normality was small and could be explained in terms of a few outlying points. Therefore, the actual mean regional duration in days of a phenological stage should be very close to the value calculated for Waddington. Moreover, if the Waddington estimate of the duration of a phase lasting 1000 °Cd was applied to the whole region, rather than interpolating the data and calculating for each 10×10 km grid square, the error introduced would be less than one week in more than 90% of the region. Since climate varies systematically, errors in the estimate of regional mean temperature will tend to result in errors in predicted yield that are relatively constant from year to year. They can, therefore, be taken account of using a correction factor. Problems with estimating the mean duration of a phase are only likely to be significant when temperatures are close to the base temperature for long periods, such as those occurring in the maritime climate of north-west Europe during winter.

Crop growth rate is often well correlated with solar radiation receipts (Monteith, 1977). Solar radiation is measured at fewer stations than temperature and modellers frequently have to estimate it using regressions with hours of sunshine. For example, only six stations in the eastern England region record solar radiation. Analysis of these records suggests that the March-to-June insolation is normally distributed across the region with a mean close to the estimated mean for Waddington and a coefficient of variation of 10%.

Other variables, such as soil temperature, wind speed and minimum daily temperature, are less easy to interpolate accurately because they are influenced by field-scale topography. Satellite observations of surface temperature show considerable heterogeneity caused by differences in soil type and topography. The mean 24-hour temperature 25 mm below the soil surface, which influences the rate of development in the early stages of crop growth, is not a unique function of the temperature in the Stevenson screen at the local meteorological station but also depends, *inter alia*, on soil texture, drainage status, slope, aspect and degree of shelter. When several weather variables are used together, such as in the computation of the soil-water balance or the identification of suitable periods for agricultural operations, the problems are compounded, especially when correlations between the variables are ignored.

Models are increasingly being run using synthetic weather data generated either stochastically (see Marshall, Crawford & Porter, this volume) or from a general circulation model (GCM). Topographical features such as coastlines and mountain ranges, which can have a significant effect on the actual weather experienced, are poorly represented in the current generation of GCMs. The significance of these omissions depends on the resolution of the weather model and the topographic heterogeneity of the region being considered.

Conclusions

Errors are introduced whenever models are used at scales for which they were not developed. It is, therefore, important for modellers to specify the scale at which their model was designed to operate. When models are used at a new scale, they need to be tested to see whether they have to be re-parameterised. This should be carried out both by re-validating the model and by undertaking an analysis of the sensitivity of the model to input and parameter values. The particular problems of scaling will vary with circumstances and some sources of error may have a smaller effect on the accuracy of regional yield estimates than expected. Researchers sometimes ask what is the best scale at which to develop a model. The answer depends on the ultimate objectives of the modelling exercise. In practice, however, the resolution of the model is often limited by the constraints of computing power and data availability. It is, therefore, important to be aware of any errors caused by the choice of scale and to think how the available data could be scaled-up or scaled-down. The increasing availability of GIS (Geographical Information Systems) data sets means that this type of information may be preferred even though it may be of an inappropriate resolution and may not include all the attributes that are required. The solution to the latter problem is to use the available data sources as surrogates of what is actually required or to convert them using correlation or more complex transfer rules (King, 1990). The technology is available with the current generation of parallel computers to solve the problems associated with insufficient computing power. In countries where there is an insufficient coverage of appropriate digital information, the most effective means of obtaining the information needed to run a model for forecasting regional yield may be to randomly sample the region of interest rather than to map it. The model can then produce a statistically valid estimate of mean yield and confidence limits.

The biggest problem with the current methods of forecasting regional yield is that they have evolved on an *ad hoc* basis from existing methods used for other purposes. Projects such as the MARS project of the European Union (Vossen & Meyer-Roux, 1995) are showing how a range of methodologies can be integrated to improve forecasts.

Acknowledgements

The following individuals and organisations provided us with data: Chris Dawson, Massey Ferguson, Moyra Farquhar (Scottish Agricultural College), Roger Sylvester-Bradley (co-ordinator of the Home-Grown Cereals Authority development project), Peter Burnhill and Joan Fairgrieve (The British Survey of Fertiliser Practice). Allied Distillers Ltd assisted with the purchase of the canopy analyser; some of the ideas were developed during discussions with Kees van Diepen (Winand Staring Centre) and Anne Burrill (Joint Research Centre, Ispra). John Porter's comments suggested useful developments of the initial ideas. Linda Sharp helped with the analysis and drew the figures. Thank you all very much.

References

Anon (1926). A census of an acre of barley. *Journal of the Royal Agricultural Society of England*, 87, 103–123.

Beckett, P.H.T. & Webster, R. (1971). Soil variability: a review. *Soils and Fertilizers*, 34, 1–15.

Blackmore, S. (1995). *Precision Farming, An Introduction.* World Wide Web URL http://www.cranfield.ac.uk/safe/cpf/papers/precfarm.htm

Bouman, B.A.M. (1994). A framework to deal with uncertainty in soil and management parameters in crop yield simulation: a case study for rice. *Agricultural Systems*, 46, 1–17.

Brisson, N., Bona, S. & Bouniols, A. (1989). SOYGRO, un modèle de simulation de la culture du soja; adaptation à des variétés cultivées dans le sud de l'Europe et validation. *Agronomie*, 9, 27–36.

Burnhill, P., Chalmers, A. & Fairgrieve, J. (1994). *The British Survey of Fertiliser Practice. Fertiliser Use on Farm Crops, 1993.* London: HMSO.

de Koning, G.H.J. & van Diepen, C.A. (1993). *Crop Production Potential of Rural Areas within the European Communities. IV: Potential, Water-limited and Actual Crop Production. Report W68.* The Hague: Netherlands Scientific Council for Government Policy.

Goudriaan, J. with contributions from Hunt, L.A., Ingram, J. *et al.* (1996). Predicting crop yields under global change. In *Global Change and Terrestrial Ecosystems*, ed. B.H. Walker & W.L. Steffens. Cambridge: Cambridge University Press.

Hay, R.K.M., Galashan, S. & Russell, G. (1986). The yields of arable crops in Scotland 1987–92: actual and potential yields of cereals. *Research and Development in Agriculture*, 3, 159–164.

Jones, C.A., Dyke, P.T., Williams, J.R., Kiniry, J.R., Benson, V.W. & Griggs, R.H. (1991). EPIC: an operational model for evaluation of agricultural sustainability. *Agricultural Systems*, 37, 341–350.

King, D. (1990) The available water capacity map compiled from the European Community soils map at scale one to one million. In *The Application of Remote Sensing to Agricultural Statistics*, ed. F. Toselli & J. Meyer-Roux, pp. 235–242. CEC: Luxembourg.

Kocabas, Z., Mitchell, R.A.C., Craigon, J. & Perry, J.N. (1993). Sensitivity analysis of the ARCHWHEAT1 crop model: the effect of changes in radiation and temperature. *Journal of Agricultural Science*, 120, 149–158.

Moen, T.N., Kaiser, H.M. & Riho, S.J. (1994). Regional yield estimation using a crop simulation model: concepts, methods and validation. *Agricultural Systems*, 4, 79–92.

Monteith, J.L. (1977). Climate and the efficiency of crop production in Britain. *Philosophical Transactions of the Royal Society of London, Series B*, 281, 277–294.

Nonhebel, S. (1994). The effect of use of average instead of daily weather data in crop growth simulation models. *Agricultural Systems*, 44, 377–396.

NSCGP (1992). *Ground for Choices: Four Perspectives for the Rural Areas in the European Community. Report 42.* The Hague: Netherlands Scientific Council for Government Policy.

Porter, J.R., Kirby, E.J.M, Day, W. *et al.* (1987). An analysis of morphological development stages in Avalon winter wheat with differing sowing dates and at ten sites in England and Wales. *Journal of Agricultural Science*, 109, 107–121.

Ragg, J.M. & Henderson, R. (1980). A reappraisal of soil mapping in an area of southern Scotland. Part I. The reliability of four soil mapping units and the morphological variability of their dominant taxa. *Journal of Soil Science*, 31, 559–572.

Russell, G. & Wilson, G.W. (1994). *An Agro-pedo-climatological Knowledge-base of Wheat in Europe*. Luxembourg: CEC/JRC.

Russell, G., Jarvis, P.G. & Monteith, J.L. (1989). Absorption of radiation by canopies and stand growth. In *Plant Canopies: Their Growth, Form and Function*, ed. G. Russell, B. Marshall & P.G. Jarvis, pp. 21–39. Cambridge: Cambridge University Press.

Searcy, S.W., Schueller, J.K., Bae, Y.H., Borgelt, S.C. & Stout, B.A. (1989). Mapping of spatially variable yield during grain combining. *Transactions of the American Society of Agricultural Engineers*, 32, 826–829.

Slafer, G.A. & Savin, R. (1994). Source–sink relationships and grain mass at different positions within the spike in wheat. *Field Crops Research*, 37, 39–49.

Thran, P. & Broekhuizen, S. (1965). *Agroecological Atlas of Cereal Growing in Europe*, Vol. I. *Agroclimatic Atlas of Europe*. Wageningen: PUDOC.

van Diepen, C.A., Wolf, J., van Keulen, H.C.A. & Rappoldt, C. (1989). WOFOST: a simulation model of crop production. *Soil Use and Management*, 5, 16–24.

van Lanen, H.A.J., van Diepen, C.A., Reinds, G.J., de Koning, G.H.J., Bulens, J.D. & Bregt, A.K. (1992). Physical land evaluation methods and GIS to explore the crop growth potential and its effects within the European Communities. *Agricultural Systems*, 39, 307–328.

Vossen, P. & Meyer-Roux, J. (1995). Crop monitoring and yield forecasting activities of the MARS project. In *European Land Information Systems for Agro-environmental Monitoring*, ed. D. King, R.J.A. Jones & A.J Thomasson, pp. 11–19. Luxembourg: European Commission.

Weir, A.H., Bragg, P.L., Porter, J.R. & Rayner, J.H. (1984). A winter wheat crop simulation model without water or nutrient limitation. *Journal of Agricultural Science*, 102, 371–382.

Wolf, J. (1993). Effects of climate change on wheat and maize production potential in the EC. In *The Effect of Climate Change on Agricultural and Horticultural Potential in Europe*, Research Report No. 2, ed. G.J. Kenny, P.A. Harrison & M.L. Parry, pp. 93–119. Oxford: Environmental Change Unit.

Zadoks, J.C., Chang, T.T. & Konzak, C.F. (1974). A decimal code for the growth stages of cereals. *Weed Research*, 14, 415–421.

R.B. LAMMERS, L.E. BAND and C.L. TAGUE

Scaling behaviour of watershed processes

Introduction

This chapter describes simulations of forest ecosystem processes over a 3000 km^2 watershed, used to investigate scaling-up effects on sampling and representing land surface attributes. Specifically, the development and control of bias in simulated carbon and water exchange processes are explored as both scale and resolution of the landscape change. An approach is outlined in which the watershed is partitioned into functional units, including hillslopes and stream channels, and the surface heterogeneity is captured as variance between land units and within land units. It is shown that an order of magnitude resolution change of the original land data sets for topography and vegetation cover can produce similar results in the carbon and water flux processes as long as the joint distribution function describing the significant surface attributes is preserved. Scaling-up is accomplished by aggregating land units into larger, more complex units, while incorporating the increased within-unit heterogeneity as subgrid variability.

In recent years, it has become apparent that knowledge of ecosystem processes at regional to global scales is required to address a number of pressing environmental problems. The interaction of atmospheric processes with land surface processes has significant implications for ecological and hydrological systems over a range of length scales. Unfortunately, much of the current knowledge of surface biophysical processes and feedbacks is derived from experiments conducted at laboratory or carefully controlled field-plot levels. Soil–vegetation–atmosphere transfer (SVAT) models represent a formalisation of the current theories and hypotheses generated from these experiments and are now being extended to operate over much larger spatial domains than those for which they were originally developed.

Initial attempts to extend these models over larger land areas implicitly assumed that the surface could be treated as a spatially exhaustive set of homogeneous areas, acting independently and in

parallel. However, recent work has demonstrated that surface heterogeneity is both strongly expressed at all regional scales and not simply
averaged in a functional sense (Avissar, 1992; Band, 1993). The non-
linear response of water and carbon flux processes to available soil
water, meteorological variables and certain vegetation canopy attributes
commonly results in significant bias when computing areal averaged
flux using mean or average surface conditions:

$$E(p(\mathbf{x})) \neq p(E(\mathbf{x})) \tag{1}$$

where $\mathbf{x}$ is a surface parameter vector, and p is a point or small area
process model that predicts some flux quantity (for example, evapotranspiration). An unbiased estimate of the areal averaged flux can be
estimated:

$$E(p(\mathbf{x})) = \int_{\mathbf{x}} p(\mathbf{x})\, f(\mathbf{x})\, d\mathbf{x} \tag{2}$$

where f is the joint distribution function of $\mathbf{x}$. While there are other
methods of estimating $E(p(\mathbf{x}))$ without explicitly integrating over the
full distribution function (for example, Bresler & Dagan, 1988; Band
et al., 1991; Rastetter *et al.*, 1992), some distributional information
is generally required if the key driving processes are non-linear. In
scaling-up or aggregating a biophysical model over progressively larger
areas, a key problem is estimation of this distributional information as
direct sampling becomes unfeasible.

Any observation made with a finite length scale will average lower
length-scale variations in surface attributes, leading to some quantifiable
information loss, and potentially biasing the sampled $f(\mathbf{x})$. This indicates that the use of fixed scale maps as information sources tends to
truncate higher-frequency information. As attention and simulation
efforts are shifted to progressively larger areas, a standard approach has
been to use more highly generalised image and cartographic products
to estimate $f(\mathbf{x})$, which has the effect of narrowing distributional information towards mean or modal values. This runs the risk of building
well-expressed scale effects into prediction of ecosystem behaviour,
reflecting the bias inherent in Equation (1). The alternative approach,
following Equation (2), indicates that methods must be developed both
to estimate subgrid resolution statistics of the distribution of surface
attributes and to incorporate this information into the simulation structure. This chapter describes a strategy that has been adopted for partitioning surface heterogeneity as variance between the spatial units
used for simulation (grid cells or irregular land features) and variance

within the units. As the land surface is progressively aggregated into larger more complex units, the overall level of heterogeneity is shifted from the between-unit to the within-unit variance using the Regional HydroEcological Simulation System (RHESSys), a data and modelling system designed for scaling-up ecosystem process simulations. It has been demonstrated that the aggregation procedure should be guided by an understanding of the significant sources of heterogeneity in the landscape (for example, topography, canopy, microclimate) and the sensitivity of the model to these driving surface and climatic factors. Different strategies for spatial aggregation are explored to optimise this process and the approach and procedures are illustrated for a set of watersheds in western Montana.

Approach and system description

The RHESSys project has been a collaborative effort between research groups at the University of Toronto, the University of Montana and the National Aeronautics and Space Adminstration (NASA)/Ames Research Center that has focused on developing methods to extend ecosystem process computations from plot to regional levels (for example, Running *et al.*, 1989; Band *et al.*, 1991; 1993; Running & Hunt, 1993). The dominant approach and philosophy of this project has been to develop a modular simulation system integrating the key interacting ecological, hydrological and atmospheric processes and factors that determine stand-to-landscape level carbon and water budgets. Representation of these processes has focused on intermediate levels of complexity that strike a balance between maintaining the key biophysical behaviour and feedbacks within the system while using levels that are feasible to determine parameters over heterogeneous landscapes. A greater level of biophysical realism could easily be put into the process model components to match the current level of understanding of individual processes; however, the inability to provide the information necessary to create parameters properly for the models negates the potential benefits of the additional levels of physical realism and may lead to a false sense of precision. Beven (1989) discussed the changing paradigm in hydrology regarding the perceptions and use of such over-parameterised approaches.

Figure 1 shows a data-flow diagram of RHESSys from data preparation through model parameterisation, simulation and visualisation, while Fig. 2 shows a more detailed rendition of the biophysical process models. The spatial framework for simulation is designed to partition the landscape into units that exist along a major environmental gradient

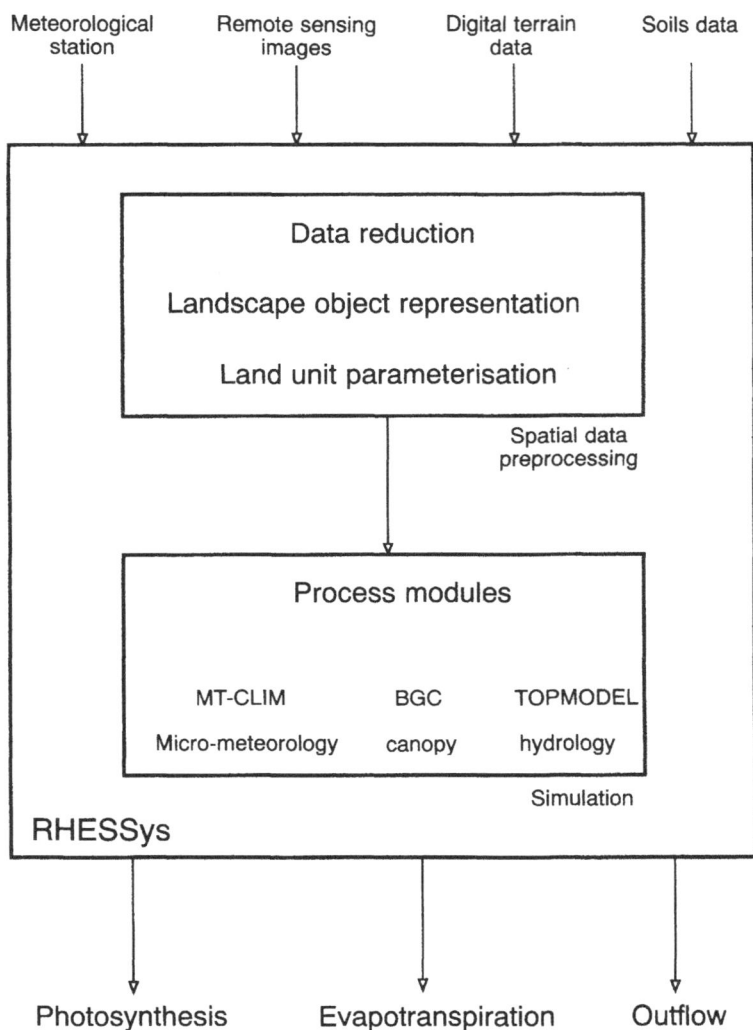

Fig. 1. RHESSys data-flow diagram showing primary input data entering at the top and selected output variables at the bottom.

over which the biophysical model shows significant variation in response. In mountainous terrain with strong differences in the radiation regime, watersheds are disaggregated into component hillslopes or valley sides, which have a low variance in exposure. In flatter terrain,

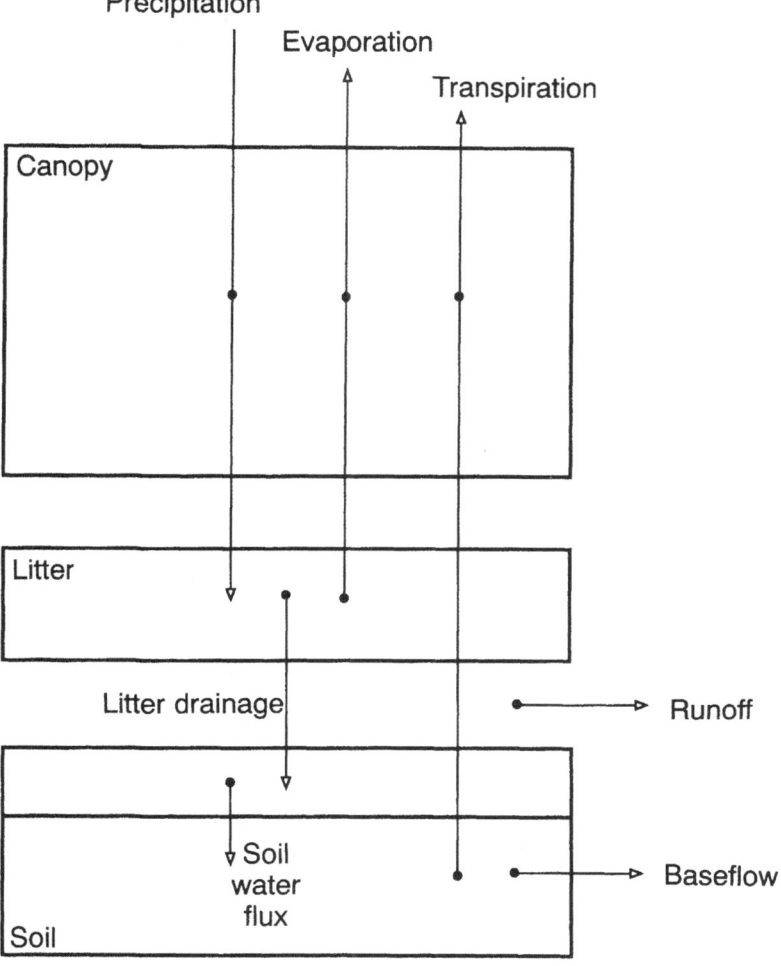

Fig. 2. RHESSys process modules showing water flows between components.

land cover may be a more important stratification. An important attribute of the approach is the incorporation of methods to process the effects of within-unit heterogeneity by stratifying the surface along an important environmental gradient. This gradient can be available soil water in water-limited areas, elevation in mountainous terrain or other variables or combination of variables that show significant variance within simulation units. For each simulation unit, the process modules are integrated over the within-unit gradient then summed to give unit-wide flux and state quantities. Incorporation of within-unit gradients is only necessary if there is sufficient variance and non-linearity of response over that variance, as discussed above. As the number of strata within a landscape unit rises rapidly and more gradients are included, it is important to choose carefully those gradients that explain the largest portion of the surface variance, and to construct efficient landscape partition methods that maximise the information captured between units.

The processes shown in Fig. 2 are called for each land surface unit and then integrated over the internal gradient. Initially, base station meteorological information is used to produce the insolation field, using the surface unit elevation, slope and aspect. Temperature, humidity and precipitation are adjusted by elevation, following the procedures in MT-CLIM (Running, Nemani & Hungerford, 1987); this can be done once for each unit if there is a limited elevation range, or as part of the internal unit integration if the elevation range is sufficient to produce significant variance in meteorological and snowpack conditions. All vertical (surface–atmosphere) exchange processes, including canopy interception, infiltration, evaporation, transpiration, photosynthesis, respiration and runoff production are computed for each within-unit stratum. With the exception of the soil hydrology, the canopy processes largely follow the methods described by Running & Coughlan (1988).

In water-limited ecosystems, the distribution of available water is controlled by a combination of soil characteristics and topographic position. In RHESSys, the landscape distribution of soil water is incorporated using basic principles of hillslope hydrology, as developed in TOP-MODEL (Beven & Kirkby, 1979). An approach to scale-up a land-unit average soil-water deficit to local deficits is used:

$$S = \bar{S} - m\left[\ln\left(\frac{aT}{T_e \tan\beta}\right) - \lambda\right] \tag{3}$$

where a is the upslope area drained per unit contour width; β is the local gradient; T and T_e are the local and unit composite soil transmissivity, respectively; $\ln(aT/T_e \tan\beta)$ is referred to as a hydrologic similarity or wetness index; S is a local saturation; $(\bar{S})$ is the land-unit average saturation deficit; λ is the areal average wetness index and m is a parameter related to the reduction in the soil saturated hydraulic conductivity with depth from the surface (but is more often treated as a calibration parameter). When $S < 0$ the soil column is saturated, while $S > 0$ indicates a deficit below saturation. The value of S is solved for each $\ln(aT/T_e \tan \beta)$ interval, along with an unsaturated zone soil-water content, which are used to control canopy transpiration. Baseflow drainage is calculated from the land unit as an exponential function of both λ and $\bar{S}$ following Beven (1986). A full land-unit water balance is computed each day, with daily flux of transpiration, runoff production and soil-water recharge (from precipitation and snowmelt) and saturation zone recharge composited from interval specific processes to update S. Additional details are given in Band *et al.* (1993).

This scheme produces spatial heterogeneity in water and carbon flux dependent on both between-unit differences and variance within land units. The surface heterogeneity is partitioned between and within units with a land-unit template, which is used to construct statistical information from the land attribute files. The land attribute files are raster images including images of leaf area index (LAI), life form, terrain gradient, aspect and elevation, contributing drainage area and soil properties. The LAI and terrain information are developed from remote sensing imagery (for example, Nemani *et al.*, 1993) and digital elevation models (DEMs) (Band, 1989; Lammers & Band, 1990), while the soils information is generally derived either from digitised soil maps or from knowledge of soil catenae along topographic gradients. The former soil data sources are typically heavily generalised and have a fixed cartographic scale, reducing information content, while the latter soil data sources are now under development to help alleviate these limitations (Moore *et al.*, 1993*a,c*; Zhu, 1994).

As remote sensing and DEM resolution decreases (grid sizes get larger), the information content of the resulting spatial data sets drop, as values are smoothed and approach the landscape mean. While the landscape mean values should be closely preserved (depending on the methods of resolution reduction), the attribute variances tend to drop and the co-variances may not be stable. This may have the effect of biasing the estimate of $f(\mathbf{x})$, and the resulting distribution and areal expectation of the model flux predictions. Higher-order information that is derived from surface patterns, such as the gradient, aspect and contributing drainage area, are also known to show regular shifts across resolution changes (Moore *et al.*, 1993*c*; Moore, Lewis & Gallant, 1993*b*; Wolock & Price, 1994; Band, Vertessy & Lammers, 1995).

Study area and simulation results

To investigate some of the effects discussed above and to illustrate the methodology of progressive land surface aggregation, RHESSys is applied to an area in the northern Rocky Mountains. The watersheds are mountainous conifer-dominated basins with steep gradients and sub-stantial vertical relief. Soils are generally thin but increase in thickness downslope in response to colluviation and the distribution of till deposits. The South Fork of the Flathead River is a 3000 km^2 watershed above the Hungry Horse Reservoir in north-western Montana. This area is south of Glacier National Park with the western side of the basin defined by the Swan mountain range and the eastern side by the Flat-head range. The overthrusted Precambrian bedrock that forms this part of the Rocky Mountains is a predominant topographic structure aligned in the north-west–south-east direction with over 1500 m vertical relief. The basin represents a large portion of the Bob Marshall Wilderness and, therefore, direct anthropogenic disturbance to the forest has been minimal since the 1950s. The hydrology of the area is snowmelt domi-nated with dry summers. There are substantial orographic gradients in precipitation with average annual amounts on the order of 600 mm at the lower elevations.

Inputs

Two sets of topographic data are used, 100×100 m grid cells interpolated from Defense Mapping Agency 3 arc second data and 1×1 km grid cells. Additionally, two sets of satellite sensor data are used: Thematic Mapper (TM) that has been re-sampled from 30 m to 100 m data and Advanced Very High Resolution Radiometer (AVHRR) that has been registered to the 1 km data set. A third data set was created by aggregating the 100 m elevation and satellite sensor data to 1 km resolution. This allows us to test differences between different resolutions while keeping the data source constant (comparison between the 100 m and aggregated 100 m) and to compare differences between data sources while keeping the resolution the same (comparison between the aggregated 100 m and the 1 km data). All secondary data sets, such as LAI, gradient and wetness index were computed from the three data sets.

Terrain and satellite sensor data form the basis for many of the inputs into the simulation model. In addition to elevation, the terrain model contributes other parameters, including gradient, aspect and the wetness index. Figure 3 shows the wetness index for the 100 m, aggregated 100 m and the 1 km data sets. It can be seen that the overall spatial characteristics appear to be preserved throughout the basin. It is import-ant to note that the wetness indices give relative values, and the actual wetness magnitude is scaled-up according to Equation (2). An important effect of reducing DEM resolution is the smoothing of the topography, resulting in a drop of surface gradients. This tends to remove the steeper

(a)   (b)   (c)

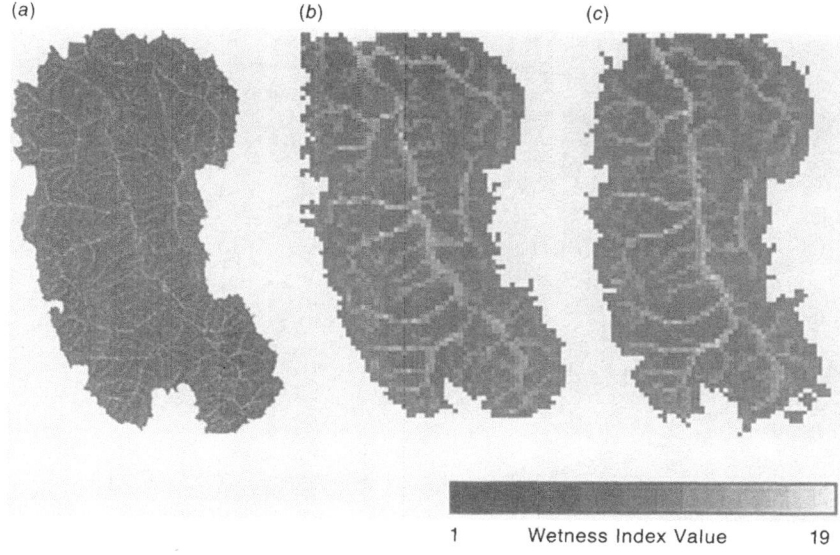

1   Wetness Index Value   19

Fig. 3. Wetness index for (*a*) 100 m, (*b*) aggregated 100 m and (*c*) 1 km data sets for South Fork, Flathead River, Montana. Light areas represent larger wetness index values and these areas have a greater propensity to become saturated.

tail of the gradient distribution over the landscape, removing some of the extreme differences of insolation.

The satellite sensor data are used to supply a measure of the surface cover of the vegetation across the landscape. The LAI is an important input to the model and, as will be seen below, the model is very sensitive to these values. Figure 4 shows the LAI surface across the landscape for the 100 m, aggregated 100 m and the 1 km data sets. Based on visual inspection of the images, the aggregation from 100 m to 1 km appears to have changed the spatial characteristics of the LAI distribution very little; however, the AVHRR 1 km data set produces a higher range of LAI than the TM-derived values. While the LAI patterns in the main valleys that are dominated by closed-canopy conifer stands are broadly similar between the data sources, areas at higher elevations over which the canopy is more open or becomes dominated by alpine meadows show significantly higher AVHRR-derived LAI. Spanner *et al.* (1990) and Nemani *et al.* (1993) have pointed out that the higher infrared reflectance of broadleaf understorey tends to produce overestimates of conifer canopy LAI. The methods of Nemani *et al.* (1993)

(a)   (b)   (c)

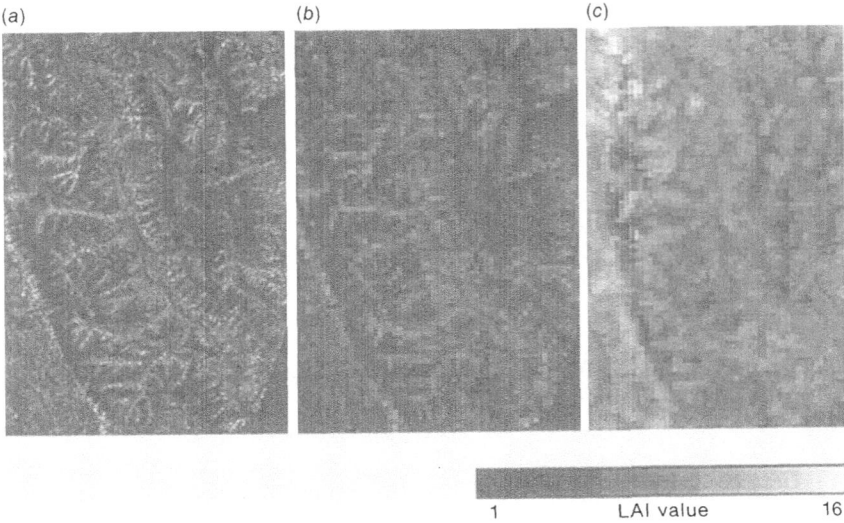

1   LAI value   16

Fig. 4. LAI images for South Fork, Flathead River, Montana for (*a*)
100 m, (*b*) aggregated 100 m and (*c*) 1 km data sets. Light areas are
high LAI values. Note the larger LAI values for the AVHRR (1 km)
data set (*c*) showing the overestimated values relative to the two TM-
derived surfaces (*a, b*).

have been used to control for the effects of canopy closure using TM
band 5; the lack of a similar band on AVHRR precludes this correction.
The ground LAI calibration data set for the imagery was largely located
in relatively closed canopies (although not exclusively), and the
AVHRR estimates (using the calibration presented by Running *et al.*,
1989) appear to produce a notable bias in LAI estimates. Therefore, the
effect of light partitioning among different life forms in the same scene
and the same pixel can be seen to have important effects both in the
estimation of surface properties and in simulated ecological processes.

Surface partition

To run the model over these large areas, the region is subdivided into
physically based modelling units that are based on the streams and
hillslopes of the basin. By changing the number of units over the basin,
a range of partition images can be produced to provide multiple representations of the landscape from few hillslopes to many (see Fig. 5).
It is important to recognise this shift in the number of modelling units

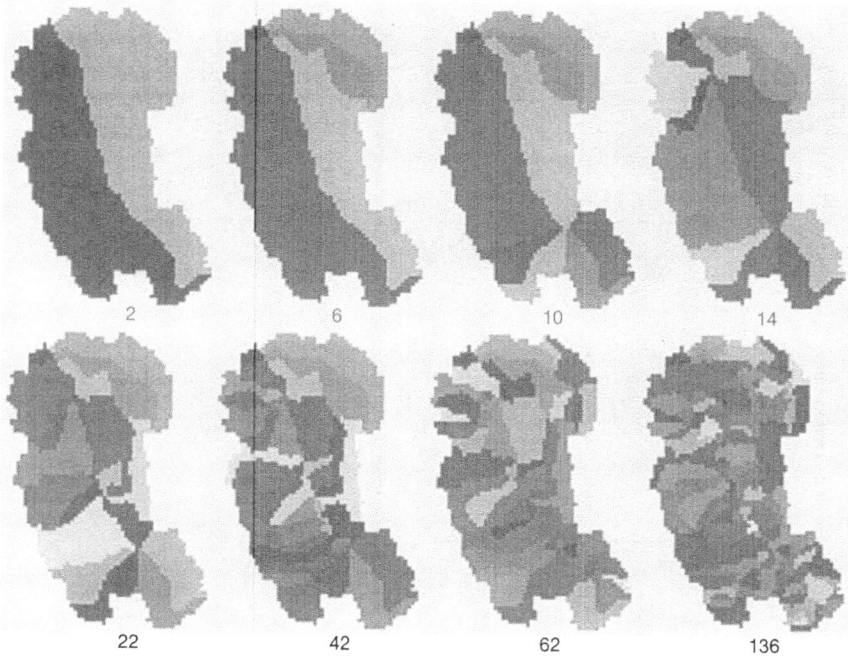

Fig. 5. The range of partitions for South Fork, Flathead River aggregated 100 m data set. Partition detail ranges from 2 hillslopes to 136 hillslopes.

as one method of scaling-up. The range in partition levels allows us to look at the effects of changing the landscape representation while keeping the resolution and the source of the input data set constant. By increasing the number of simulation units within the drainage basin there is an exchange of within-unit variance for between-unit variance.

To gain an understanding of the differences in the input data sets to the model, the means and variances for the three data sets (100 m, aggregated 100 m and 1 km) are plotted across a range of landscape partitions. Four variables, elevation, gradient, LAI and the wetness index (λ), have been chosen for closer inspection (Fig. 6). The primary topographic variables of elevation, gradient and aspect are generalised within RHESSys to the level of the land unit such that the landscape is represented as an exhaustive partition of hillslope facets and stream or valley reaches. Despite the differences between the different data sets, the means for elevation are fairly conservative (the small relative shift

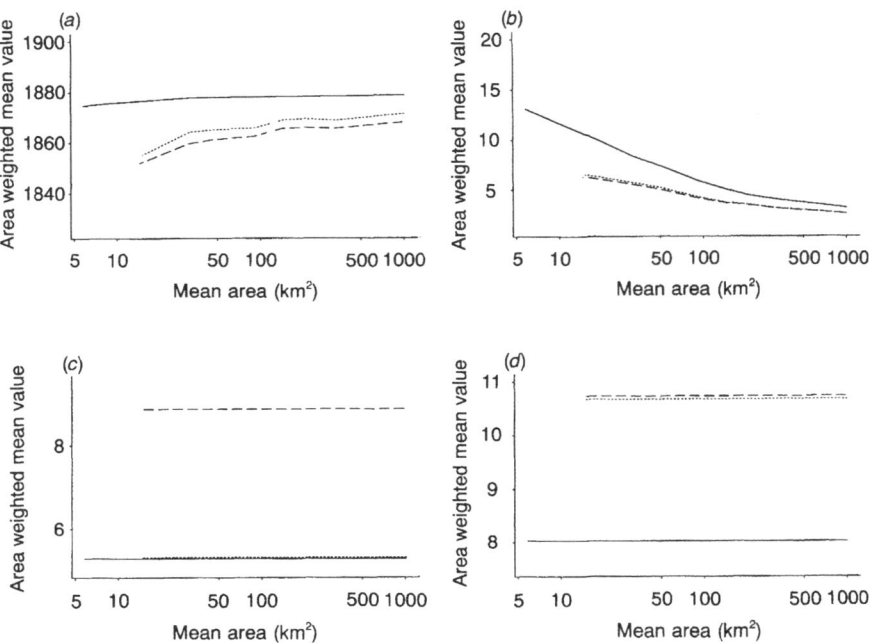

Fig. 6. Area weighted basin mean values of the input data to the simulation model for four variables: (*a*) elevation; (*b*) gradient; (*c*) LAI; and (*d*) wetness index. The data sets are: 100 m (—); aggregated 100 m (· · · ·); and 1 km (– – –). Mean areas on the *x*-axis represent the mean land unit size for a given partition.

of mean elevation is assumed to be a computing round-off effect). It can be seen that moving to coarser partitions of the landscape decreases the gradient. This is a result of the hillslope normal vectors, from which gradient is determined, tending towards zenith (the surface is flatter). The spherical variance of the surface normal shows an asymptotic trend (Fig. 7) as the number of units is increased and the spherical distribution of the terrain is better sampled.

The large difference in LAI, as noted above, is a result of the two sources of the satellite sensor data and the different methods from which the LAI values were derived. The calibration data set used to relate the normalised difference vegetation index (NDVI) to LAI was collected over the divide to the west of the present study area. Since the South Fork of the Flathead River tends to be somewhat drier, there are more open canopies and, therefore, a stronger deciduous understorey signal; consequently, there is greater bias, especially as elevation increases. It

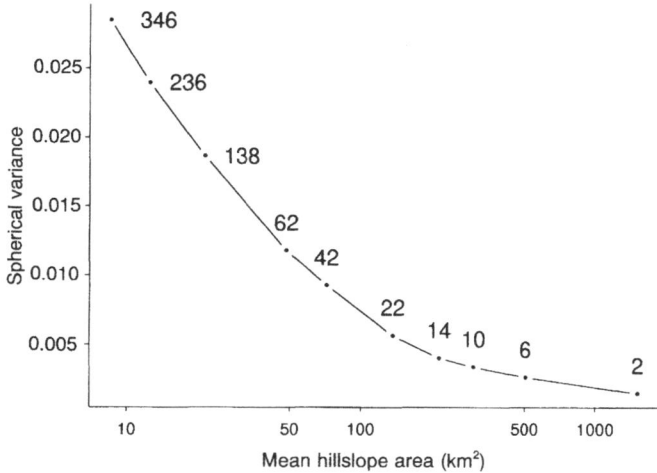

Fig. 7. Spherical variance (South Fork, Flathead River, Montana) across the range of partitions for the 100 m data set. The values on the curve indicate the number of hillslopes.

should be noted that this finding presents a serious problem for any modelling effort given that, by moving over the divide from the calibration sites, the climate gradients were strong enough to affect the results.

The mean wetness index shows a large difference as a result of the resolution of the data set. Wetness index values are much higher for the two 1 km resolution data sets because of the dependence of the wetness index on gradient and the drainage area per unit contour length (dependent on cell dimension). Gradient is strongly influenced by the resolution because of the direct functional relationship on a length scale. However, an inspection of the histograms for each data set suggests the resolution shifts are largely restricted to the distribution location but with little shift in shape. According to Equation (2), this conserves the distribution of soil water relative to the mean, although the rate of base-flow drainage is dependent on the wetness index.

In similar plots (Fig. 8), it can be seen that the between-unit variance decreases as the resolution of landscape representation is decreased. In the case of gradient and aspect, within-unit distributional information has not been incorporated so that the loss of between-unit variance is not compensated by increasing the within-unit variance. As with the plots of the means (Fig. 6), the LAI differences are driven by data

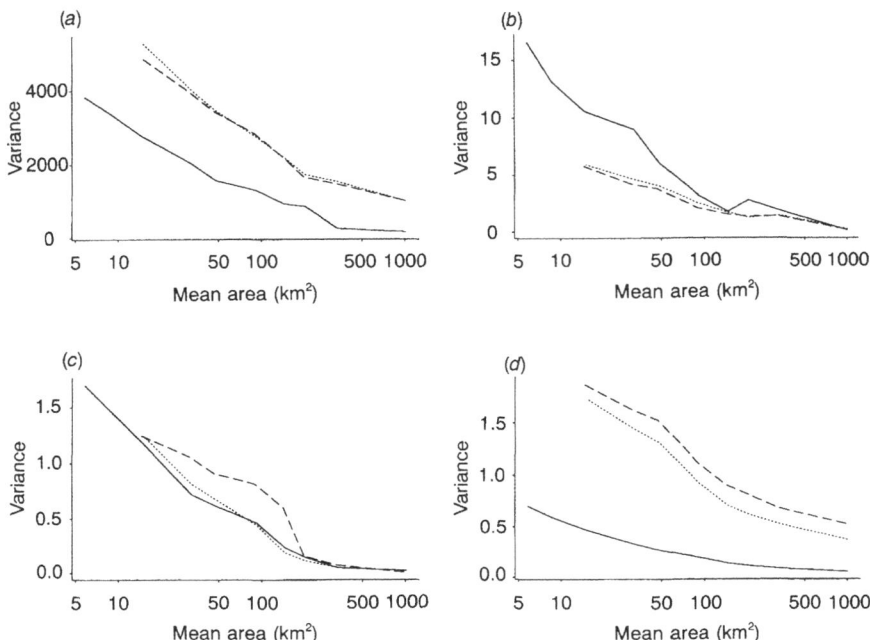

Fig. 8. Variances of the simulation model input data for four variables: (*a*) elevation; (*b*) gradient; (*c*) LAI; and (*d*) wetness index. The data sets are: 100 m (—); aggregated 100 m ($\cdots$); and 1 km (– – –). Mean areas on the *x*-axis represent the mean land unit size for a given partition.

source (discrepancies between the TM- and AVHRR-derived LAI images). Similarly, for the wetness index, the differences result from the underlying resolution of the data sets.

Output

The outlet of the drainage basin was selected to coincide with a United States Geological Survey (USGS) monitored gauge so that outflows could be directly compared. Figure 9 shows the observed outflow for 1988 along with the precipitation used for the simulation. Included in this figure are three model-derived basin outflow results from the 100 m, aggregated 100 m and 1 km data sets. As can be seen, there is variation in the outflows depending upon which data set is chosen. These hydrographs have not been calibrated or tuned but have been directly parameterised by the available geographic data and by transferring the

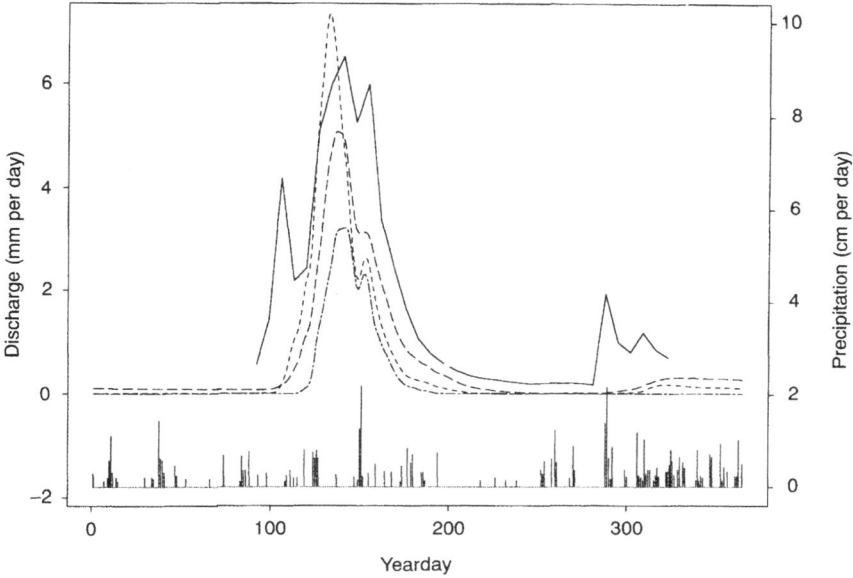

Fig. 9. Observed and modelled outflow for 1988. Simulated results are for landscape partitions with 42 hillslopes for the 100 m (– – –); aggregated 100 m (- - -); and the 1 km (– · – ·) data sets and for the observed discharge (—). Observed discharge data prior to yearday 91 and after yearday 334 were not available. The bottom trace is precipitation.

parameter *m* in Equation (2) from a much smaller calibrated basin just over the watershed divide to the west of the South Fork (Band *et al.*, 1993). It was decided not to tune the model because the primary goal was to look at the effects of scaling-up between simulation runs and any calibration would invariably fit the curves more closely for the wrong reasons. Calibration is particularly significant in this case, since the climate file is derived from three widely dispersed meteorological stations and it is possible that a few significant storms may have been missed.

It is suspected that a significant source of error in the outflow computations relative to the observed values resides in the magnitude and distribution of the snowpack, for which initialisation data did not exist. The observed hydrograph is shown without tuning the model for outflows simply to compare form. At two times during the year, no correspondence is shown between the observed and the simulated hydrographs. An early outflow event caused by snowmelt is not being picked up by the model. This appears to be a result of the soils absorbing the early snowmelt. This may correspond to snowmelt over bare rock or thin soil areas that not adequately parameterised, or potentially the impact of frozen soil early in the melt season. Another outflow event occurs near the end of the summer. The year 1988 represented the last year in a drought that lasted several years and finally broke at the end of the summer. It can be seen from the climate file that there was precipitation for several days, which produced a signal in the observed outflow. This suggests the variability in soil-water capacity has not been adequately incorporated. The difference in baseflow between resolutions is caused by differences in wetness indices, which directly affects baseflow and then, by enhanced drainage of the soil water, the generation of surface flow. If the wetness index could be corrected for resolution dependency, the baseflow problem may be correctable.

The simulated differences of net canopy photosynthesis (PSN) between the 100 m and the aggregated 100 m data are very small relative to the 1 km data set simulations (Fig. 10a). The large difference between the 100 m and the 1 km data sets is directly related to the differences noted earlier in LAI. The larger AVHRR-derived LAI values cause much larger photosynthetic activity. Similarly, when evapotranspiration is plotted it can be seen that there is a very close correspondence between the 100 m and the aggregated 100 m data sets (Fig. 10b), while AVHRR-derived LAI shows larger amounts of evapotranspiration taking place.

Landscape partitions

The effect of changing the landscape representation on model results while holding the data source and the input data layer resolution constant was investigated. Figure 11 shows a three-dimensional plot of evapotranspiration on the z-axis with time and number of hillslopes on the horizontal axes. The number of hillslopes is represented by the different landscape partitions for the 100 m simulations. In all eight simulations, the input data layers are identical but the basin is partitioned with different numbers of simulation units, ranging from 2 hillslopes to 138 hillslopes (Fig. 5). For most of the year evapotranspiration remains reasonably constant across all partition levels, but at other times of the year, around yearday 150 for example, there is some minor variation. It is encouraging to see such stability across this range of aggregation for a parameter such as evapotranspiration.

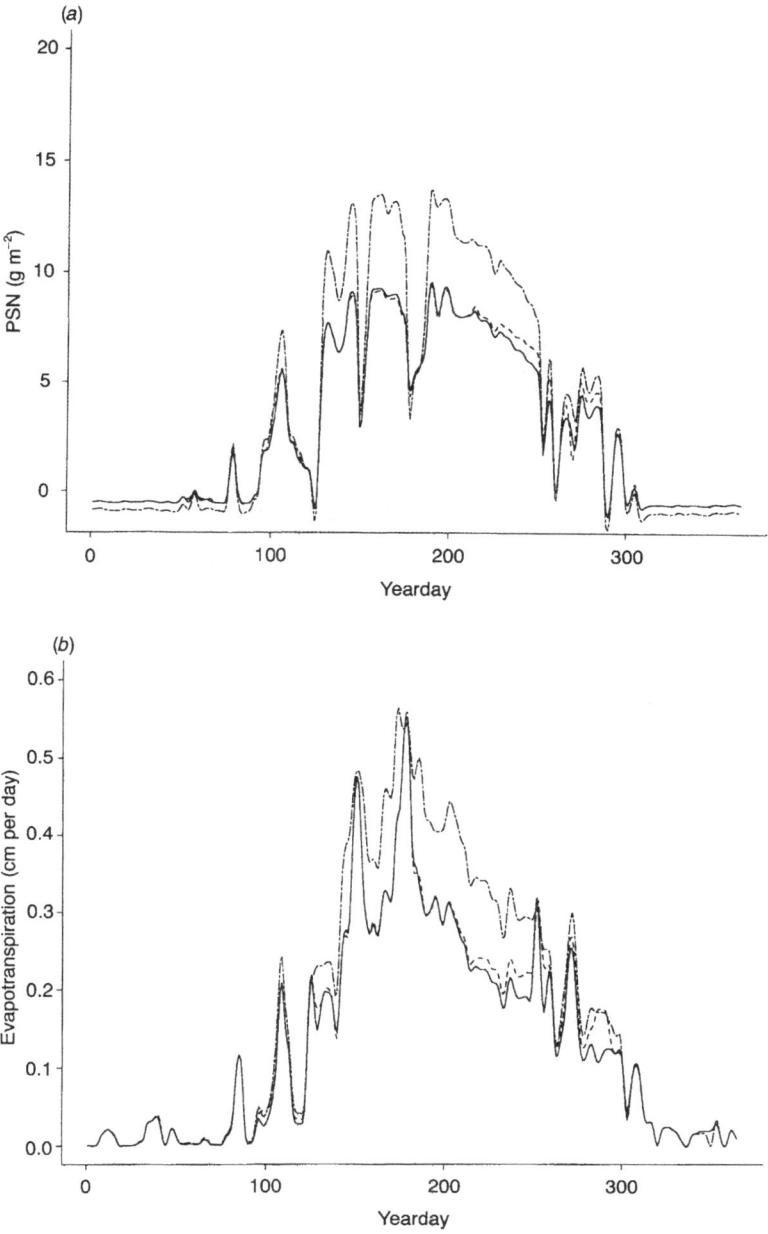

Fig. 10. Simulated net photosynthesis (PSN) (*a*) and evapotranspiration (*b*) for the three data sets using landscape partitions with 42
hillslopes: 100 m (——); aggregated 100 m (- - -); and 1 km (– · – ·).

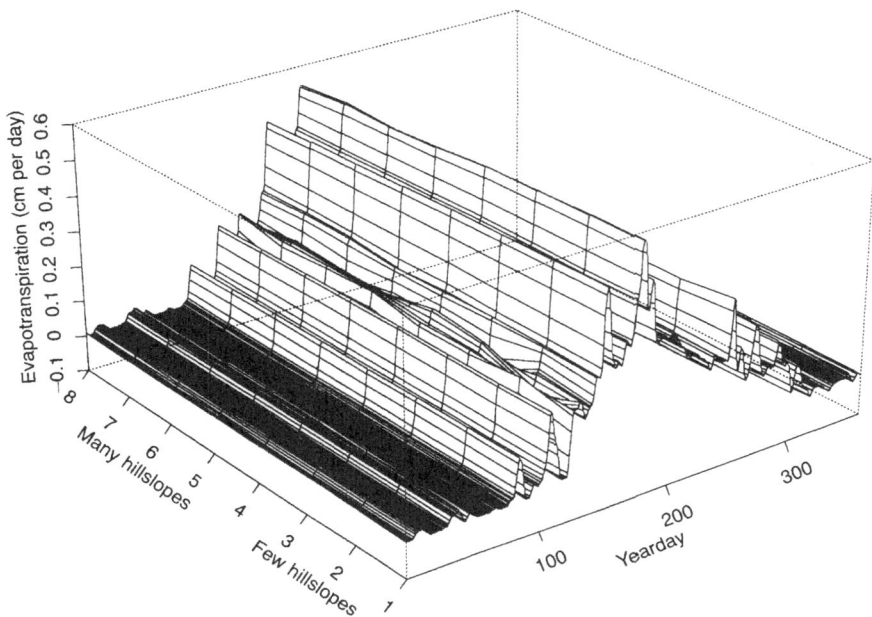

Fig. 11. A three-dimensional plot showing evapotranspiration through the year for all eight partition levels of the 100 m data set. Partitions range from 2 hillslopes to 138 hillslopes.

Representing elevation gradients

In steep mountainous terrain, such as the study area, the elevation gradient within the valley-side units can contribute substantial variation to ecosystem and hydrological processes both within individual units and integrated over the entire basin. The method described above does not incorporate this effect as mean elevations are used for each unit. Further partitioning of the hillslopes has been performed to include elevation bands, which can be adjusted to provide a variable degree of resolution. Selecting a large elevation band that covers the entire hillslope with a single elevation interval duplicates the results described above. Figure 12 shows the variation of (a) snowpack and (b) outflow over the set of 200 m elevation intervals over a single valley-side unit while Fig. 13 shows outflow production over the full South Fork Flathead River basin. Additional simulations suggest that a fair portion of this marginal increase in correspondence between observed and predicted runoff can be gained with much larger elevation intervals, reducing extra compu-

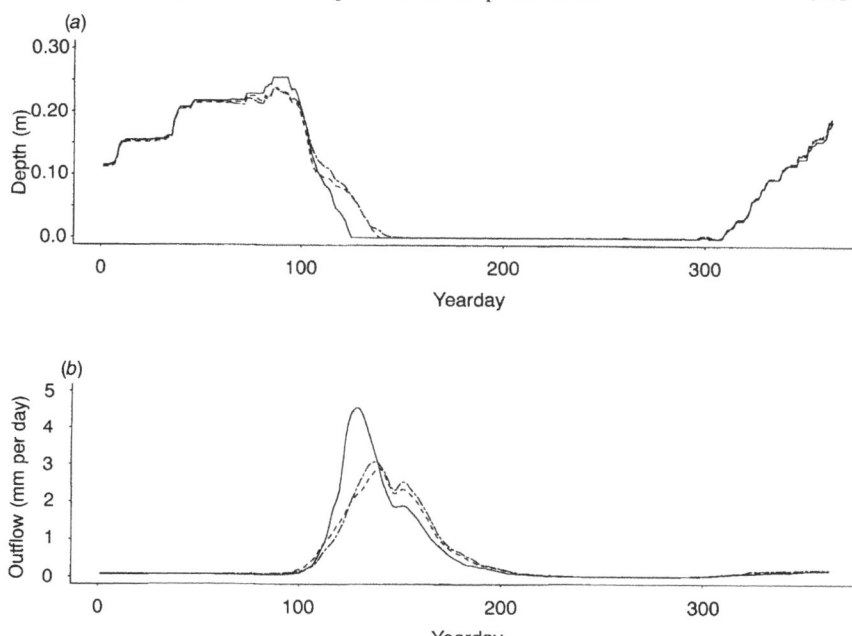

Fig. 12. Simulation results of (*a*) snowpack and (*b*) outflow for a single hillslope (number 3) from the 100 m data set using the partition with 42 hillslopes. Multiple elevation bands were used: one elevation band (—); two elevation bands (- - -); and eight elevation bands (– · – ·). For this hillslope, using 200 m intervals gives eight elevation bands.

tational demands. Otherwise the extra detail of surface description may not add substantially to the surface distribution of elevation and related variables. Experiments over different landscapes will be used to explore these marginal benefits.

Conclusions

The simulations over the South Fork of the Flathead River have suggested several hypotheses. The unifying theme that emerges is the need to sample and represent surface attribute information in a manner that adequately estimates the joint distribution function of the key model parameters. The most obvious point is that care must be taken about the use of calibrated functions that define variables such as LAI, which still requires significant research for estimation at coarse satellite sensor

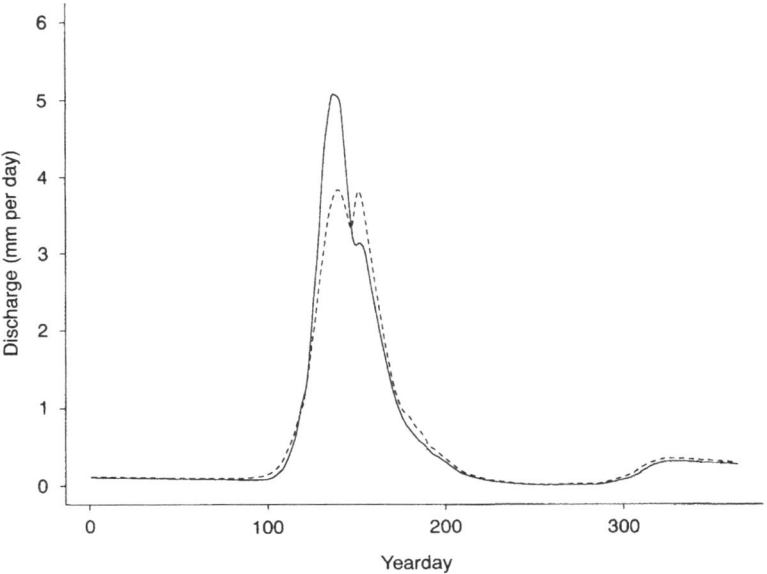

Fig. 13. Simulation results for basin aggregated discharge from the 100 m data set using the partition with 42 hillslopes and resolving elevation bands with 200 m intervals (- - -); single elevation band (—).

(AVHRR) resolution. The effects of mixing a spectrally bright forest floor under an open canopy and the effect of light partitioning between different life forms produces problems both for interpreting spectral reflectance for LAI estimation and for subsequent simulation of canopy processes. Another conclusion is shown by those measures dependent on a length scale, such as gradient and the wetness index. There are fairly strong differences in gradient and the wetness index across resolution for which controls are needed. These differences contribute strongly to the variation of magnitude seen in the simulated hydrographs. However, the similar distribution shapes of the 100 m and 1 km data sets suggest that there may exist some form of topographic similarity across these resolutions, which would be very useful to exploit for large area simulation. Wolock and Price (1994) have shown that the wetness index appears to show a log-linear scaling with DEM resolution, suggesting this effect may be easily controlled for. Although the distributions are sampled on very different supports, the close correspondence of the canopy processes and to a lesser extent the runoff indicates the large resolution differences may be controlled for as long as the variance of soil water and other controlling surface variables are incorporated. Testing of this hypothesis in other landscapes is in order. In the case of gradient, subgrid distributions may have to be incorporated to preserve the values.

There is also an effect across scales when all the input data layers are held constant but the partitioning resolution is changed. For both types of resolution and landscape partition shift, some *a priori* information on landscape grain, or spatial statistics of the terrain and land cover may be required in order to regularise distributional information estimated from one level (particularly from coarse resolution sampling). In many cases, the precise spatial patterns of attribute co-occurrence do not need to be maintained at all regional scales. There may not exist a single resolution at which all parameters need to be measured, as many vary over different length scales. Those variables that show the largest length scales can be used to form the major landscape partition units, while variables showing higher frequency variations can be described statistically as within-unit variations. Finally, the mix of length scales and the sensitivity of the modelled carbon and water fluxes to each of the significant landscape attributes will vary across different landscapes, so the methods shown here need to be sufficiently flexible to allow stratification over different types of gradient, both between and within units.

Acknowledgements

The research presented in this paper was supported by NASA grant NAGW-952, and a NOAA grant to Colorado State University. The RHESSys project has been a collaborative effort between our group at the University of Toronto and groups led by Steve Running and Ramakrishna Nemani at the University of Montana and David Peterson at NASA/Ames Research Center.

References

Avissar, R. (1992). Conceptual aspects of a statistical dynamical approach to represent landscape subgrid-scale heterogeneities in atmospheric models. *Journal of Geophysical Research*, 97, 2729–2742.

Band, L.E. (1989). A terrain based watershed information system. *Hydrological Processes*, 3, 151–162.

Band, L.E. (1993). Effect of land surface representation on forest water and carbon budgets. *Journal of Hydrology*, 150, 749–772.

Band, L.E., Peterson, D.L., Running, S.W. *et al.* (1991). Forest ecosystem processes at the watershed scale: basis for distributed simulation. *Ecological Modelling*, 56, 171–196.

Band, L.E., Patterson, P., Nemani, R. & Running, S.W. (1993). Forest ecosystem processes at the watershed scale: incorporating hillslope hydrology. *Agricultural and Forest Meteorology*, 63, 93–126.

Band, L.E., Vertessy, R. & Lammers, R.B. (1995). The effect of different terrain representations and resolution on simulated watershed processes. *Zeitschrift für Geomorphologie*, 101, 187–199.

Beven, K. (1986). Runoff production and flood frequency in catchments of order *n*: an alternative approach. In *Scale Problems in Hydrology*, ed. V.K. Gupta, L. Rodriguez-Iturbe & E.F. Wood, pp. 107–131. Dordrecht: Reidel.

Beven, K. (1989). Changing ideas in hydrology: the case of physically based models. *Journal of Hydrology*, 105, 157–162.

Beven, K. & Kirkby, M. (1979). A physically based, variable contributing area model of basin hydrology. *Hydrological Science Bulletin*, 24, 43–69.

Bresler, E. & Dagan, G. (1988). Variability of yield of an irrigated crop and its causes. 1. Statement of the problem and methodology. *Water Resources Research*, 24, 381–388.

Lammers, R.B. & Band, L.E. (1990). Automated object description of drainage basins. *Computers and Geosciences*, 16, 787–810.

Moore, I.D., Gessler, P.E., Nielsen, G.A. & Peterson, G.A. (1993*a*). Soil attribute prediction using terrain analysis. *Soil Science Society of America Journal*, 57, 443–452.

Moore, I.D., Lewis, A. & Gallant, J.C. (1993*b*). Terrain attributes: estimation methods and scale effects. In *Modelling Change in Environmental Systems*, ed. A.J. Jakeman, M.B. Beck & M.J. McAleer, pp. 189–214. New York: Wiley.

Moore, I.D., Turner, A.K., Wilson, J.P., Jenson, S.K. & Band, L.E. (1993*c*). GIS and land surface–subsurface process modelling. In *Environmental Modeling and GIS*, ed. M.F. Goodchild, B.O. Parks & L.T. Steyaert, pp. 196–230. New York: Oxford University Press.

Nemani, R., Pierce, L.L., Running, S.W. & Band, L.E. (1993). Forest ecosystem processes at the watershed scale: sensitivity to remotely-sensed leaf area index estimates. *International Journal of Remote Sensing*, 14, 2519–2534.

Rastetter, E.B., King, A.W., Cosby, B.J., Homberger, G.M., O'Neill, R.V. & Hobbe, J.E. (1992). Aggregating fine-scale ecological knowledge to model coarser-scale attributes of ecosystems. *Ecological Applications*, 2, 55–70.

Running, S.W. & Coughlan, J.C. (1988). A general model of forest ecosystem processes for regional applications. I. Hydrologic balance, canopy gas exchange and primary production processes. *Ecological Modelling*, 42, 125–154.

Running, S.W. & Hunt, R. (1993). Generalization of a forest ecosystem model for other biomes, BIOME-BGC, and an application to global scale models. In *Scaling Physiological Processes: Leaf to Globe*, ed. J.R. Ehleringer & C.B. Field, pp. 141–158. San Diego: Academic Press.

Running, S.W., Nemani, R. & Hungerford, R.D. (1987). Extrapolation of synoptic meteorological data in mountainous terrain, and its use for simulating forest evapotranspiration and photosynthesis. *Canadian Journal of Forest Research*, 17, 472–483.

Running, S.W., Nemani, R., Peterson, D.L. *et al.* (1989). Mapping regional forest evapotranspiration and photosynthesis by coupling satellite data with ecosystem simulation. *Ecology*, 70, 1090–1101.

Spanner, M.A., Pierce, L.L., Peterson, D.L. & Running, S.W. (1990). Remote sensing of temperate coniferous forest leaf area index: the influence of canopy closure, understory vegetation and background reflectance. *International Journal of Remote Sensing*, 11, 95–111.

Wolock, D.M. & Price, C.V. (1994). Effects of digital elevation model map scale and data resolution on a topography-based watershed model. *Water Resources Research*, 30, 3041–3052.

Zhu, A. (1994). SOLIM: soil inference model using fuzzy logic. PhD thesis, University of Toronto.

R.J. GURNEY and I.J. SEWELL

Observation and simulation of energy budgets at the surface of a prairie grassland

Introduction

Energy and water exchanges at the land surface are important parts of the global energy and water cycles. The models for these exchanges have been developed using point data and need to be scaled-up for inclusion in general circulation models (GCMs). The processes do not scale linearly and so both observational and modelling developments are required. Remote sensing from aircraft and spacecraft makes radiation observations at a scale intermediate between point observations and the spatial scales required for climate models. Such observations can probably help with scaling problems but the methods to use are not clear. Further, the remote sensing observations are sparse in time.

In his 1992 co-ordinator's report for the Hadley Centre for Climate Prediction and Research, Bennetts (1992) reviewed the dependence of climate model development on the availability and analysis of observational data. He included a list of key observations, including surface flux and soil moisture data, which required 'continuing effort'. This chapter presents results and conclusions from a detailed analysis of data of the type described by Bennetts. Model developments that are intended to implement, test and refine the conclusions are also presented. The modelling work focuses on the parameterisation of the surface conductance term used in land surface models to represent both physiological and physical controls on moisture and heat fluxes between a vegetated land surface and the base of the planetary boundary layer.

Energy budgets at the surface of a prairie grassland: FIFE

The First ISLSCP (International Satellite Land Surface Climatology Programme) Field Experiment (FIFE) resulted from recognition within ISLSCP that new experimental work was required to obtain a clearer understanding of the links among surface states (such as biomass, cover type, temperature), surface processes (such as transpiration, photosynthesis) and their radiative properties, as observed from satellites (Sellers *et al.*, 1992). As part of the experiment, five intensive field campaigns (IFCs) were mounted, four in 1987 and one in 1989.

Details of the broad range of ground-based and remote observations during these IFCs are given, for example, in Sellers and Hall (1987) and Sellers *et al.* (1992). A major part of the observations was of surface moisture and heat fluxes, which were measured at 22 flux measurement stations (FMSs), distributed across the 15 by 15 km area of Kansas prairie grassland, the Konza Prairie, which served as the FIFE site (Fig. 1).

The FMS data have provided a valuable opportunity for improving our understanding of the nature and implications of variations in energy partition at the Earth's surface. In particular, such data are an important

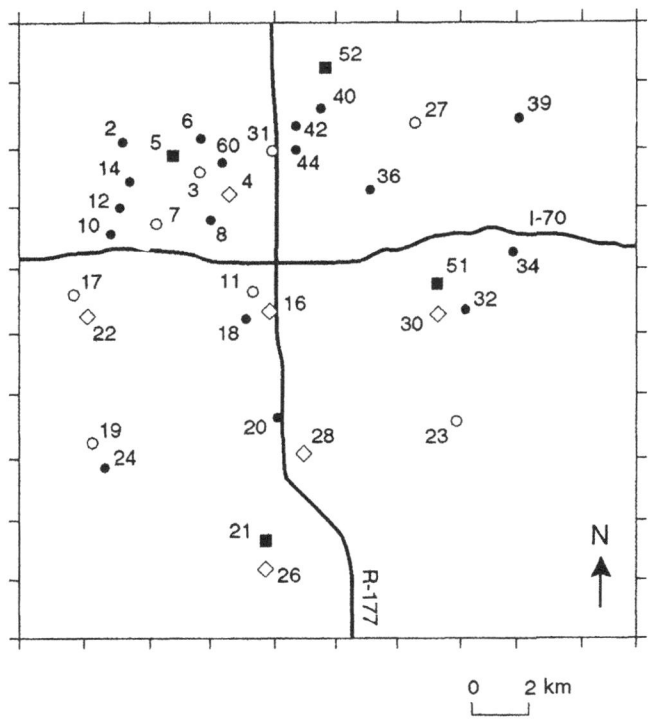

Fig. 1. Distribution of flux measurement sites across the FIFE site in 1987 (IFCs 1–4): PAM (○); SAM (super PAM (■); Bowen ratio flux measurement (●); and Eddy correlation flux measurement (◇).

resource in the development of models that are physically realistic, supported by observations and computationally efficient.

By using FMS data in model development work, it is possible to contribute to two of the objectives of the FIFE experiment, which stretch well beyond the data collection and quality control objectives that have already, to a large degree, been fulfilled:

1. To obtain knowledge of the detailed variations in sensible and latent heating of the atmosphere, which will lead to a better understanding of variations (and the controlling factors) in land surface energy budgets (Crosson & Smith, 1992)
2. To use the data to facilitate the development and testing of simplified parameterisation schemes for modelling studies of surface fluxes (Smith *et al.*, 1992).

The analysis of FMS data described below uses the evaporative fraction (EF), the proportion of available energy that is used for evaporation and transpiration, as a diagnostic. This focus on partitioning of available energy recognises the strong relationship between energy partition and the availability of water for evaporation, and the powerful feedbacks between energy partition and the state variables of the lower atmosphere (Brutsaert, 1982).

The EF also has the advantages of simplicity in computational terms and of clear physical relevance as a descriptor of partition of available energy at the land surface. EF, defined in Equation (1), is also closely related to the Bowen ratio (β) and to the Priestley–Taylor parameter (α) (Priestley & Taylor, 1972) (Equations (2) and (3)) both often used in data analyses and modelling work.

$$\text{EF} = \lambda E / (\lambda E + H) \equiv \lambda E / (Q^* - G) \tag{1}$$

$$\text{EF} = 1 / (1 + \beta) \tag{2}$$

$$\text{EF} \approx \alpha [\Delta / (\Delta + \gamma)] \tag{3}$$

where Q^* is net radiation; H is the sensible heat flux; G is the soil heat flux; λe is the latent heat flux; Δ is the slope of the saturation vapour pressure versus temperature curve; and γ is the psychrometric constant.

The use of EF values has been shown to be very conservative at the FIFE site during daylight hours (Brutsaert & Sugita, 1992) under a wide range of environmental conditions (Fig. 2). Its formulation means that it is significantly more conservative than the Bowen ratio (Fritschen & Qian, 1992). Accordingly, it may prove a highly attractive component

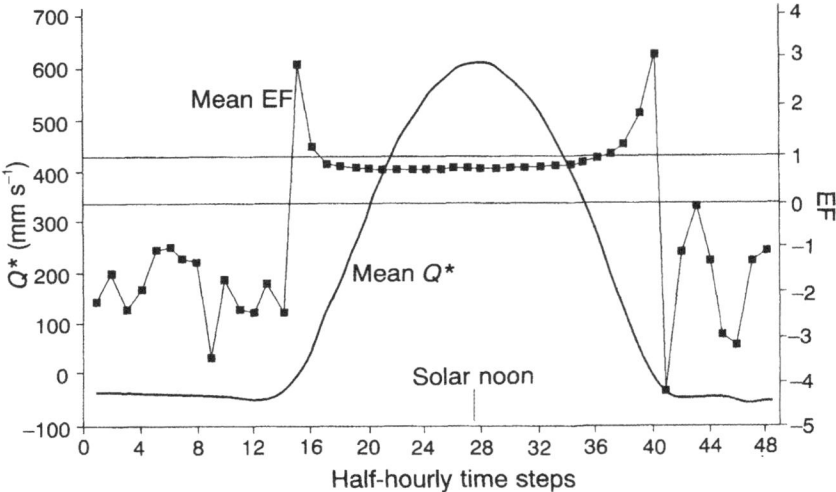

Fig. 2.Mean EF and Q^* for JD 227, a golden day during IFC3, for the FIFE site. The key EF range (0–1) is delineated.

of research into simplified parameterisations of surface energy budgets. The analysis presented below seeks to investigate this possibility.

During most of the FIFE measurement days, evaporation during daylight hours made up 80–90% of the total evaporation in each 24-hour period (Brutsaert & Sugita, 1992). In recognition of this fact and in order to avoid the complex measurement difficulties and data-reliability problems during night-time hours, only daytime FMS data (recorded between 08:00 and 18:00 hours) were included in the analysis.

Studies of the flux measurement systems used during the FIFE experiment have underlined the need for cautious use of the FIFE flux data. Any identified variation must be examined in the light of the varied systems used to collect flux data during FIFE. The instrumentation at individual FMS sites was not consistent. In most cases, site leaders deployed either an Eddy correlation (EC) or a Bowen ratio (BR) system. In just two cases, EC and BR systems were co-located for extended periods.

Although well established, both approaches (EC and BR) may only be reliable under certain conditions, leading to noise in the FMS data set. EC systems' performance is complicated in non-neutral conditions by the need for stability corrections. Although not unduly affected by stability problems, the BR method becomes indeterminate when available energy $(Q^* - G)$ tends to zero. It is also notoriously unstable at times of rapid change in temperature and vapour pressure gradients in the near surface layer, such as at dawn and at dusk. These rapid large percentage changes can make time averaging undesirable (Monteith & Unsworth, 1990) and the half-hourly FIFE flux data recorded around sunset and sunrise are, therefore, prone to unrealistic perturbations. The BR method also assumes the absence of horizontal temperature and vapour pressure gradients, the existence of a homogeneous land surface (sinks and sources of heat, moisture and momentum are not distinguished) and steady-state conditions (Fritschen & Qian, 1992).

The differences that result from the choice of flux measurement technique are complicated by the differing performance of the specific instruments used. Several workers have, for example, investigated the consequences of the variety of net radiometers used in FIFE. Seven different radiometer designs from five manufacturers were used for net radiation measurement. This range of instrumentation caused substantial instrument-related differences (Field *et al.*, 1992).

Using a mobile BR system at 20 of the 22 FMS sites, Nie *et al.* (1992) have underlined the need for caution. Although the work is largely limited to instantaneous measurements and half-hourly integrations (rather than comparison of longer time series), they identify highly significant maximum differences ranging from $\pm 10\%$ (instantaneous Q^* measurements) to $\pm 30\%$ (instantaneous BR measurements). These results have prompted Nie *et al.* (1992, p. 18 724) to issue a 'health warning' to all FIFE data users: 'The users of [FIFE] flux data are strongly recommended to take these differences into account if they are to examine the spatial variation of fluxes'.

Differences between daytime BR- and EC-derived EF data have, therefore, been examined in two ways:

1. Pairwise comparison of EF data recorded at co-located FMSs using the different techniques; this allows an assessment of the relative performance of EC and BR systems in recorded energy partition (rather than absolute flux values)
2. Comparison of the average EF recorded by stations grouped by technique; this approach to the assessment of EC–BR differences tests the hypothesis that EC–BR differences will not significantly reduce the value of energy partition records in the FMS data to the meso-scale and GCM modelling communities.

Figures 3 and 4 show typical results from these comparisons. The results have confirmed the widely accepted assessment that 'less than

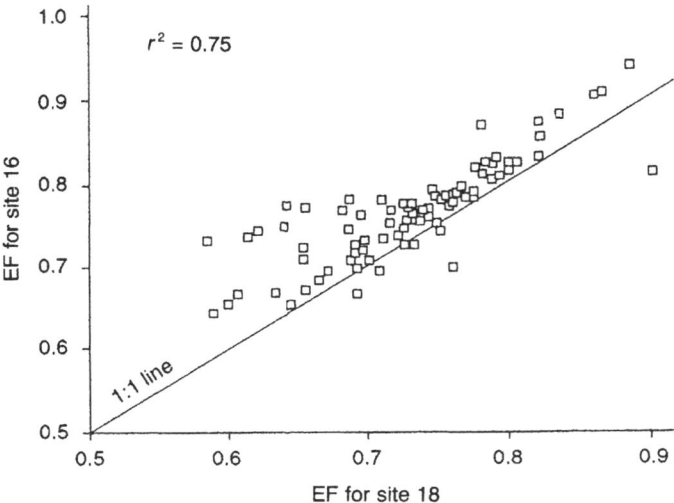

Fig. 3. FIFE 87/IFC1: correlation of daytime 08:00–16:00 hours EF data from sites 16 and 18.

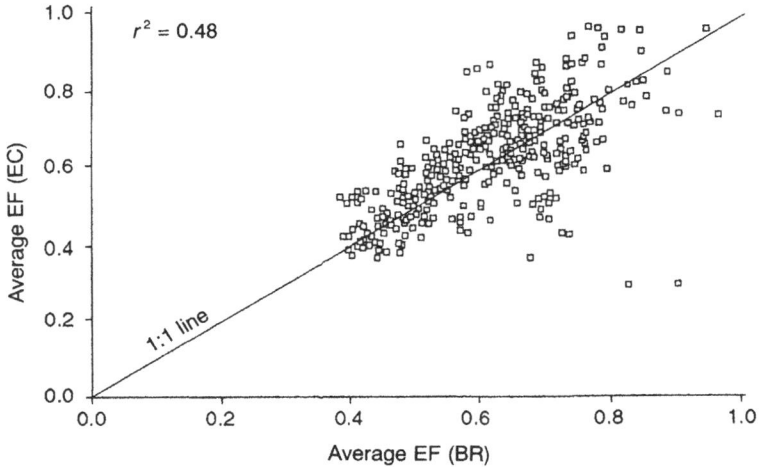

Fig. 4. FIFE 89/IFC5: correlation of mean daytime 08:00–18:00 hours EF data recorded by EC and BR flux measurement systems.

optimal' (Sellers & Hall, 1993) implementation during FIFE of some measurement techniques has introduced some problems. Differences between flux data recorded at EC and BR sites may be seen in several surface flux studies.

Implications of technique-derived noise

When evapotranspiration proceeds at or near the potential rate, scatter and bias such as that in Figs. 3 and 4 seem, in large part, the result of technique-derived variation. Previous analyses of FIFE FMS data have also emphasised the possibility of serious discrepancies in EC and BR flux measurements in complex terrain (Fritschen *et al.*, 1992). Whenever net radiation fluctuates sharply over short distances, the half-hourly flux data and the derived EF values recorded at co-located EC and BR systems can differ significantly. Local or non-dimensional modelling studies, particularly those examining short-term trends such as the development of the mixed layer during any one day, may, therefore, be susceptible to inaccuracies if based on data from a small number of sites using a mixture of EC and BR systems.

However, pairwise comparison does not help the assessment of EC–BR differences when using the full FMS data set as a basis for characterisation of energy budgets across the entire 15 × 15 km FIFE area: a current goal for both the FIFE experiment and for the meso-scale modelling community as a whole. Comparison of grouped EC and BR data begins to address this question. Throughout the five IFCs, characterisation of energy partition over the entire FIFE site (using the EF) is not affected by the choice of data group. Once grouped, EC- and BR-derived data produce very similar results (Fig. 5). Local technique-derived differences are not maintained for the FIFE site as a whole.

This result of aggregating or averaging data is very important for research using the FIFE data. The data volume available for the 15 × 15 km FIFE area is relatively small as a basis for studies seeking to characterise land surface conditions for the entire area. Data dropouts mean that, at some steps, there are fewer than a dozen full flux samples for the entire area. In the light of results such as those shown in Fig. 5, the flux data can, therefore, be treated as a whole in meso-scale modelling work, but caution must be observed in subdividing the FIFE data set. This approach is further supported by analyses of flux data by research teams who deployed both EC and BR systems during FIFE. They too have suggested that, when considering the entire FIFE area, there are no indications of any serious problems with measurement bias

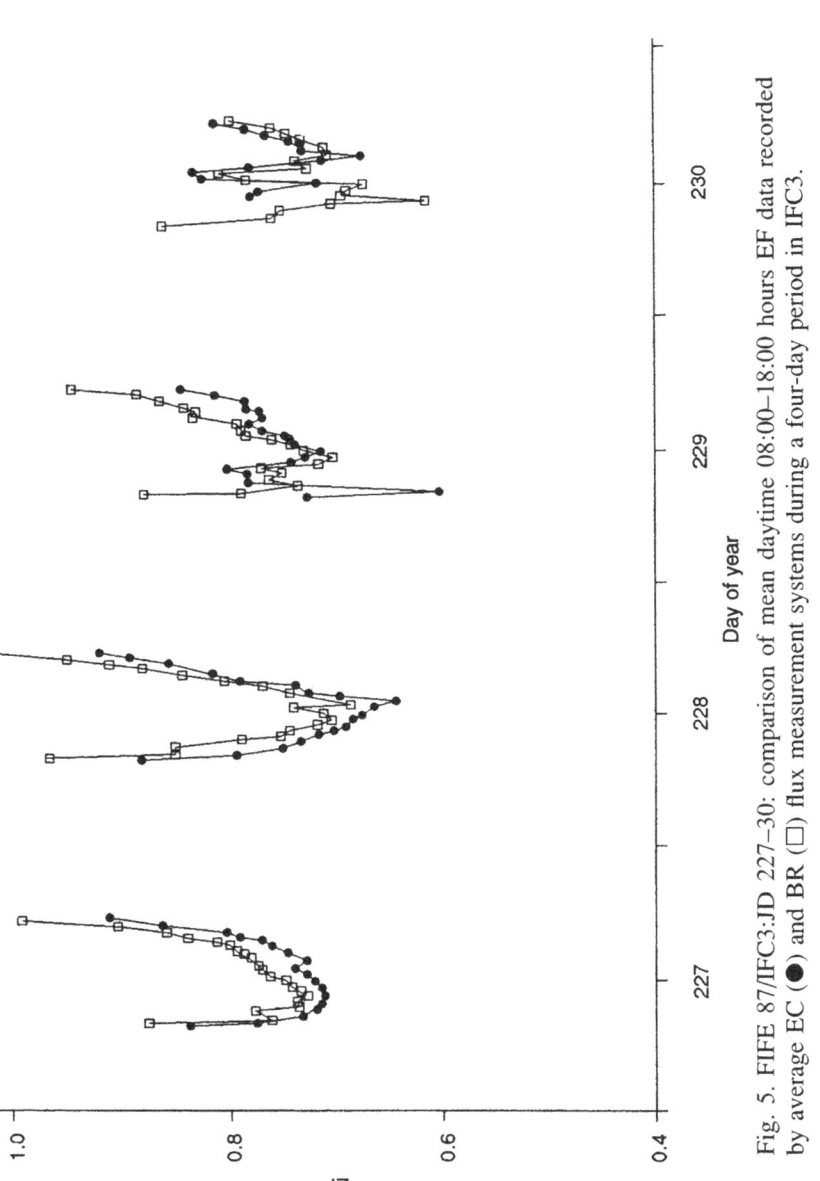

Fig. 5. FIFE 87/IFC3:JD 227–30: comparison of mean daytime 08:00–18:00 hours EF data recorded by average EC (●) and BR (□) flux measurement systems during a four-day period in IFC3.

with the possible exception of data recorded over senescent vegetation (Smith *et al.*, 1992).

Temporal variation in energy partition

The apparent conservatism of EF values during daylight hours suggests that EF could provide a useful link (for modelling purposes) between one-off 'snapshots', as provided by remote sensing instruments, and the course of the surface energy budget during daylight hours. In addition to aiding model development, an understanding of the temporal variation of daytime energy partition is, therefore, central to any attempt to assess the potential of remote sensing data as a surrogate, in meso-scale and GCM modelling, for extensive and often impractical flux measurement work on the ground.

To test whether the EF trends could provide a link between modelling and remote sensing data, regression statistics have been generated from correlations of single EF values and the overall daytime mean. Figure 6a shows an example from the analysis. For each of three days during IFC1, the noontime EF value at every FMS is compared with the average EF measured at that station during the 08:00–18:00 hours period. The regression gives r^2 values of 0.84, 0.95 and 0.97 (all significant at the 0.1% level). These strong correlations suggest a powerful predictive role for a single noontime measure of EF.

To enlarge the data population and, therefore, the statistical significance of any correlation, this approach has also been applied to the entire FIFE data set (Fig. 6b). At each time step, the arithmetic mean EF across the 22 FMS sites was calculated. The regression result is a reduced but more significant (0.01% level) r^2 value of 0.84, again indicating a significant correlation. However, the correlation does show a small, consistent offset above the 1:1 line, indicating that use of the mid-day figure would lead to a small underestimate in predicting the daytime mean EF.

This offset is caused by the slight upward concavity of the daytime EF curve (Fig. 2); in the case of IFC4, the offset is amplified by the shorter winter daylight period. The underestimation is in line with the figure of 1.5% calculated by Shuttleworth *et al.* (1989) and can be avoided by accounting for seasonal variations in the daylight period. For example, Fig. 6c is a repeat of Fig. 6a with the daytime period restricted to 10:00–16:00 hours. Despite the senescent vegetation, IFC4 data now also falls on or close to the 1:1 line.

As a final component in this temporal analysis of EF data (and its implications for the use of remote sensing data), the same comparison

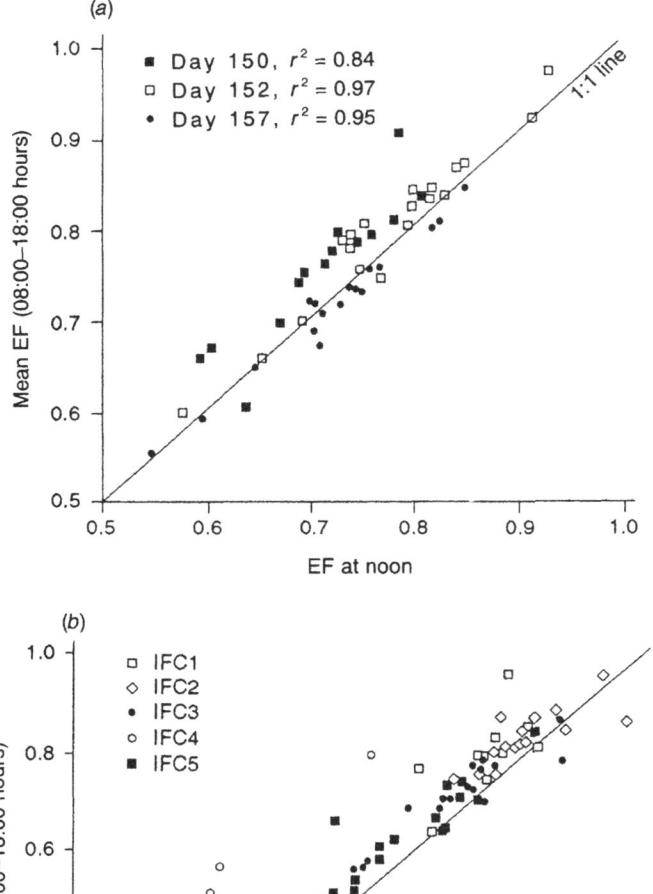

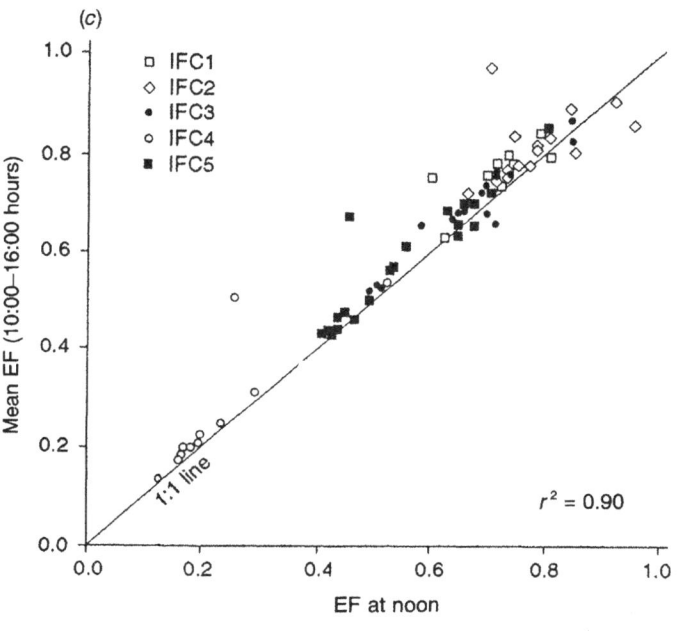

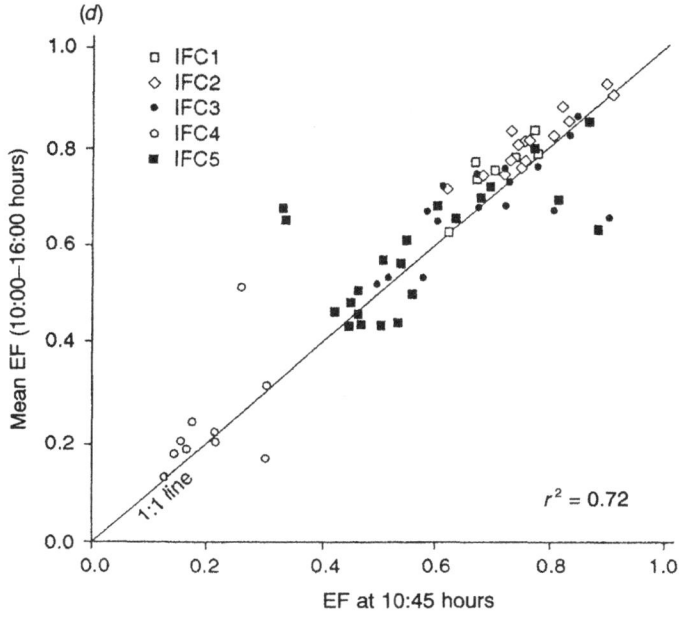

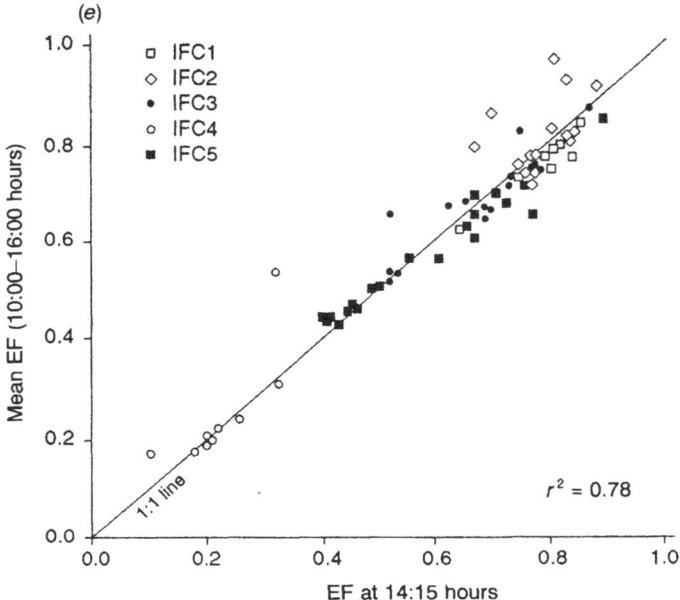

Fig. 6. Regression statistics for correlations of single EF values with overall daytime means. (*a*) FIFE 87/IFC3: correlation of noon versus daytime (08:00–18:00 hours) values for three days during IFC1. (*b*) FIFE 87 and 89/all IFCs: correlation of noon versus daytime (08:00–18:00 hours) values. (*c*) FIFE 87 and 89/all IFCs: correlation of noon versus daytime (10:00–16:00 hours) values. Note improved correlation results produced by reduced daytime period. (*d*) FIFE 87 and 89/all IFCs: correlation of 10:45 versus daytime (10:00–16:00 hours) values. (*e*) FIFE 87 and 89/all IFCs: correlation of 14:15 versus daytime (10:00–16:00 hours) values.

of single EF records with the daylight mean was carried out using different single points as predictors of the daytime mean. Figures 6*d* and 6*e* show the results for 10:45 and 14:15, respectively.

Previous studies of EF trends have not provided the assessment given here of the predictive power of EF snapshots. The 'single' EF values used by Shuttleworth *et al.* (1989) and Verma *et al.* (1992) in correlations with daytime means are in fact EF means for a 2-hour period, usually 1 hour either side of the solar noon. The 2-hour mean was then compared with a daylight mean. Accordingly, earlier work did not allow for an assessment of the true predictive power of a single, instantaneous remotely sensed value. Use of single EF records increases the power of any assessment, although even single EF records from the FMS data are half-hour integrations (chosen to avoid the problems of non-equilibrium measurements).

Spatial variation in EF data

Simple pairwise comparisons of data from distributed FMS sites show a wide range of results. EF data from sites well apart from each other often exhibit correlations that match or exceed those calculated for co-located sites. No clear spatial structure is apparent.

This result has been confirmed by the construction of correlation matrices for the 22 FMS sites. Figure 7a–c show the results for IFCs 2, 4 and 5 (chosen to show 'wet', 'dry'/senescent and 'dry-down' conditions, respectively). In each case, correlation coefficients have been computed for all possible FMS site pairs. With the exception of IFC4, the correlations do not show any decreasing trend across the distances sampled by the FMS sites (that is, up to *c*. 13 km). A decrease in the correlation with distance from *c*. 0.7 to *c*. 0.4 may be seen in the results for IFC4. Spatial consistency of EF is also shown by the analyses shown in Fig. 6, in which the scatter of the data away from the 1:1 line is remarkably limited. The clustering was investigated by means of an analysis of residuals. In the expectation that there may be biases, the *Y* residuals (+ and −) in Fig. 6c were compared with a range of site characteristics. Example results are shown in Fig. 8. The residuals show no bias for aspect, elevation, slope, location or site number (that is, instrumentation and/or technique). None of these spatial factors appears to exert any systematic influence over the power of a noontime measurement to predict the course of daytime EF values.

The rather limited, pairwise approach given in Fig. 7 produces little evidence of the strong spatial autocorrelation structures that might have been expected at the strongly undulating FIFE site. Despite significant slope angles and elevation differences, the EF values across the area are very consistent. In energy partition terms, it is reasonable to treat the area as a single modelling unit with the energy partition assigned to it being the arithmetic mean of available FMS records.

This conclusion is re-inforced by both the analysis of residuals and multiple correlation results. None of the admittedly coarse characterisations of site conditions and positions (such as aspect, management, elevation) causes a consistent offset or bias in the residuals drawn from correlations of FMS data.

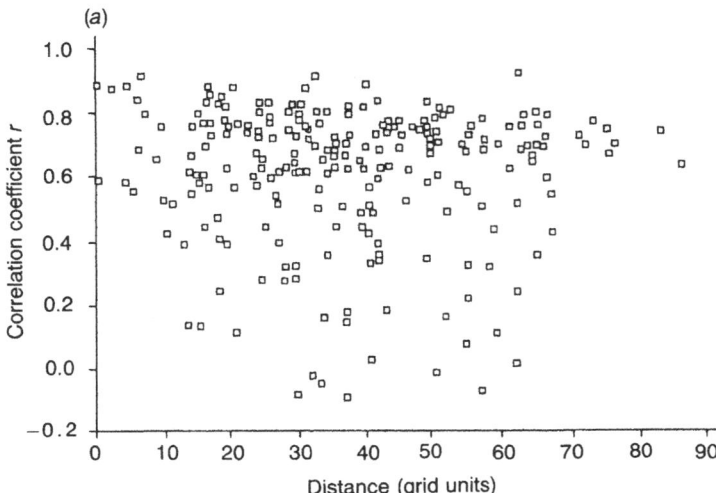

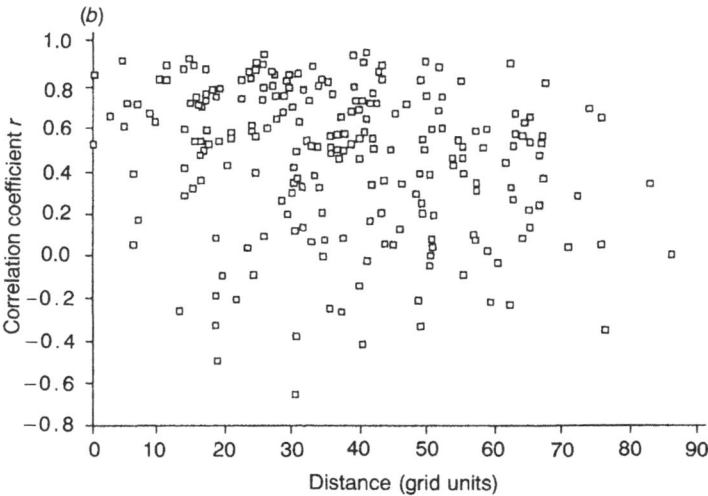

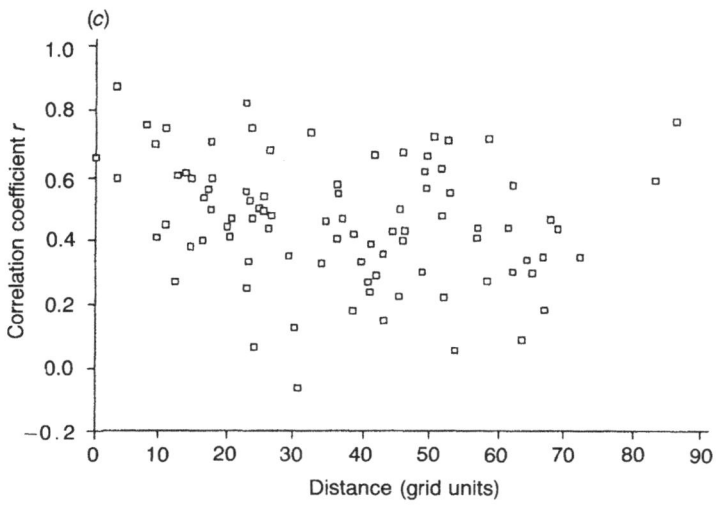

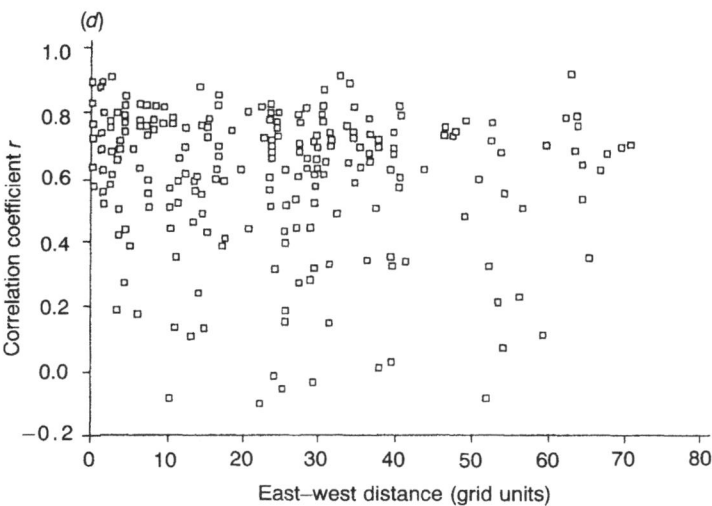

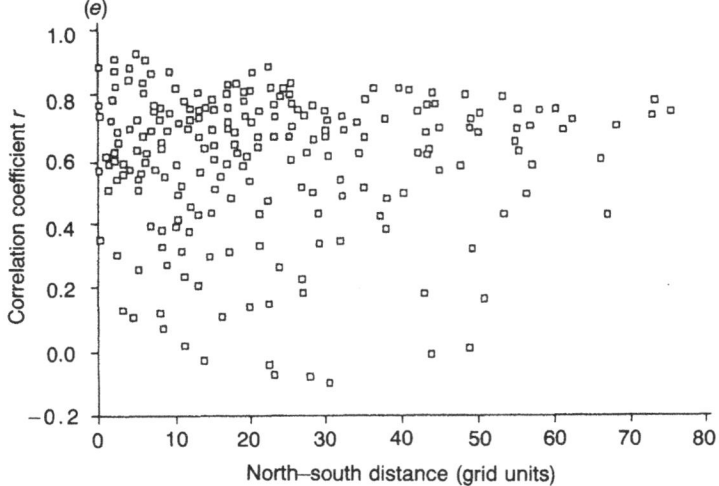

Fig. 7. Scattergram to show pairwise correlation coefficients for EF values versus distance (in units of the 99 × 99 FIFE site grid) between all possible pairings of FMSs. (*a*) FIFE 87/IFC2: wet conditions; (*b*) FIFE 87/IFC4: 'dry'/senescent conditions; (*c*) FIFE 89/IFC5: 'dry-down' conditions; (*d*) FIFE 87/IFC2: distance calculated parallel to the prevailing wind direction (east–west); (*e*) FIFE 87/IFC2: distance calculated perpendicular to the prevailing wind direction (north–south).

Data analysis conclusions

This detailed analysis of EF variation, made possible by the FIFE data, has highlighted some important characteristics of prairie grassland surface energy budgets:

1. The FMS data confirm the suggested daytime conservatism of EF values
2. The analysis reveals a remarkable degree of consistency across the FIFE site, despite its significant variations in elevation, soil depth, aspect, slope angle and vegetation cover
3. The observation of conservative EF values in IFC4, when the prairie vegetation was largely senescent, suggests that daytime variation in energy partition is a rather weaker function of stomatal control than has been implicit in many modelling studies to date.

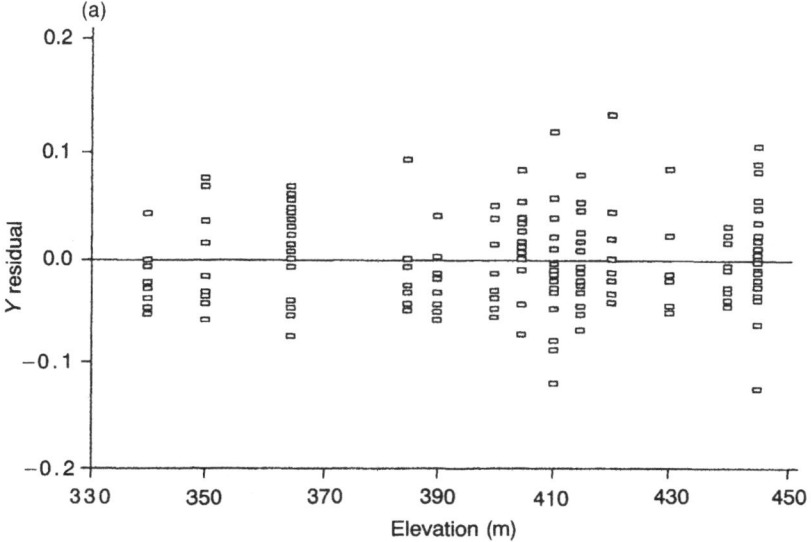

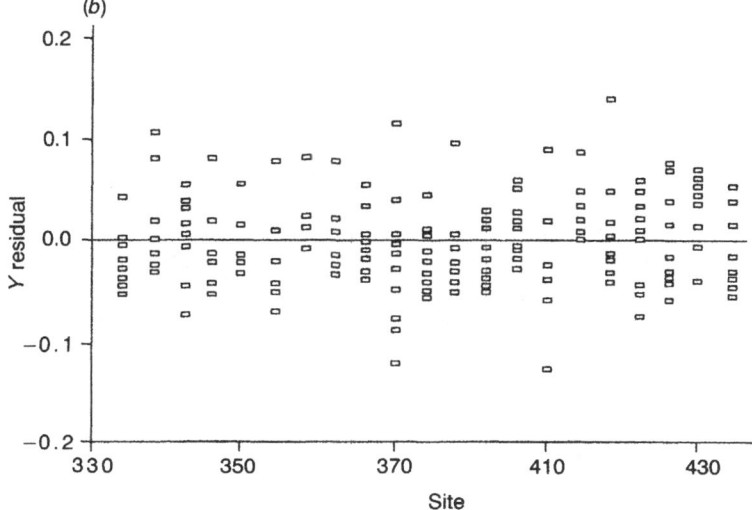

Fig. 8. FIFE 87/IFC1: analysis of residuals. *Y* residuals away from the 1:1 line (see Fig. 6c) plotted against (a) site elevation; (b) site.

Simulation of energy partition

As they have been shown to be consistent, the FMS data may be used for development of new soil–vegetation–atmosphere transfer (SVAT) schemes.

The UK Meteorological Office maintains a single column version of its unified general circulation model. The single column model (SCM) uses the same physics routines as the full GCM (Lean, 1992) and is used for parameterisation development work. Using the SCM, it is possible to assess the performance of new land surface parameterisations (LSPs) by using FIFE data. Use of this model allows the atmospheric boundary layer to interact with the free atmosphere in the model, rather than being fixed as in most SVAT schemes. This allows the effect of the bottom surface boundary on the atmosphere to be assessed.

Land surface parameterisations in GCMs

Recent attempts to simulate the heat, momentum and moisture fluxes at the land surface have focused on SVAT models, many of which assume simple diffusion terms for flux calculations. The land surface components of GCMs use similar schemes and are, therefore, based on the assumption that the unit area vapour flux can be calculated using a so-called Ohm's law analogue model. For instance, the evaporation flux may be modelled as

$$E = \rho \, \delta q / (r_{s} + r_{a}) \tag{4}$$

where E is the evaporation rate, ρ is the density of dry air, δq is the specific humidity deficit in the lowest layer of the atmosphere, r_{s} is the surface or canopy resistance to vapour flux and r_{a} is the aerodynamic resistance between the canopy and the atmosphere (Brutsaert, 1982). The effective implementation of such models requires that, at each time step in a model run, the physiological and physical controls on evapotranspiration, which are represented through the parameterisation of r_{s}, are accurately represented and varied.

In GCMs, the need for an accurate, physical representation has to be balanced against computational considerations. Until recently, these considerations and the lack of demonstrably superior models have meant that the r_{s} term in Equation (4) has been held constant for each biome and at all time steps. This constant conductance approach has the merits of computational simplicity. When soil moisture conditions are non-limiting and evapotranspiration takes place at or near the potential rate, it produces acceptable flux values, but with no attempt made to parameterise the complex controls on canopy conductance under subpotential conditions. It is, therefore, important to develop parameterisations that may produce more realistic results for subpotential conditions. Such resistance models also represent an important step in introducing fully interactive biospheres in GCMs.

Following the work of Jarvis (1976) and others, the use of a constant conductance (inverse of resistance) approach has been replaced with interactive conductance models that use a range of environmental control factors to modify the conductance at each time step:

$$g_s = \left\{ g_{max} \cdot f(\delta g) \cdot f(Q^*) \cdot f(T_s) \cdot f(\theta) \right\} \tag{5}$$

where g_{max} is the maximum conductance of the vegetated surface and $f(\delta q)$, $f(Q^*)$, $f(T_s)$ and $f(\theta)$ are modelled as independent functions of vapour pressure deficit, total downward shortwave (solar) radiation, surface temperature and soil moisture content, respectively. Each function varies between 0 and 1 and constants in each function are best obtained using multiple optimisation techniques (Stewart & Verma, 1992). Just such a parameterisation has been included in the simulation trials described below.

Though relatively simple in computational terms, multiplicative models have only a weak physical basis. This is particularly true of $f(\delta q)$, which dominates modelled conductance variation. No physiological mechanism has been established for a link between vapour pressure deficit and stomatal aperture. The simple multiplication of environmental control factors also assumes that the controls of conductance act independently. The FIFE data suggest that this is far from the truth. Indeed, the use of a simple, independent soil moisture function (a so-called β function) runs counter to the conclusions of the data analysis described above. The exact strength of atmospheric controls on energy partition is often a strong function of subsurface conditions, and the consequent inappropriateness of an independent soil moisture-availability term has led Sellers (1992) to suggest that 'the use of the β-function ... is probably the most damaging of all the flaws in the early [conductance] models'.

Abandoning the vapour pressure deficit–conductance relationship

Mott & Parkhurst (1991) have shown that stomatal response in *Vicia faba* leaves may be the result of changes in transpiration rate rather than a response to changes in humidity in the surrounding air. Such a response has also been suggested by results from a model of photosynthesis which assumes that plants optimise conditions within leaves for photosynthesis (Friend, 1991). Following these suggestions and a comprehensive re-analysis of existing observations of canopy processes and fluxes, both in the laboratory and in the field, Monteith (1995a, b) has proposed an alternative interpretation of conductance for vegetated surfaces. In place of an explicit and dominant relationship with the vapour pressure deficit, conductance is related to the evapotranspiration rate. This approach is founded on the linear relationship between evapotranspiration and conductance, seen in almost all the data re-analysed by Monteith (1995a, b) (Fig. 9). The gradient of the g–E relationship (and the value of the intercepts g_{max} and E_{max}) was postulated by Monteith (1995a, b) to be a function of temperature, soil moisture and CO_2 concentration.

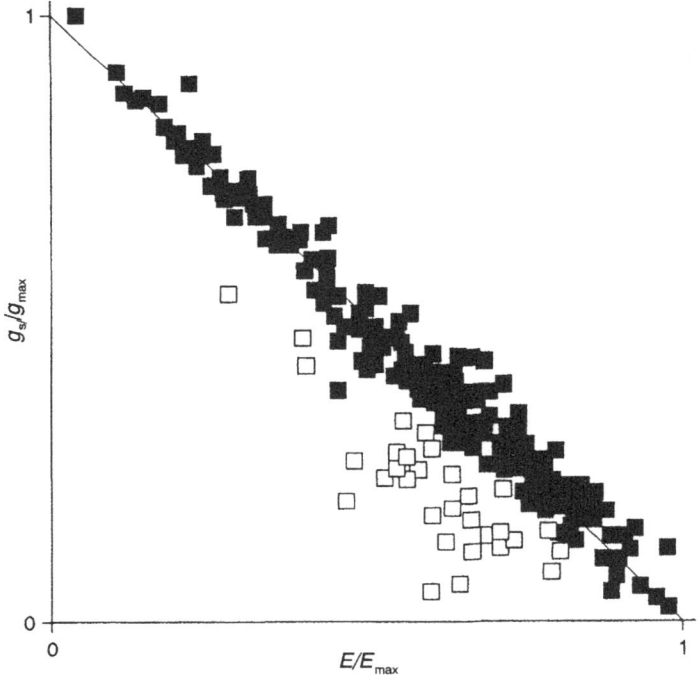

Fig. 9. The apparent dependence of leaf stomatal conductance (g_l) on transpiration rate (E) from laboratory measurements on single leaves. (Open squares are values that may indicate anomalous 'patchy' closure of stomata, as suggested in Mott, Cardon & Berry (1993).) Graph from Monteith (1994a).

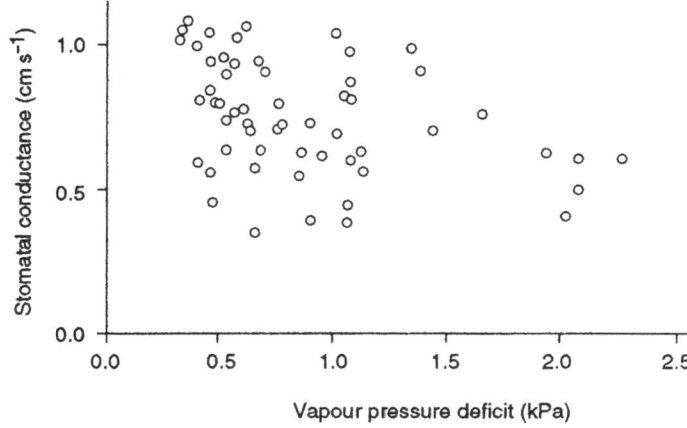

Fig. 10. Stomatal conductance at the top layer of an oak canopy versus vapour pressure deficit. All points are for solar radiation >200 W m⁻². Graph from Dolman & van den Burg (1988).

Existence of a linear relationship between conductance and evapotranspiration prompts the use of an alternative land surface parameterisation in GCMs. The highly non-linear relationship observed between vapour pressure deficit and conductance (for example, Fig. 10, taken from Dolman & van den Burg, 1988) can be abandoned. This alternative conductance model ('the Monteith model') has been implemented in the SCM as follows:

$$g_s = g_{max} \{1 - (E / E_{max})\} \tag{6}$$

where
$$g_{max} = C \cdot f(T_s) \cdot f(Q^*) \tag{7}$$

and
$$E_{max} = (\Psi_s - \Psi_{lc}) / r_{sl} \tag{8}$$

where C is a constant, Ψ_s is the soil–water potential, Ψ_{lc} is the critical (or wilting) leaf–water potential and r_{sl} is a soil–leaf resistance.

Running the SCM for the FIFE site: observational forcing

During the FIFE experiment, there was a programme of radiosonde ascents (Kelly, 1992). The observations of temperature, humidity and winds from the sonde provided model inputs for the boundary layer. Specifically, the sondes give values for the lowest seven of the twenty pressure levels in the SCM's model atmosphere. The lowest five of these layers represent the boundary layer.

In adapting the SCM to run for the FIFE site, the statistical forcing of average values, which is usually used for model development, has been replaced with forcing based on the actual observations from the sondes over time.

The SCM has been run for a three-day period in both the August (IFC3) and October (IFC4) intensive field campaigns. For each run, the atmosphere is initialised using values from a contemporary sonde ascent. Using a modified version of the standard model, the SCM then uses data from all sonde ascents during the three-day run. Modelled values of temperature, humidity and wind are relaxed towards the observed values using a user-defined forcing parameter, τ:

$$A_{model} = A_{model} - \{(A_{model} - A_{sonde})/\tau\} \qquad (9)$$

where A is either temperature, humidity or wind speed. Simulation runs can be allowed to proceed unforced after initialisation by using a large τ value. Alternatively, by reducing τ, the atmosphere overlying the model vegetation becomes increasingly close to that which existed during FIFE. Since modelling of advection is neglected in the SCM, this method of forcing means that each land surface parameterisation under investigation operates beneath a realistic atmosphere.

All simulation results presented here are for the periods 15–17 August in IFC3 and 6–8 October in IFC4. There was little or no cloud on all six days and no rainfall. August 13 saw continuous heavy rainfall (average of 83 mm). In all SCM runs the value of τ has been set to 5. Unless otherwise indicated, the constant conductance used in these trials is that for the SCM's type 5 vegetation (pasture): $g_s = 1.27$ cm s^{-1} ($r_s = 79$ s m^{-1}).

Figure 11 shows a comparison of results for the EF for IFC3 from all three parameterisations (constant, multiplicative and Monteith) and the FIFE observations (displayed as an 'all-site' average). Figure 12 shows a comparison for IFC4 using three constant conductance values. Figure 13 shows the result of varying the key r_{sl} parameter in the Monteith model.

As a whole, Figs. 11–13 demonstrate two key results from the model runs:

1. Using the SCM in its observational forcing mode combined with accurate initialisation of key state variables and surface/subsurface parameters, it is possible to produce reasonable simulations of surface energy budgets
2. In non-water-limited conditions, the form of the conductance parameterisation can have little impact on modelled

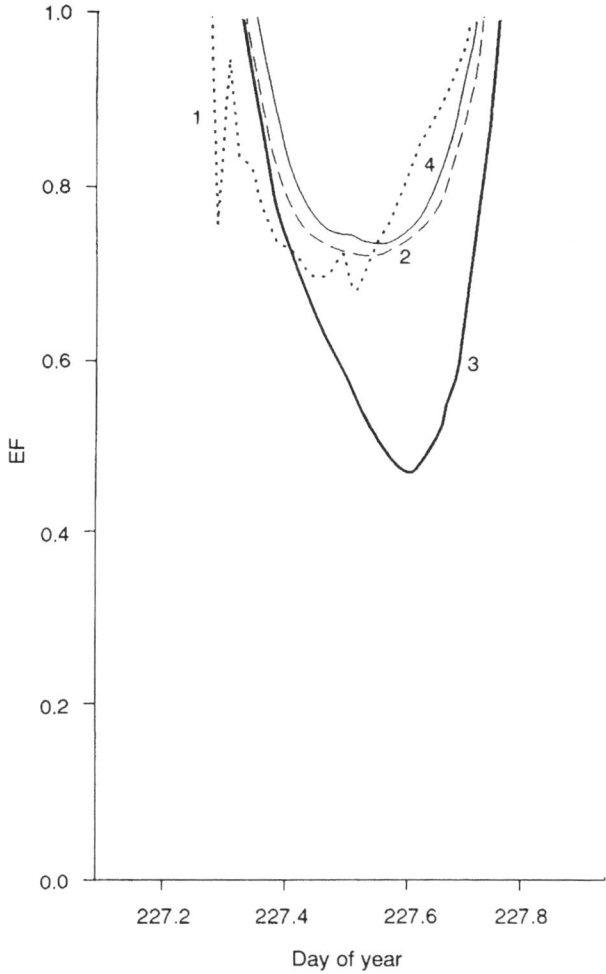

Fig. 11. Comparison of simulations using all three conductance models (see text) and FIFE observations during a three-day period in IFC3. 1, observations; 2, constant conductance; 3, multiplicative; 4, Monteith.

flux values (see, for example, the results from the Monteith and constant conductance runs in Fig. 11).

The greatest difference between the multiplicative and the Monteith model occurs in the afternoon. As the humidity deficit increases, the

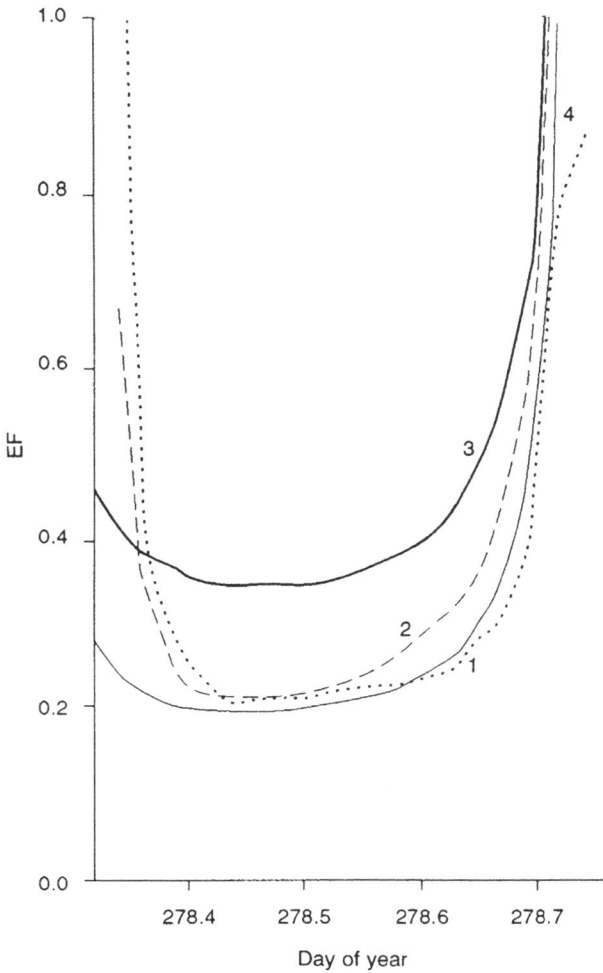

Fig. 12. Comparison of simulations using different constant conductance values and the FIFE observations. 1, observations; 2, constant
constant conductance; 3, multiplicative; 4, Monteith.

multiplicative model reduces the surface conductance, even in unlimited
water conditions. The latent heat flux is reduced to an unrealistic level
and the EF, therefore, drops down well below observed values. In contrast, the Monteith model relates soil moisture conditions to E_{max} and,
provided that the soil–water potential remains above the wilting poten-

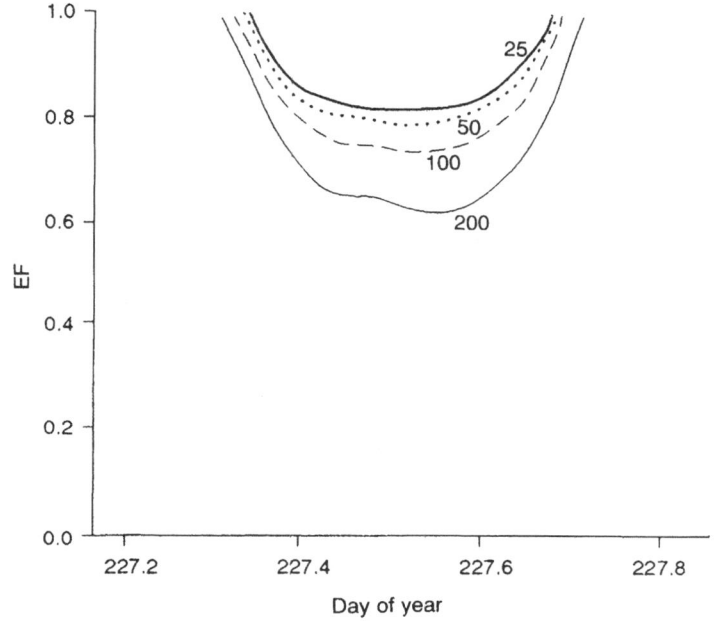

Fig. 13. Comparison of simulations using different r_{sl} values in the Monteith model and the FIFE observations: r_{sl} values of 25, 50, 100 and 200 MPa m^2 smol^{-1} H$_2$O.

tial, the conductance is held high well into the afternoon. For a clear-sky daylight period, the Monteith model, therefore, performs very much like the constant conductance model (Fig. 12). Both models produce EF results that are realistic in both value and daytime conservatism.

The soil–leaf resistance in the Monteith model is a powerful determinant of the model's performance (Fig. 13). This sensitivity to r_{sl}, which is intended to represent the controls on transpiration exerted by the root and xylem system, concurs with the FIFE data analysis above. Subsurface conditions, in particular the inter-dependence of subsurface and surface–atmospheric controls on conductance, have not been modelled realistically in previous models and the use of r_{sl} begins to address this problem.

The preliminary simulation work discussed here must be developed to produce a more rigorous assessment of model performance in a wide range of environments and under a wide range of atmospheric conditions. In particular, it is not yet possible to demonstrate whether the constant conductance and multiplicative models could be modified to produce comparable results to the Monteith model and its empirical g–E relationship. It may, for instance, be possible to use an effective constant conductance whilst retaining constant conductance for each day. Certainly, care is required before deciding that a more complicated, physiologically based model will necessary give better estimates of the surface energy partition.

Acknowledgements

The authors are indebted to John Monteith for his constant advice and comments, and to Peter Cox for initiating, directing and informing use of the Single Column Model. Ian Sewell was funded by a NERC studentship (GT4/91/GS/114) and the work is part of NERC's TIGER (Terrestrial Initiative in Global Environmental Research) programme (GST/02/603). The authors are also indebted to all the FIFE investigators, and particularly to Ann Hsu for the quality control work she has done on the FIFE flux data.

References

Bennetts, D.A. (1992). *A Review of the Science Base Underpinning Climate Prediction Research: Second Research Coordinator's Report*. Bracknell, UK: Hadley Centre for Climate Prediction and Research.

Brutsaert, W.H. (1982). *Evaporation into the Atmosphere*. Dordrecht: D. Reidel.

Brutsaert, W.H. & Sugita, M. (1992). Regional surface fluxes under nonuniform soil moisture conditions during drying. *Water Resources Research*, 28, 1669–1674.

Crosson, W.L. & Smith, E.A. (1992). Estimation of surface heat and moisture fluxes over a prairie grassland 2. Two-dimensional time filtering and site variability. *Journal of Geophysical Research*, 97, 18 583–18 598.

Dolman, A.J. & van den Burg, G.J. (1988). Stomatal behaviour in an oak canopy. *Agricultural and Forest Meteorology*, 43, 99–108.

Field, R.T., Fritschen, L.J., Kanemasu, E.T. *et al.* (1988). Calibration, comparison and correction of net radiation instruments used during FIFE. *Journal of Geophysical Research*, 97, 18 681–18 695.

Friend, A.D. (1991). Use of a model of photosynthesis and leaf microenvironment to predict optimal stomatal conductance and leaf nitrogen partitioning. *Plant, Cell and Environment*, 14, 895–905.

Fritschen, L.J. & Qian, P. (1992). Variation in energy balance components from six sites in a native prairie for three years. *Journal of Geophysical Research*, 97, 18 651–18 661.

Fritschen, L.J., Qian, P., Kanemasu, E.T. *et al.* (1992). Comparisons of surface flux measurement systems used in FIFE 1989. *Journal of Geophysical Research*, 97, 18 697–18 713.

Jarvis, P.G. (1976). The interpretation of variations in leaf water potential and stomatal conductance found in canopies in the field. *Philosophical Transactions of the Royal Society of London, Series B*, 273, 593–610.

Kelly, R.D. (1992). Atmospheric boundary layer studies in FIFE: challenges and advances. *Journal of Geophysical Research*, 97, 18 373–18 376.

Lean, J. (1992). *A Guide to the UK Meteorological Office Single Column Model*. Bracknell, UK: Hadley Centre for Climate Prediction and Research.

Monteith, J.L. (1995*a*). Accommodation between transpiring vegetation and the convective boundary layer. *Journal of Hydrology*, 166, 251–263.

Monteith, J.L. (1995*b*). A reinterpretation of stomatal responses to humidity. *Plant, Cell and Environment*, 18, 357–364.

Monteith, J.L. & Unsworth, M.H. (1990). *Principles of Environmental Physics*, 2nd edn. London: Hodder and Stoughton.

Mott, K.A. & Parkhurst, D.F. (1991). Stomatal responses to humidity in air and helox. *Plant, Cell and Environment*, 14, 509–515.

Mott, K.A., Cardon, Z.G. & Berry, J.A. (1993). Asymmetric patchy stomatal closure for the surfaces of *Xanthium strumarium* leaves at low humidity. *Plant, Cell and Environment*, 16, 25–34.

Nie, D., Kanemasu, E.T., Fritschen, L.J. *et al.* (1993). An intercomparison of surface energy flux measurement systems used during FIFE 1987. *Journal of Geophysical Research*, 97, 18 715–18 724.

Priestley, C.H.B. & Taylor, R.J. (1972). On the assessment of surface heat flux and evaporation using large-scale parameters. *Monthly Weather Review*, 106, 81–92.

Sellers, P.J. (1992). Biophysical models of land surface processes. In *Climate System Modelling*, ed. K.E. Trenberth, pp. 451–490. Cambridge: Cambridge University Press.

Sellers, P.J. & Hall, F.G. (1987). *FIFE: Experiment Plan, International Satellite Land Surface Climatology Project*. Greenbelt, MD: NASA-GSFC.

Sellers, P.J. & Hall, F.G. (1993). *FIFE-89: Experiment Operations, Preliminary Results and Project Documentation, International Satellite Land Surface Climatology Project*. Greenbelt, MD: NASA-GSFC.

Sellers, P.J., Hall, F.G., Asrar, G., Strebel, D.E. & Murphy, R.E. (1992). An overview of the First International Satellite Land Surface Climatology Project (ISLSCP) Field Experiment (FIFE). *Journal of Geophysical Research*, 97, 18 345–18 371.

Shuttleworth, W.J., Gurney, R.J., Hsu, A.Y. & Ormsby, P.J. (1989).

FIFE: the variation in energy partition at surface flux sites. In *Remote Sensing and Large-scale Global Processes*. Baltimore, MD: IAHS Press.

Smith, E.A., Hsu, A.Y., Crosson, W.L. *et al.* (1992). Area-averaged surface fluxes and their time–space variability over the FIFE experimental domain. *Journal of Geophysical Research*, 97, 18 599– 18 622.

Stewart, J.B. & Verma, S.B. (1992). Comparison of surface fluxes and conductances at two contrasting sites within the FIFE area. *Journal of Geophysical Research*, 97, 18 623–18 628.

Verma, S.B., Kim, J. & Clement, R.J. (1992). Momentum, water vapour and carbon dioxide exchange at a centrally located prairie site during FIFE. *Journal of Geophysical Research*, 97, 18 629– 18 639.

J.A. BERRY, G.J. COLLATZ, A.S. DENNING,
G.D. COLELLO, W. FU, C. GRIVET,
D.A. RANDALL and P.J. SELLERS

SiB2, a model for simulation of biological processes within a climate model

Introduction

Atmospheric general circulation models (GCMs) are widely used for weather forecasting and for climate sensitivity studies. Early studies (e.g. Charney *et al.*, 1977; Shukla & Mintz, 1982; Sud, Shukla & Mintz, 1988) established that interaction of the land surface with the atmosphere can strongly influence climate over the continents. Recognition of the need to understand and predict the implications of global change has generated demand for increased accuracy and realism in all aspects of climate modelling. At the very least, a land surface parameterisation (LSP) should provide the climate model with the immediate fluxes of radiation, heat, water vapour and momentum between the surface and the atmosphere. These fluxes can then be integrated over extended model runs to keep track of the thermal and moisture balance of the land surface. Up to now, the focus has been on calculating surface energy and water budgets in LSPs, but a need for realistic representation of the carbon cycle in GCMs is emerging. Surface properties controlling these fluxes (albedo, the surface roughness, the effective resistance of the soil and vegetation to evaporation of water and the physiological properties of the vegetation) can vary dramatically at all scales and may change substantially with season and year by year. Part of this variation is a result of feedbacks between the land biosphere and climate. To depict the dynamics and spatial variability of these processes on a global scale requires vast quantities of information on the properties of the land surfaces of the planet and a parameterisation of surface processes that can work in all of the major climatic regions. At the same time, the models should be robust and computationally efficient.

Early approaches to modelling biosphere–atmosphere interactions prescribed the properties of the surface and used a simple bucket hydrology (Budyko, 1974). Dickinson (1984) and Sellers *et al.* (1986) introduced models with a more realistic soil–plant–atmosphere structure. In these models, roughness and albedo were specified as a function of soil reflectance, vegetation type and leaf area index (LAI). Maps of vegetation type and soils were constructed from ground-based classification work of Matthews (1985) and others, and seasonal variation in green LAI was specified from data gathered from the ecological literature (see Dorman & Sellers, 1989). In a fundamental departure from previous approaches, the canopy conductance was made a dynamic function of environmental conditions provided by the climate model using an empirical formulation based on Jarvis (1976), which was a function of vegetation type, LAI, light, temperature, humidity, wind speed and precipitation. This introduced feedback between the biosphere and the atmosphere.

These more complete models have been shown to generate more realistic continental climates. For example, Sato *et al.* (1989) found in parallel GCM runs, one with the Simple Biosphere Model of Sellers *et al.* (1986) (SiB) and one with a conventional bucket model, that SiB gave consistently lower evaporation and more realistic continental precipitation fields, mainly because of the more realistic treatment of the surface resistance to water vapour transport. Beljaars *et al.* (1996) recently showed how the implementation of a more realistic soil moisture and surface-resistance parameterisation in a numerical weather prediction model was responsible for a greatly improved simulation of the precipitation anomaly that gave rise to the mid-western floods in the USA in the summer of 1993. These results and others, reviewed by Garratt (1993), show that improved land surface models have helped to improve GCM simulations, numerical weather forecasting and studies of climate sensitivity to hypothetical land cover changes.

Application of these models is, nevertheless, hampered by a shortage of appropriate data to characterise the density and phenology of terrestrial vegetation on a global scale, and by unstable model behaviour in certain vegetation types and climatic zones. In recent years, there has been rapid progress in two areas that offers promise of improved model performance. First, as anticipated by the treatment of Sellers (1985; 1987), optical remote sensing of the land surface has advanced to the point that processed satellite observations can be directly used to define the time–space variations in the fraction of incident photosynthetically active (solar) radiation (PAR) absorbed by vegetation on a global scale (Sellers *et al.*, 1996a,b). Second, advances in plant physiology (e.g. Woodrow & Berry, 1988; Collatz *et al.*, 1991; Sellers *et al.*, 1992a) provide new insight into the mechanisms of photosynthesis and how this fundamental process is tied to stomatal function on the one hand and to the use of absorbed PAR and optical properties of the canopy on the other. An interdisciplinary team (Asrar & Dokken, 1993) was organised to modify SiB to take advantage of the improved mechanistic understanding of the coupling between optical remote sensing, light interception and physiology, to incorporate this model in an advanced GCM and to develop appropriate global data sets from satellite observations to drive the model. This chapter provides an overview of this work with special emphasis on issues of scaling-up from detailed biochemical and physiological processes of leaves; it also reviews progress towards the longer-term goal of including an interactive carbon cycle in climate models.

Model overview

SiB2 is largely based on an earlier model, SiB, of Sellers *et al.* (1986) and a detailed description of the new model is given by Sellers *et al.* (1996b). A schematic diagram of the model is shown in Fig. 1.

General

Atmospheric boundary conditions include wind speed (U_m), temperature (T_m) and concentrations of CO_2 (C_m) and water vapour (e_m) at a reference level above the canopy, ($\bar{z}_m$) and the shortwave ($S\downarrow$) and longwave ($L\downarrow$) components of incident radiation and precipitation (P). The last is divided into convective and large-scale components, which are distributed differently within the grid area. The exchanges of radiation ($L\uparrow$, $S\uparrow$), momentum, sensible heat ($H_m\uparrow$), latent heat ($\lambda E_m\uparrow$) and CO_2 ($a_m\downarrow$) between the surface and the atmosphere are simulated.

The canopy is considered as a single layer of one of nine vegetation types with associated morphological parameters (canopy height, leaf dimensions, etc.), optical properties and physiological constants. The vegetation type and physiological type (C_3/C_4) of each grid cell is obtained from maps. Time-varying properties of the vegetation (fractional absorptance of PAR (FPAR), LAI, greenness fraction and surface roughness length) are derived from satellite data. The soil is considered as three layers: a surface layer subject to direct water loss by evaporation, a layer that supplies water for plant transpiration and a subsoil layer. The albedo and physical properties of the soil are obtained from other global data sets (Zobler, 1986; Harrison *et al.*, 1990).

A two-stream radiative transfer scheme (Sellers, 1985; 1987) is used to describe interception, reflection, transmission and absorption of radiation by the vegetation and soils and to calculate their net radiation balance. The build-up of snow on the canopy and soil surface also affects the reflectance and aerodynamic properties of the surface, and

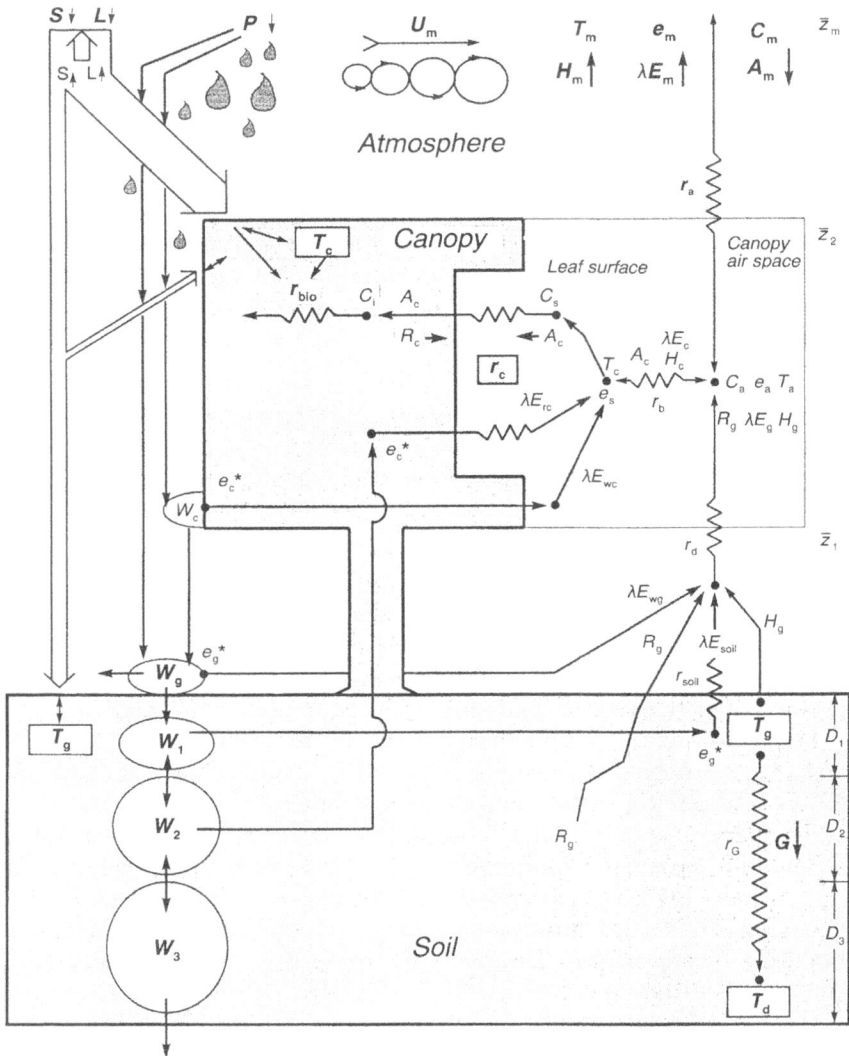

thin snow cover is assumed to become patchy to avoid drastic changes in albedo and temperature as snow depth approaches zero.

The aerodynamic transport of SiB2 follows a first-order closure scheme as modified by Sellers *et al.* (1989) with 'resistance' elements associated with the boundary layer of leaves (r_b), transport from the soil to the canopy air (r_d) and the turbulent transport between the canopy air and the mixed layer (r_a). Precipitation, less surface runoff and interception loss from liquid water (or snow) stores on the canopy and the ground surface infiltrated into the soil. Moisture transport in the soil is a function of gravitational drainage and hydraulic diffusion between horizontal layers. Runoff can occur from the surface during intense precipitation or by bottom-drainage of the soil profile. Heat transport in the soil is described by the force-restore model of Deardorff (1977). Conductance of the soil surface to diffusion of water vapour is described by an empirical function that accounts for drying of the surface layer (Sellers, Heiser & Hall, 1992*b*). Water transport from the soil to the canopy is not explicitly treated in the model. Rather, physiological water stress is parameterised as a function of the available water stored in the soil.

SiB2 has nine prognostic (time-stepped) state variables: three temperatures (canopy temperature, soil surface and deep ground temperature), two intercepted precipitation stores (on the canopy and

---

Fig. 1. Schematic diagram depicting most of the processes modelled in SiB2. The soil is divided into three layers (D_1–D_3) as is the canopy–atmosphere space (z_1, z_2, z_m). Bold symbols represent meteorological inputs (incoming solar radiation, $S\downarrow$; incoming longwave radiation, $L\downarrow$; precipitation, P; mixed layer temperature, T_m; humidity, e_m; wind speed, U_m; CO_2 concentration, C_m) or prognostic variables, which are temperatures (canopy, T_c; ground surface, T_g; deep soil, T_d), water stores (intercepted precipitation by canopy, W_c, and soil, W_g; soil water, W_1–W_3) and canopy resistance, r_c. The resistances to mass transfer through the atmosphere include aerodynamic (r_a) and canopy and soil boundary layer (r_b, r_{soil}, respectively) resistance. Biochemical limitations on photosynthesis are represented by r_{bio}. Other diagnostic variables include fluxes of heat (sensible, $H_m\uparrow$; latent, $\lambda E_m\uparrow$; soil, $G\downarrow$) and CO_2 (A_c, $A_m\downarrow$, R_g) and the concentrations of water vapour (e, e^*); and CO_2 (C). Subscripts m, a, c, s and g refer to mixed layer, canopy air space, canopy, leaf surface and ground, respectively. The superscript, *, denotes saturation with respect to water vapour at the respective temperature.

the soil surface), three soil moisture values and a time-stepped stomatal conductance. The energy balance is solved iteratively because of the non-neutral treatment of the aerodynamics. Stomatal adjustments interact with photosynthesis, the energy balance and the aerodynamic conductance. Therefore, a full solution with interactive stomatal adjustment to the leaf environment would require at least two levels of nested iterations. This is eliminated by taking advantage of the fact that the time step of the model (0.1 hour) is similar to the relaxation time for stomatal movements. Hence it is possible to treat stomatal conductance as a prognostic variable. This more realistic treatment of stomata adds stability to the calculations and improves the computational efficiency of the model.

The coupled photosynthesis–stomatal conductance model

Photosynthesis and the transpiration of water vapour in SiB2 are handled by coupled photosynthesis–stomatal conductance models for leaves with C_3 (Collatz *et al.*, 1991) or C_4 (Collatz, Ribas-Carbo & Berry, 1992) photosynthetic systems. The photosynthesis models are derived from the model of Farquhar, von Caemmerer & Berry (1980). The primary adjustable parameter that may change from place to place is the amount (V_{max}) of the CO_2 fixing enzyme rubisco in the cells. Adjusting V_{max} changes the value of absorbed PAR at which photosynthesis becomes light saturated. Other parameter values and their response to temperature are constrained by independent biochemical and physiological experiments. Water stress and extremes of temperature attenuate the V_{max} from its unstressed value. When applied in the coupled photosynthesis–stomatal conductance model, this empirical approximation results in linear decreases in photosynthesis and stomatal conductance as stress develops. This pattern is generally observed when stress develops under natural conditions (Björkman, 1989).

The photosynthetic parameterisation used in SiB2 is based on a mechanistic understanding of the biochemistry of photosynthesis. However, this knowledge would be of little use without a means of simultaneously predicting the stomatal conductance of leaves. At the present time, there is no satisfactory basis for constructing a mechanistic parameterisation of stomatal conductance. A number of empirical parameterisations are available. Most of these are based on Jarvis (1976); Ball, Woodrow & Berry (1987) or a theoretical treatment of optimum stomatal regulation (Cowan & Farquhar, 1977). These empirical models of stomatal conductance are generally calibrated from gas-exchange measurements of leaf conductance over a range of environmental conditions. Some of these models have been discussed by Collatz *et al.* (1991) (see also Lloyd, 1991), who pointed out that most available models could be parameterised to give quantitatively similar responses under restricted conditions. The correct form for a stomatal parameterisation is an area of active debate at the moment with several forms of empirical equation under consideration (see Aphalo & Jarvis, 1993a; Amthor, 1994; de Pury *et al.*, 1995; Leuning, 1995). At the same time others (e.g. Friend, 1991; Deware, 1995; Monteith, 1995) have proposed approaches based on mechanistic hypotheses. We have not resolved these issues, and for the present, we are using the original parameterisation of Ball *et al.* (1987). This has been calibrated for the major vegetation types, and the computational stability of the coupled photosynthesis–stomatal conductance formulation has been extensively tested in GCM simulations.

Several recent papers have attached considerable significance to the ability of various forms of empirical equation to fit gas-exchange data. Table 1 shows a comparison of the residual standard error obtained by analysis of four large data sets ($n = 42, 93, 97$ and 123). The Ball parameterisation is compared with alternative forms recently suggested by several other workers. Table 1 shows that all of the models account for at least 80% of the variance in g_s (stomatal conductance to water vapour). The formulation of Leuning (1995) provides a marginally better fit, but it also uses an additional adjustable parameter. Other criteria such as the ability to simulate a wide range of environmental conditions and numerical stability in combination with a photosynthesis model also need to be considered.

It may be useful to review briefly what is required of a stomatal parameterisation. The stomata of leaves respond over a wide dynamic range from values that strongly restrict diffusive exchange between the interior of the leaf and its environment to values that essentially allow transpiration to proceed at the potential rate. At the same time, stomata, by restricting the diffusion of CO_2, affect the rate of photosynthesis. These responses have been correlated with factors of the environment including light, temperature, CO_2 concentration, atmospheric humidity and the availability of soil water. Furthermore, empirical studies indicate that the rate of photosynthesis of a leaf influences the conductance of its stomata (apart from the effect of stomata on photosynthesis noted above). All of these change, some rapidly, during the course of natural growing seasons, and a very wide range of conditions is encountered in global-scale simulations of canopy processes. It is important that the parameterisation be accurate, robust and capable of representing the dynamic interactions of leaf physiology within the coupled system that

Table 1. *Comparison of the residual standard error obtained by regression for models for stomatal conductance*

Species	Aspen	Soybean	Peanut	Maize
n	42	123	97	93
g_s (mol m^{-2} s^{-1})	0.026–0.447	0.070–1.280	0.080–0.910	0.030–0.730
Residual standard error[a]				
Ball's model	0.0473	0.0563	0.1188	0.0520
Aphalo's model	0.0458	0.0585	0.1222	0.0574
Amthor's model	0.0655	0.0997	0.1705	0.0909
de Pury's model	0.0551	0.1102	0.1431	0.1365
Leuning's model	0.0417	0.0702	0.1012	0.0519

[a] The residual standard error is the square root of the sum of squared residuals, divided by number of degrees of freedom. Compare with g_s.

Ball's model (Ball, Woodrow & Berry, 1987)

$$g_s = m \frac{A_n}{c_s} h_s + g_0$$

Aphalo's model (Aphalo & Jarvis, 1993a)

$$g_s = \frac{A_n}{c_s} (k + mD_s + nT_1)$$

Amthor's model (Amthor *et al.*, 1994):

$$g_s = \frac{(A_n - R_1)}{(c_i - \Gamma)} m e^{-0.00045 D_s} + g_0$$

de Pury's model (de Pury, 1995):

$$g_s = A_n \sqrt{\frac{1.6\lambda}{(c_s - \Gamma)D_s}} + g_0$$

Leuning's model (Leuning, 1995):

$$g_s = m \frac{A_n}{(c_s - \Gamma)(1 + D_s / D_0)} + g_0$$

Abbreviations: g_s, stomatal conductance to water vapour; A_n and R_1, leaf rates of photosynthesis and respiration (mol m^{-2}s^{-1}), respectively; h_s and c_s, (decimal) relative humidity and mole fraction of CO_2 at the leaf surface, respectively; c_i, intercellular CO_2; Γ, CO_2 compensation point; T_1, leaf temperature; D_s, leaf-to-air vapour pressure gradient at the leaf surface; g_0, D_0, k, m, n and λ are adjustable parameters.

includes the leaf, its laminar boundary layer, the canopy air and the aerodynamic mixing of canopy air with the bottom layer of the atmosphere.

An experimental result presented in Fig. 2 provides a useful example

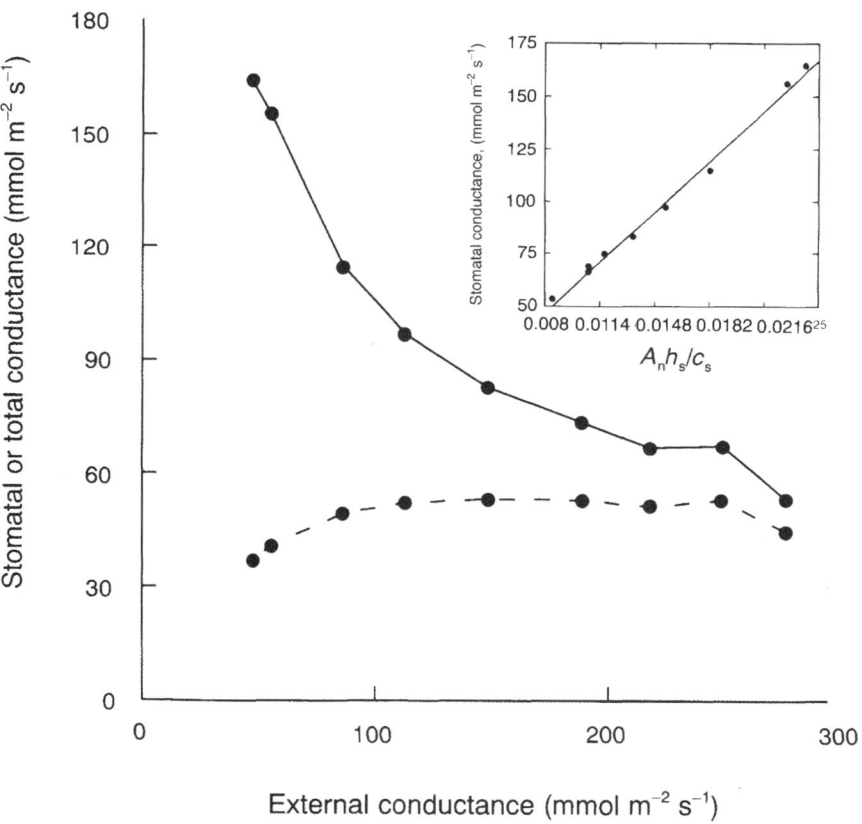

Fig. 2. Response of stomatal conductance, g_s (upper curve) and total conductance, g_t ($g_s + g_e$ in series) (lower curve) to changes in the flow rate through a leaf chamber. The flow is expressed in terms of conductance units (flow rate in mol s^{-1} divided by leaf area) and is analogous to an aerodynamic conductance, g_e. The water vapour and CO_2 partial pressures of the incoming air were held constant as were light intensity and temperature. The insert graph is a plot of measured stomatal conductance versus an index that is the basis of the empirical stomatal model described in the text. The linear regression line is also plotted.

of the dynamic coupling of leaf physiology with aerodynamics. In this experiment, a leaf of barley was enclosed in a stirred gas-exchange cuvette and stomatal conductance, surface CO_2 and H_2O concentrations and the rates of photosynthesis and transpiration were monitored as a flow of air (of constant CO_2 and H_2O concentration) was varied through the cuvette. The flow rate (mol s^{-1}) divided by the leaf area (m^2) has the dimensions of a conductance, g_s, and is plotted on the x-axis. The changes in gas composition in the vicinity of the leaf are equivalent to what would occur with changes in turbulent mixing in a natural canopy (Fig. 1, where, $g_e = 1/(r_a + r_b)$). On the y-axis are plotted the stomatal conductance, g_s (upper line) and the total conductance, g_t (g_s and g_e (environmental conductance) in series, $1/g_t = 1/g_s + 1/g_e$). Note that changes in g_e elicit a very large change in g_s. Aphalo & Jarvis (1993b) and Bunce (1985) report responses of stomata to changes in boundary layer conductance elicited by changing the wind speed over the leaf. Physiological responses similar to those illustrated in Fig. 2 are widespread, if not universal, in higher plants, and such interactive coupling of stomatal physiology with aerodynamics must be a universal feature of canopies.

An insert to Fig. 2 shows changes in the local concentrations of H_2O and CO_2 at the leaf surface (plotted as the index, $A_n h_s/c_s$; see Table 1), which are strongly correlated with the change in stomatal conductance, and these responses are accurately predicted by the empirical stomatal model of Ball and co-workers. All of the models considered in Table 1 can also be made to fit this response, but the parameter values may need to be adjusted as temperature, atmospheric humidity or CO_2 concentration are varied. Proper simulation of these interactions over the wide range of environmental conditions encountered in global simulations places a stringent requirement on the stomatal parameterisation.

Extensive tests of the combined photosynthesis–stomatal conductance models at the leaf scale are presented by Collatz *et al.* (1991; 1992). These models provide realistic responses to changes in light, temperature, CO_2 and r_b and r_a, with a single set of parameters. Interactive coupling of stomatal responses with the local environment at the surface of the leaf and with the photosynthetic processes within the leaf lead to a number of interesting system properties (e.g. mid-day stomatal closure) discussed by Collatz *et al.* (1991). Values for the adjustable physiological constants required as inputs to SiB2, such as V_{max} and the stomatal slope parameter m, have been selected for the various vegetation types (Sellers *et al.*, 1996a). Initially, these have been based on laboratory measurements. The need for studies at the field and plot scale is recognised, and an example will be presented later in this chapter.

Scaling from the leaf to the canopy

In SiB2, scaling-up from leaf-level processes to the canopy is accomplished by invoking 'economic arguments' to derive a simple scheme that integrates the performance of individual leaves to the canopy as a whole (Sellers *et al.*, 1992a). This scheme essentially relates photosynthesis and stomatal conductance of the complete canopy to that calculated for a reference leaf at the top of the canopy. The top-leaf photosynthesis and conductance are multiplied by a canopy integration factor, Π. This factor generally falls within the range 0–2 and is the ratio of the PAR absorbed by the reference leaf to that absorbed by the entire canopy (see Kruijt *et al.*, this volume). The value of Π can be calculated from a biophysical description of the canopy, or it can be inferred from optical remote sensing. It follows from this scheme that the biochemical parameters required for simulation of canopy responses can be determined from process-level studies of single leaves of species comprising the canopy.

This integration scheme assumes that non-linearities introduced by gradients in the environmental and the physiological properties of leaves within the canopy are negligible. The most important source of non-linearity is in the response of leaves to changes in the PAR flux incident at the top of the canopy. For example, if some of the leaves within the canopy were light-saturated while others were light-limited at the same time, the sum of the light responses of all leaves, taken together, would be more curved than the light response of any single leaf. More important, the total photosynthesis of the canopy would be less than it could be if the biochemical capacity of the leaves were 'tuned' to their local light regimes (see Collatz *et al.*, 1990). As reviewed by Sellers *et al.* (1992a), there is abundant evidence that leaves within a canopy acclimate to their local light regimes with adjustments in photosynthetic capacity and stomatal conductance. If this tuning were perfect, all leaves would respond in synchrony to the external light field, and the sum of photosynthesis and stomtal conductance of all of the leaves of the canopy would scale linearly with the reference leaf. This tuning tends to improve the efficiency with which limiting resources (water, light and nutrients) are used for photosynthesis, and these 'economic arguments' lead, conveniently, to a simplified scheme for canopy integration. It is not realistic to expect that this tuning will be perfect, nor that actual canopy integration will be free of non-linear effects owing to other factors such as temperature or humidity gradients in the canopy. This would result in greater 'co-limitation' of photosynthesis by light and biochemical capacity at the canopy scale than is observed at the leaf scale (Collatz *et al.*, 1990). In our view, these, 'real world effects' can be handled by adjustments to constants in the photosynthesis model that specify the amount of co-limitation.

Respiration

Net CO_2 flux to the atmosphere is given by $A_m = A_c - R_g$, where A_c is the net canopy exchange (including leaf respiration) and R_g is the respiratory flux from the rest of the ecosystem. At the present time, we do not attempt to differentiate between plant respiration and decomposition of soil organic matter. The value of R_g is calculated as a non-linear function of the soil moisture and temperature. For simulation of micrometeorological measurements we use an empirical parameterisation developed by Norman, Garcia & Verma (1992) that is calibrated using chamber measurements made at the site. For GCM simulation of global-scale carbon cycling, we follow the approach of Fung, Tucker & Prentice (1987). First, we calculate the seasonal pattern of respiration assuming that the net exchange of CO_2 at each grid point sums to zero over a year. Source or sink activity can then be arbitrarily imposed. To obtain seasonally balanced CO_2 exchange, we calculate a linear term, R^*, that reflects the non-linear responses of respiration to soil moisture and temperature (Raich *et al.*, 1991), and running totals of R^* and A_c are accumulated. At any time,

$$R_g(t) = R^*(t) \frac{\sum_{1 \text{ year}} A_c(t) \Delta t}{\sum_{1 \text{ year}} R^*(t) \Delta t}$$

where the summations are for the previous year. Net CO_2 flux to the atmosphere from the terrestrial biosphere modelled by SiB is mixed in the GCM with prescribed fields of net CO_2 exchange from fossil fuels, tropical deforestation and net exchange with the ocean to simulate the concentration of CO_2 at grid points close to the stations used by atmospheric sampling programmes. This permits the model to be used in a 'pseudo inverse' mode to examine the plausibility of various scenarios for the strength and location of terrestrial and ocean sinks for CO_2.

Model results and tests

GCM results

Several multiyear runs of the Colorado State University (CSU) GCM with SiB2 have now been completed. Including SiB2 increases the execution time of the GCM by about 10%, relative to that required for a simple bucket hydrology. Climate statistics indicate that the SiB2 surface parameterisation yields a realistic continental climate, and the new model runs faster and is more stable than the prior version of SiB. Details of these results are presented by Randall *et al.* (1996). It is of interest here to focus on some of the carbon cycle results. Figure 3 shows the predicted annual gross CO_2 assimilation field. The simulated total annual gross primary productivity (GPP) is 143 Gt. This value is very sensitive to the simulated climate and varies between 125 and 150 Gt per year, depending on the versionn of the GCM used. Assuming that global GPP should be 2–2.5 times net primary productivity (NPP), this value is consistent with most recent estimates of NPP.

With these results it is possible to analyse the impact of the carbon

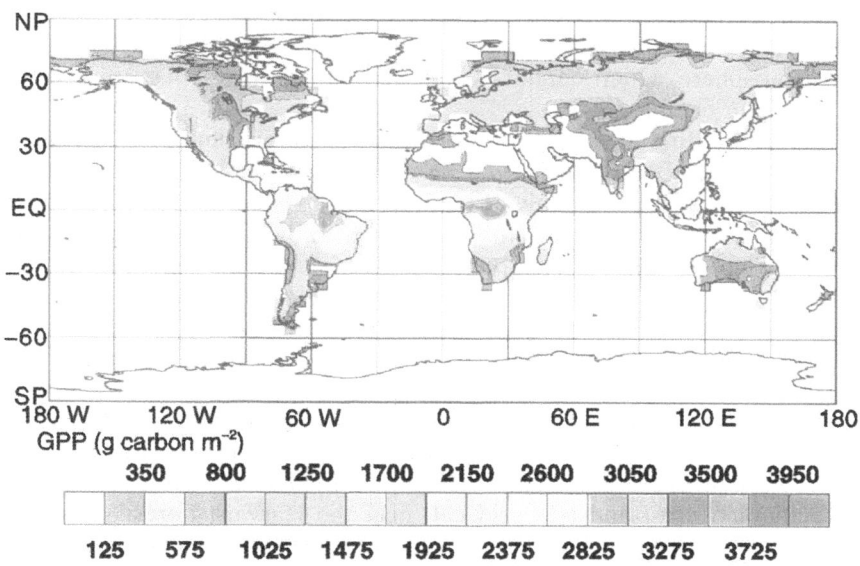

Fig. 3. A map of the distribution of annual GPP predicted by the SiB2 GCM model at a $4° \times 5°$ resolution.

A colour version of this figure is available for download from
www.cambridge.org/9780521471091

cycle (as simulated by SiB2 in the CSU GCM) on the spatial and temporal variation in the CO_2 concentration of the atmosphere. Figure 4*a* illustrates the simulated net CO_2 flux to the atmosphere (plotted by latitudinal zones versus month of the year). The fluxes are most strongly seasonal at temperate northern latitudes, with strong net CO_2 uptake in the summer balanced by net CO_2 efflux during the spring and autumn. The strong seasonality of net CO_2 exchange is responsible for a large seasonal change in atmospheric CO_2 observed at these latitudes.

Figure 4*b* presents the seasonal cycle of atmospheric CO_2 simulated using SiB2 at four observing stations operated by the National Oceanic and Atmospheric Administration (NOAA) (Conway *et al.*, 1994). Although the simulated concentrations reflect only the activity of the terrestrial biota (no fossil fuel emissions or air–sea fluxes were included, for example), the model captures much of the spatial variation in the phase and amplitude of the observed seasonal cycle. These results indicate that the seasonality and spatial distribution of net CO_2 fluxes simulated by SiB2 are approximately correct. One unique feature of these simulations is that the carbon exchange with the biosphere is simulated in step with the simulation of atmospheric mixing. Other analyses have used monthly summaries of net CO_2 exchange and atmospheric wind fields to mix CO_2 'off-line'. However, there is substantial co-variation between atmospheric circulation patterns and net CO_2 exchange with the biosphere which cannot be resolved in these off-line simulations. These are captured here.

A key finding is that a substantial meridianal gradient of CO_2 concentration would be established in the northern hemisphere by a biosphere with no net sources or sinks (Fig. 5; also see Denning, Fung & Randall, 1995). This occurs because of differences in the vertical distribution of CO_2 in the atmosphere. Respired CO_2 tends to stay close to the surface because there is little convective mixing when this process prevails (at night and in the winter), while active convective mixing during mid-day (peaking in the summer) ensures that net photosynthetic uptake of CO_2 draws from a larger volume of the atmosphere.

Figure 6 presents the seasonal course of photosynthesis (A_c), respiration (R_g) and net ecosystem production (A_m) for a single grid cell selected from a GCM run. This cell is specified as a mixed evergreen–deciduous forest and is a grid cell near the Harvard Forest site where Steven Wofsy and co-workers have conducted long-term studies of CO_2 flux by eddy correlation (Wofsy *et al.*, 1993). By taking advantage of the separation of photosynthesis and respiration in the diurnal cycle (photosynthesis and respiration occur together in the day and respiration occurs alone at night), they were able to obtain an estimate of the yearly

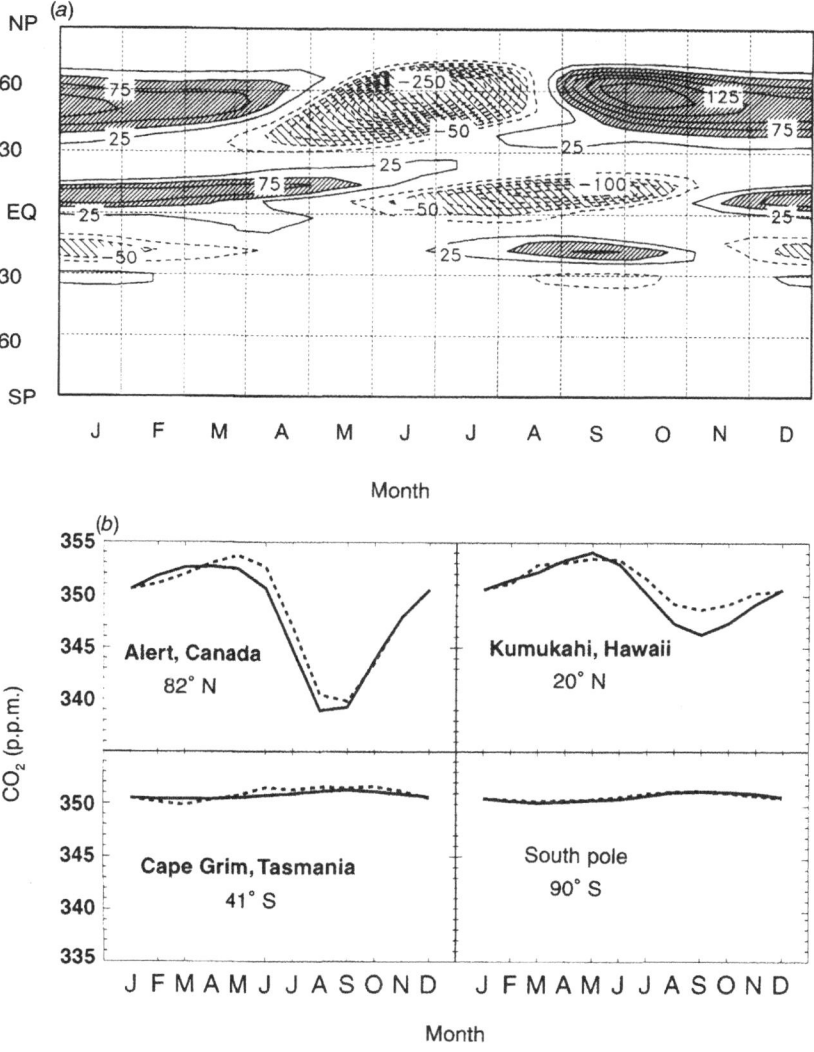

Fig. 4. (*a*) Zonal total flux of carbon from the terrestirla biosphere to the atmosphere as predicted by SiB2. The contour interval is 25×10^9 kg carbon per 4° latitude zone per month. Absolute values greater than 50×10^9 kg per month are shaded. (*b*) Monthly mean CO_2 concentration as measured by the NOAA/CMDL Carbon Cycle Group at four flask-sampling stations (solid curves), and as predicted for the same locations by the CSU GCM using SiB2 (dashed curves). The flask data are represented by an annual cycle of four harmonics fitted to the data from all available years at each station. The secular trend owing to fossil fuel emissions has been removed (Denning, 1994).

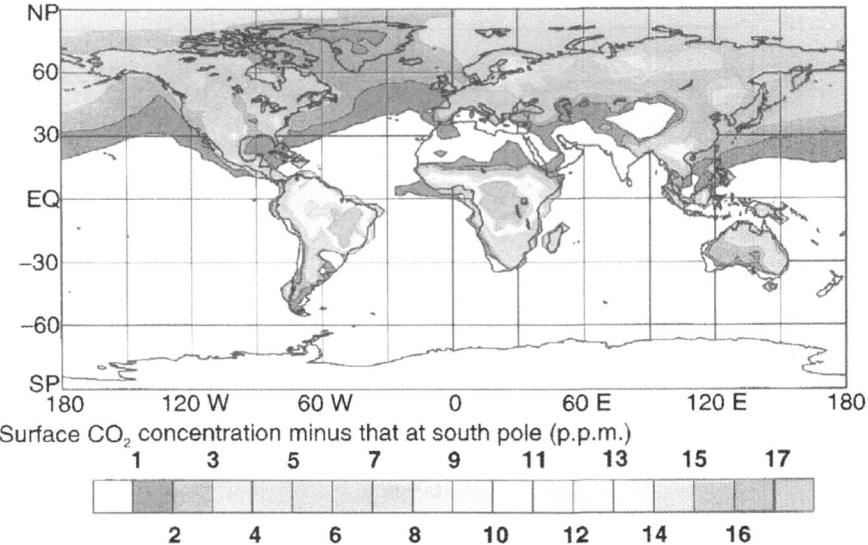

Fig. 5. Annual mean surface concentration (p.p.m.) of CO_2 simulated using SiB2 in the CSU GCM (SiB2 contribution *only*). The model was initialised with a globally uniform concentration and integrated for four years to reach equilibrium between surface fluxes and atmospheric transport. Only the effects of the annually balanced exchanges between the land surface and atmosphere are represented here. The annual net flux was close to zero at every model grid point. The concentration at the south pole has been subtracted.

A colour version of this figure is available for download from
www.cambridge.org/9780521471091

integrated A_c, which was 1110 g carbon m^{-2} per year from that forested site. This is close to the simulated output from that region in the model (1260 g carbon m^{-2} per year). They observed a net seasonal accumulation of carbon (S) of 370 g carbon m^{-2} per year, and this was used to constrain our calculation of respiration in Fig. 6. No other tuning was done. The simulated seasonal pattern of A_m is close to the observed. Simulated seasonal courses of A_c, R_g and A_m for other ecosystems are plausible, but long-term CO_2 exchange measurements are not available for comparison.

Local-scale simulations

Off-line simulations using SiB2 have been conducted for comparison with daily courses of CO_2 and energy flux measurements made as part of the FIFE (first ISLSCP field experiment) (Verma, Kim & Clement,

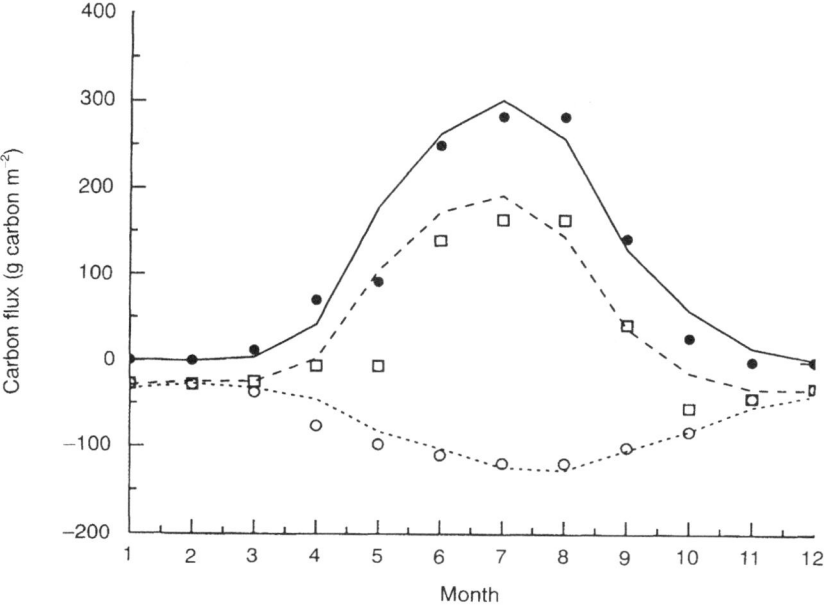

Fig. 6. Carbon fluxes simulated by SiB2 GCM for a grid cell corresponding to mixed evergreen–deciduous forest. Symbols show measurements taken from Wofsy *et al.*, 1993, for the year 1991 and lines show results of simulations. Closed circles and the solid line represent gross primary production (A_c), open circles and the dotted line are ecosystem respiration (R_g) and the squares and dashed line are net ecosystem CO_2 exchange (A_m).

1992). FIFE was conducted in a tall-grass prairie ecosystem with a canopy composed of a mixture of C_3 and C_4 species. Figure 7 shows a comparison of simulations driven by independent meteorological observations (dashed line) and flux observations shown as points. Thirteen periods (of 1–3 days) were examined spread over the season, when eddy correlation data were available. These spanned three episodes of dry down with severe water stress and two periods of recovery following precipitation events. Plots comparing observed and simulated fluxes (Fig. 8) show that SiB2 produced reasonable simulation of energy and net site CO_2 flux, as modulated by water stress. The V_{max} value of rubisco for top leaves of this canopy under unstressed conditions is the primary adjustable variable in these simulations. Collatz *et al.* (1992) report a value of 35 μmol m^{-2} s^{-1} fits their studies of *Zea mays*, and the

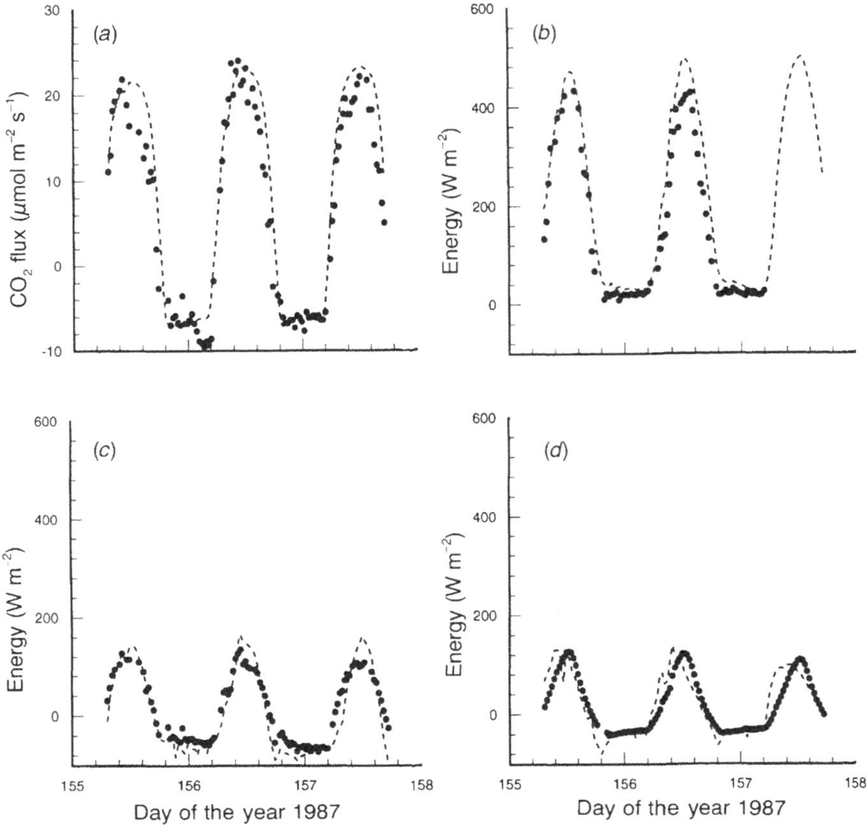

Fig. 7. An example of diurnal carbon and energy fluxes from a prairie grassland as measured (symbols) and simulated by SiB2 (dotted line) using meteorological inputs. (*a*) A_m; (*b*) λE_m; (*c*) H_m; and (*d*) G (see Fig. 1).

canopy simulations (Fig. 7) fit best to a value of 29 μmol m^{-2} s^{-1}. For GCM runs, a value of V_{max} of 30 μmol m^{-2} s^{-1} is currently used for C$_4$ vegetation. This study illustrates that small-scale experiments can be useful for calibrating the model for canopy- and larger-scale simulations.

Conclusions

Tests conducted so far seem to indicate that the simulations of climate driven by exchange of energy and momentum at the land surface and

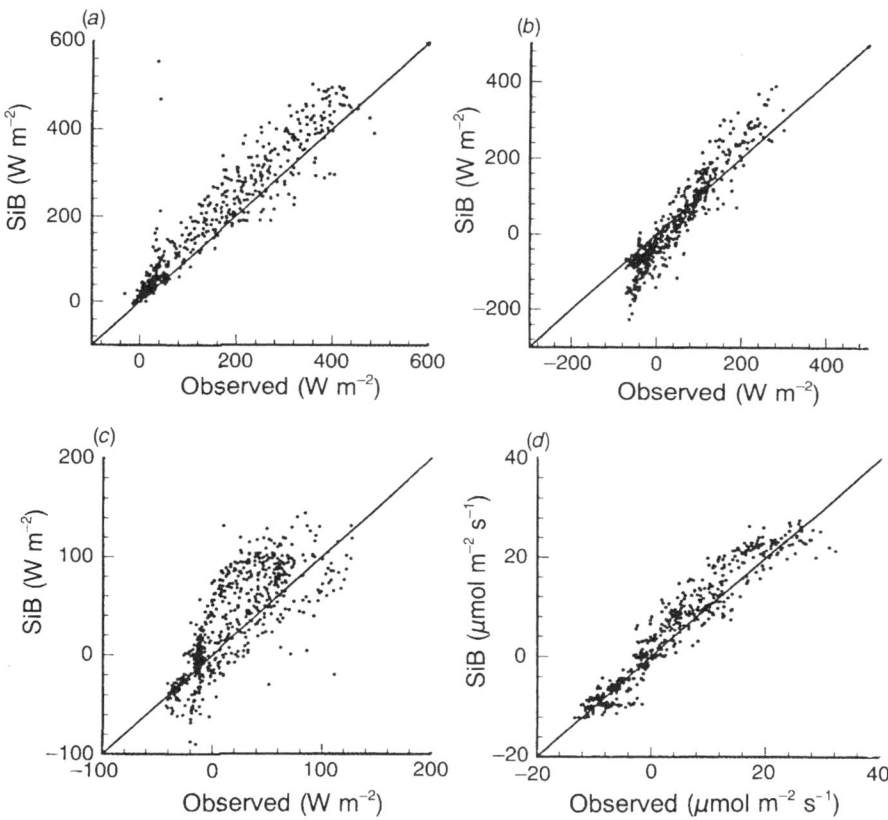

Fig. 8. Plots of carbon and heat fluxes predicted by SiB2 compared with measurements during the growing season in a prairie grassland. (*a*) λE_m; (*b*) H_m; (*c*) G; and (*d*) A_m (see Fig. 1).

the carbon cycle using SiB2 have worked reasonably well. It is important to note, however, that these tests have not yet been particularly rigorous or extensive – much more work is needed. Work in progress will test the application of SiB2 to other ecosystems, and it is our long-term goal to integrate modelling of the climate system discussed here with the longer-term biogeochemical processes that link the carbon cycle to the cycling of nitrogen and other elements.

Acknowledgements

Flask station data were provided by Pieter Tans and Thomas Conway, NOAA Climate Monitoring and Diagnostics Laboratory (CMDL)

Carbon Cycle Group. Support for this project was provided by NASA Earth Observing System (EOS) funds through the Sellers–Mooney Interdisciplinary Science Project. Additional support was provided by the Mellon Foundation and the Carnegie Institution of Washington. This is CIW-DPB Publication Number 1227.

References

Amthor, J.S. (1994). Scaling CO_2–photosynthesis relationships from the leaf to the canopy. *Photosynthesis Research*, 39, 321–350.

Aphalo, P.J. & Jarvis, P.G. (1993a). An analysis of Ball's empirical model of stomatal conductance. *Annals of Botany*, 72, 321–327.

Aphalo, P.J. & Jarvis, P.G. (1993b). The boundary layer and the apparent responses of stomatal conductance to wind speed and the mole fractions of CO_2 and water vapour in the air. *Plant, Cell and Environment*, 16, 771–783.

Asrar, G. & Dokken, D.J. (1993). *EOS Reference Handbook*. Washington, DC: NASA.

Ball, J.T., Woodrow, I.E. & Berry, J.A. (1987). A model predicting stomatal conductance and its contribution to the control of photosynthesis under different environmental conditions. In *Progress in Photosynthesis Research*, Vol. 4, ed. J. Biggins, pp. 221–224. Nijhoff: Dordrecht.

Beljaars, A.C.M., Viterbo, P., Miller, M.J. & Betts, A.K. (1996). The anomalous rainfall over the USA during July 1993: sensitivity to land surface parameterization and soil moisture anomalies. *Monthly Weather Review*, 124, 362–383.

Björkman, O.E (1989). Some viewpoints on photosynthetic response and adaptation to environmental stress. In *Photosynthesis*, ed. W.R. Briggs, pp. 45–58. New York: Alan R. Liss.

Budyko, M.I. (1974). *Climate and Life*. New York: Academic Press.

Bunce, J.A. (1985). Effect of boundary layer conductance on the response of stomata to humidity. *Plant, Cell and Environment*, 8, 55–57.

Charney, J.G., Wuirk, W.J., Chow, S.H. & Kornfield, J. (1977). A comparative study of the effects of albedo change on drought in semi-arid regions. *Journal of Atmospheric Sciences*, 34, 1366–1385.

Collatz, G.J., Berry, J.A., Farquhar, G.D. & Pierce, J. (1990). The relationship between the rubisco reaction mechanism and models of photosynthesis. *Plant, Cell and Environment*,13, 219–225.

Collatz, G.J., Ball, J.T., Grivet, C. & Berry, J.A. (1991). Physiological and environmental regulation of stomatal conductance, photosynthesis and transpiration: a model that includes a laminar boundary layer. *Agricultural and Forest Meteorology*, 54, 107–136.

Collatz, G.J., Ribas-Carbo, M. & Berry, J.A. (1992). Coupled photosynthesis–stomatal conductance model for leaves of C_4 plants. *Australian Journal of Plant Physology*, 19, 519–538.

Conway, T.J., Tans, P.P., Waterman, L.S. *et al.* (1994). Evidence for interannual variability of the carbon cycle from the NOAA/CMDL global air sampling network. *Journal of Geophysical Research*, 99, 22 831–22 855.

Cowan, I.R. & Farquhar, G.D. (1977). Stomatal function in relation to leaf metabolism and environment. *Symposium of the Society for Experimental Biology*, 31, 471–505.

Deardorff, J.W. (1977). Efficient prediction of ground surface temperature and moisture with inclusion of a layer of vegetation. *Journal of Geophysical Research*, 83, 1899–1903.

Denning, A.S. (1994). Investigations of the transport, sources, and sinks of atmospheric CO_2 using a general circulation model. *Atmospheric Science Paper No. 564*. Fort Collins: Colorado State University.

Denning, A.S., Fung, I.Y. & Randall, D.A. (1995). Latitudinal gradient of atmospheric CO_2 due to seasonal exchange with land biota. *Nature*, 376, 240–243.

de Pury, D. (1995). Scaling, photosynthesis and water use from leaves to paddocks. PhD thesis, The Australian National University, Canberra, ACT, Australia.

Deware, R.C. (1995). Interpretation of an empirical model for stomatal conductance in terms of guard cell function. *Plant, Cell and Environment*, 18, 365–372.

Dickinson, R.E. (1984). Modeling evapotranspiration for three-dimensional global climate models. In *Geophysical Monograph of the American Geophysical Union 29, Climate Processes and Climate Sensitivity*, ed. J.E. Hanson & T. Takahashi, pp. 58–72. Washington, DC: American Geophysical Union.

Dorman, J.L. & Sellers, P.J. (1989). A global climatology of albedo, roughness length and stomatal resistance for atmospheric general circulation models as represented by the simple biosphere model (SiB). *Journal of Applied Meteorology*, 28, 833–855.

Farquhar, G.D., von Caemmerer, S. & Berry, J.A. (1980). A biochemical model of photosynthetic CO_2 assimilation in leaves of C_3 plants. *Planta*, 149, 78–90.

Friend, A.D. (1991). Use of a model of photosynthesis and leaf microenvironment to predict optimal stomatal conductance and leaf nitrogen partitioning. *Plant, Cell and Environment*, 14, 895–905.

Fung, I.Y., Tucker, C.J. & Prentice, K.C. (1987). Application of very high resolution radiometer vegetation index to study atmosphere–biosphere exchange of CO_2. *Journal of Geophysical Research*, 92, 2999–3015.

Garratt, J.R. (1993). Sensitivity of climate simulations to land-surface and atmospheric boundary-layer treatments – a review. *Journal of Climate*, 6, 419–449.

Harrison, E.F., Minnis, P., Barkstrom, B.R., Ramanathan, V., Cess, R.D. & Gibbson, G.G. (1990). Seasonal variation of cloud radative forcing derived from the Earth Radiation Budget Experiment. *Journal of Geophysical Research*, 95, 18 687–18 703.

Jarvis, P.G. (1976). The interpretations of the variation in leaf water potential and stomatal conductance found in canopies in the field. *Philosophical Transactions of the Royal Society of London, Series B*, 273, 593–610.

Leuning, R. (1995). A critical appraisal of a combined stomatal–photosynthesis model for C_3 plants. *Plant, Cell and Environment*, 18, 339–355.

Lloyd, J. (1991). Modelling stomatal responses to environment in *Macadamia integrifolia*. *Australian Journal of Plant Physiology*, 18, 649–660.

Matthews, E. (1985). Atlas of archived vegetation, land use and seasonal albedo data sets. *NASA Technical Memorandum*. Washington, DC: NASA.

Monteith, J.L. (1995). A reinterpretation of stomatal responses to humidity. *Plant, Cell and Environment*, 18, 357–364.

Norman, J.M., Garcia, R. & Verma, S.B. (1992). Soil surface CO_2 fluxes and carbon budget of a grassland. *Journal of Geophysical Research*, 97, 18 845–18 853.

Raich, J. W., Rastetter, E.B., Melillo, J.M. *et al.* (1991). Potential net primary productivity in South America: application of a global model. *Ecological Applications*, 1, 399–429.

Randall, D.R., Sellers, P.J., Berry, J.A. *et al.* (1996). A revised land-surface parameterization (SiB2) for GCMs. Part 3: the greening of the Colorado State University General Circulation Model. *Journal of Climate*, 9, 738–763.

Sato, N., Sellers, P.J., Randall, D.A. *et al.* (1989). Effects of implementing the simple biosphere model (SiB) in a GCM. *Journal of Atmospheric Sciences*, 46, 2757–2782.

Sellers, P.J. (1985).Canopy reflectance, photosynthesis and transpiration. *International Journal of Remote Sensing*, 6, 1335–1372.

Sellers, P.J. (1987). Canopy reflectance, photosynthesis and transpiration. II. The role of biophysics in the linearity of their interdependence. *Remote Sensing of Environment*, 21, 143–183.

Sellers, P.J., Mintz, Y., Sud, Y.C. & Dalcher, A. (1986). A simple biosphere model (SiB) for use within general ciruclation models. *Journal of Atmospheric Sciences*, 43, 305–331.

Sellers, P.J., Shuttleworth, J.W., Dorman, J.L., Dalcher, A. & Roberts, J.M. (1989). Calibrating the simple biosphere model (SiB) for Amazonian tropical forest using field and remote sensing data. Part 1, average calibration with field data. *Journal of Applied Meteorology*, 28, 727–759.

Sellers, P.J., Berry, J.A., Colaltz, G.J., Field, C.B. & Hall, F.G. (1992a). Canopy reflectance, photosynthesis, and transpiration. III. A reanalysis using improved leaf models and a new canopy integration scheme. *Remote Sensing of Environment*, 42, 187–216.

Sellers, P.J., Heiser, M.D. & Hall, F.G. (1992b). Relationship between surface conductance and spectral vegetation indices at intrmediate ($100 \, m^2 - 15 \, km^2$) length scales. *Journal of Geophysical Research*, 19 033–19 060.

Sellers, P.J., Los, S.O., Tucker, C.J. *et al.* (1996a). A revised land surface parameterization (SiB2) for atmospheric GCMs. Part 2: the generation of global fields of terrestrial biophysical parameters from satellite data. *Journal of Climate*, 9, 706–737.

Sellers, P.J., Randall, D.R., Collatz, G.J. *et al.* (1996b). A revised land-surface parameterization (SiB2) for GCMs. Part 1: model formulation. *Journal of Climate*, 9, 676–736.

Shukla, J. & Mintz, Y. (1982). Influence of land-surface evapotranspiration on the earth's climate. *Science*, 215, 1498–1501.

Sud, Y.C., Shukla, J. & Mintz, Y. (1988). Influence of land-surface roughness on atmospheric circulation and precipitation. A sensitivity study with a general circulation model. *Journal of Applied Meteorology*, 27, 1036–1054.

Verma, S.B., Kim, J. & Clement, R.J. (1992). Momentum, water vapor, and carbon dioxide exchange at a centrally located prairie site during FIFE. *Journal of Geophysical Research*, 97, 18 629–18 639.

Wofsy, S.C., Goulden, M.L., Munger, J.W. *et al.* (1993). Net exchange of carbon dioxide in a mid-latitude forest. *Science*, 260, 1314–1317.

Woodrow, I.E. & Berry, J.A. (1988). Enzymatic regulation of photosynthetic CO_2 fixation in C_3. *Annual Review of Plant Physiology*, 39, 533–594.

Zobler, L. (1986). A world soil file for global climate modeling. *NASA Technical Memorandum*, 87802. Washington, DC: NASA.

P.R. VAN GARDINGEN, G. RUSSELL,
G.M. FOODY and P.J. CURRAN

Science of scaling: a perspective on future challenges

Research into the biological components of global environmental change has been driven by the need to: (i) understand and predict the impact of change on whole ecosystems, food and fibre production, water resources and environmental quality; (ii) quantify the values of variables such as biomass or the rates of processes such as net primary productivity; and (iii) evaluate options such as re-forestation for their potential to mitigate environmental change. These are not easy tasks. It is now accepted that global environmental change will not occur evenly across the globe and that actions in one locality can affect conditions elsewhere. Predictions are difficult because of the inextricable links between global environmental change and such socio-economic variables as population growth and individual prosperity. Analysis is further complicated by the need to establish communication between existing scientific disciplines. These problems have required a major shift in the way that research is carried out. Traditional reductionist science, which has been so productive, requires modification to take account of the concepts of systems analysis. The key processes must be included within the boundary of the system under study, which implies at least a regional scale. Since these processes lie within the province of disciplines such as ecophysiology, soil science, geography and chemistry, studies have to be interdisciplinary. Finally, since the results of these studies may be used for policy-making, they must be integral with, or at least linkable to, the socio-economic components of the system. These issues and related questions have resulted in the evolution of new scientific approaches. The most significant change has been the development of interdisciplinary, integrative programmes addressing issues at a regional or global scale. The need for this approach was clearly identified by the Intergovernmental Panel on Climate Change (IPCC, 1990) and is exemplified by programmes such as the International Geosphere Programme (IGBP, 1994) and the World Climate Programme (WCP). These programmes have led to the successful completion of international field campaigns such as HAPEX, HAPEX-

Sahel, FIFE and BOREAS. Results from some of these campaigns have been presented in this volume (e.g. the chapters by Gurney & Sewell and Berry *et al*; and elsewhere (e.g. Stewart *et al.*, 1996).

Complex, interdisciplinary programmes consist of studies at different levels of organisation or scales within the system being studied. In any one programme, information may be collected at scales ranging from the cellular to the individual organ (for example, a single leaf) or organism, up to a region or the whole globe. The data collected may range from estimates of biochemical variables to remotely sensed estimates of system parameters (e.g. topography). All these data need, ideally, to be integrated within some type of model. This need for integration is what has generated the rapid development of the science and mathematics of scaling in recent years. The previous chapters in this volume have discussed some of the methods and current applications of scaling. We would now like to place these activities into a wider context and highlight likely future needs and possible technological developments.

Global environmental change

Global environmental change is an issue that has stimulated international activity and discussion amongst scientists, economists, sociologists and politicians since the middle of the 1980s. The demands from a rapidly growing human population for resources are undoubtedly the main factors driving change. The burning of fossil fuels, deforestation and other land-use changes have resulted in an increase in atmospheric CO_2 concentrations from preindustrial levels of around 280 p.p.m.v. (IPCC, 1990) to current levels (1994) of around 360 p.p.m.v. (Enting, Trudinger & Francey, 1995). However, land-use change and the industrialisation of agriculture have led to increasing emission rates of other greenhouses gases including methane and nitrous oxide, as well as environmental degradation, for example loss of biodiversity, desertification and salinisation, with its consequent effect on regional albedo and water balance. Biophysical processes in the global climate system have long been known to be affected by the environmental factors that have been observed to be changing. This knowledge, combined with a public perception of abnormal climatic conditions during the 1980s and 1990s, has fuelled the debate regarding global climate change. These two decades have certainly contained an abnormally large number of extreme events, such as droughts, floods, hot summers and cold winters, in many regions of the world. The observed changes in global climate are consistent with the expected effects of rising concentrations of greenhouse gases but, as yet, do not form proof positive of global warning. From an environmental standpoint, the political discussion as to whether global warming is taking place is less relevant than development of an understanding at regional to global scales and the investigation of the impacts of the observed or predicted environmental changes, in particular of CO_2 concentration and land-use change. The well-attested measurements of these last two factors prompt the following crucial questions. How will global environmental change impact on biological systems in the medium and long term? What role do biological systems and processes play in regulating the global environment and climate? What options are available for modifying biological systems to mitigate the effects of global environmental change? These questions, however, remain largely unanswered.

The biology of environmental change

The major achievement of what is now around ten years of scientific endeavour has probably been the development of methods for studying the impacts on, and options for, mitigation of environmental change on whole ecosystems. New experimental protocols and analytical techniques, often integrated through mathematical modelling, were required. Since many measurements can only be made at the organism level, there was a clear requirement for scaling information between levels. The development of more rigorous research protocols has greatly aided this process. A generalised scheme seems to be evolving that can accommodate within it the formality of deductive science:

- definition of problem or issue and its scale or extent (e.g. the effects of deforestation on regional hydrology)
- definition of objectives of the programme
- definition of evaluation of methods of investigation
- specification of the desired products of the research programme
- identification of data requirements and methods of integration
- definition of the experimental programme
- evaluation of results using independent data sets
- interpretation of results
- production of recommendations based on research findings.

This scheme of research is now starting to deliver information in a form that can be of use to policy-makers. For example, estimates of net primary productivity (Field, Randerson & Malmstrom, 1995) and regional water balance (Woodcock, Collins & Jupp, this volume) are now available. The approach has also been useful for identifying significant unknowns in the system that require future investigation. Examples include the 'missing sink' in the global carbon balance (Tans, Fung & Takahashi, 1990; Enting, Trudinger & Francey, 1995) and the role of soils in the carbon cycle (Atkinson & Fogel, this volume).

The search for components of the system likely to be strongly affected by environmental change has shown that they can occur at different scales. For example, the direct effects of increasing CO_2 concentrations are observed at the level of leaf biochemical composition (Kruijt, Ongeri & Jarvis, this volume). These direct effects, however, may be modified by secondary effects such as changes in canopy structure and resulting interactions between the canopy and the physical environment.

The challenge for scaling has been to encapsulate these system features, taking due account of the heterogeneity of the system in space and time, the existence of feedback loops and the non-linearity of many functional relationships. The scientific community has responded to these challenges and now has in place measurement and modelling programmes designed to answer the key questions about global environmental change.

New technologies for scaling

Techniques for modelling the productivity and dynamics of vegetation have developed significantly since they were first introduced in the early 1960s. Although there are now many well-validated models for predicting the yield of agricultural crop monocultures, models for predicting the effect of environmental change on natural ecosystems, managed forests or whole agro-ecosystems are less reliable. The reasons for this are partly to do with the temporal and spatial scale at which the models operate and partly to do with the quality of the input data representing the surface characteristics. Some of these problems can be addressed by developing standards for model building, analogous to those employed in software engineering, which include procedures for parameterising, driving and testing a model for all the situations for which it is to be used.

Six conclusions can be drawn from the review of recent research presented in this volume. First, modelling of environmental processes at a given range of scales is only possible if rigorous statistical sampling procedures are used to derive the values of the parameters and variables that are appropriate for that given range of scales. Second, since computing power is no longer a serious limitation to what is possible, it should be normal practice to introduce stochastic elements into both input variables (such as those that represent weather) and parameter values, which are often not known accurately. The outcome of a model should, therefore, be a probability distribution rather than a single value. Third, a major limitation to applying such models is the generally low quality of the input data. Geographical Information Systems (GIS) are commonly used to store input data for large numbers of mapping units. However, for large parts of the world, maps are only available at a scale that hides much significant variability. Fourth, long series of meteorological data are now available for many parts of the world. However, work needs to continue into how best to interpolate these data for heterogeneous topography and to increase the accuracy of synthetic weather generation for input to models. Fifth, a clearer separation needs to be made in models between the atmospheric environment (defined as those atmospheric factors that affect but are not significantly affected by the system of interest) and the local micro-climate, which has been modified by the land surface. Sixth, until recently, many of these models have been quantitative in nature and based on classical statistical techniques. However, there is an increasing interest in the use of other approaches, especially artificial intelligence methods (e.g. rule-based techniques), which may be more suited to the task. These may make fewer assumptions and may be more able to accommodate factors such as missing data values.

The role of GIS and remote sensing

The realisation that many environmental processes operate at different spatial and temporal scales may necessitate multiscale investigations and modelling. At present, investigators have to choose between acquiring environmental information for small areas frequently or for large areas infrequently. Transferring knowledge from one extreme to the other is a complex task but required if we are to understand and model environmental processes and their effects. To rise to this challenge, extensive data sets on environmental properties at a range of spatial and temporal scales will be required. The large volumes of spatially related data that will be required for this will call for the use of GIS for data storage and analysis. Furthermore, with attention focused on regional–global scale issues, greater use of remotely sensed data as a source of environmental information and for scaling environmental models is expected. Satellite remote sensing is the only feasible means of acquiring information on significant environmental properties (e.g. land cover, leaf area index (LAI), net primary productivity (NPP)) over large areas and at a relatively fine temporal resolution. Moreover, remotely sensed data are in a format conducive for relatively simple integration into a GIS. The launch of new sensing systems such as those associated with the EOS programme should also herald a new era in environmental remote sensing.

With both GIS and remote sensing there remain many scientific and technical problems to be resolved before their full potential in scaling environmental information is realised. Presently, for instance, we do not fully understand how radiation interacts with many environmental features. The techniques for extracting environmental information from remotely sensed data and using this within a multiscale analysis in a GIS also require further refinement. Some of these issues have been touched on in this book. For instance, Barnsley, Barr & Tsang (this volume) outline some of the problems associated with scaling and generalising land cover maps derived from remotely sensed data and Woodcock, Collins & Jupp (this volume) illustrate the need to consider the relationship between the spatial resolutions of a remotely sensed data set and the scale of the environmental features of interest. A major breakthrough may be made when fuller use is made of the rich spatial information provided by remote sensing. Presently, many models used in environmental investigations are not explicitly spatial yet a major aim is to understand and predict processes over different regions. Furthermore, spatial and temporal variations in the quality of products derived from remotely sensed data are required to ensure not only maximum information extraction but ultimately also the quality of later products based upon them.

Socio-economic dimensions

The problems of global environmental change are not only scientific, their solution requires the socio-economic aspects to be addressed also. Science can predict what would happen to biological productivity and biodiversity for a given global change scenario. However, the degree of social change will depend on the reaction of people to the prevailing and predicted conditions. The demands for resources vary regionally, with, for example, the demand for energy being highly concentrated in the northern hemisphere, particularly in North America and Europe. Likewise, the *per capita* emissions of CO_2 are on average ten times higher in developed than in developing countries (IPCC, 1990), emphasising the inequality between cause and impact. One of the most worrying aspects of global environmental change is that as developing countries understandably attempt to increase their standard of living to emulate that of developed countries, their impact on the global environment increases rapidly.

It follows that any solution to the problems of change must take account of social and economic dimensions including a consideration of the aspirations of the people directly affected by either impact or solution.

Policy implications

One of the tasks of the scientist should be to provide reliable information that can be used to help politicians and other policy-makers to make better decisions. For this process to be successful, the scientist needs to be able to understand the needs of the policy-maker. Information must be made available at the correct *scale*, which will often mean an administrative region, and with an indication of the probability of the predicted outcome being correct. Politicians are used to dealing with incomplete and imperfect information. Scientists must learn not to take refuge in the excuse that something has not been proved but instead to distinguish changes that are likely to happen from those that are unlikely. After all, the essence of conventional scientific approaches is that nothing is ever actually proved. All we can do is to show that hypotheses are not supported. Our conclusions at whatever scale can only ever provide us with:

> a body of statements of varying degrees of certainty – some most unsure, some nearly sure, but none absolutely certain. (Feynman, 1989, p. 245.)

Let us hope that a rigorous approach to scaling-up can increase that degree of certainty.

References

Enting, I.G., Trudinger, C.M. & Francey, R.J. (1995). A synthesis inversion of the concentration and $\delta^{13}C$ of atmospheric CO_2. *Tellus*, 478, 35–62.

Feynman, R.P. (1989). *What do you Care what Other People Think?* London: Unwin Hyman.

Field, C.B., Randerson, J.T. & Malmstrom, C.M. (1995). Global net primary production – combining ecology and remote sensing. *Remote Sensing of Environment*, 51, 74–88.

IGBP (1994). *IGBP in Action: Work Plan 1994–1998, IGBP Global Change Report No. 28*. Stockholm: IGBP.

IPCC (1990). *Climate Change. The IPCC Scientific Assessment.* Cambridge: Cambridge University Press.

Stewart, J.B., Engman, E.T., Feddes, R.A. & Kerr, Y. (eds.) (1996). *Scaling up in Hydrology using Remote Sensing.* Chichester: Wiley.

Tans, P.P., Fung, I.Y. & Takahashi, T. (1990). Observational constraints on the global atmospheric CO_2 budget. *Science*, 247, 1431–1438.

Index